A Special Issue of
Social Neuroscience

Interpersonal Sensitivity: Entering Others' Worlds

Guest Editors

Jean Decety
University of Chicago, USA

and

C. Daniel Batson
University of Kansas, USA

Routledge
Taylor & Francis Group
LONDON AND NEW YORK

First published 2007 by Psychology Press

Published 2018 by Routledge
2 Park Square, Milton Park, Abingdon, Oxon OX14 4RN
52 Vanderbilt Avenue, New York, NY 10017

First issued in paperback 2018

Routledge is an imprint of the Taylor & Francis Group, an informa business

British Library Cataloguing in Publication Data
A catalogue record for this book is available from the British Library

ISSN 1747-0909

Cover design by Design Deluxe
Typeset in Ireland by Datapage International, Dublin

ISBN 13: 978-1-138-87776-4 (pbk)
ISBN 13: 978-1-84169-838-0 (hbk)

Contents*

(continued overleaf)

* This book is also a special issue of the journal *Social Neuroscience*, and comprises pp. 151–366 of Volume 2 (2007). The page numbers are taken from the journal and so begin with p. 151.

SOCIAL NEUROSCIENCE, 2007, 2 (3–4), 151–157

Psychology Press
Taylor & Francis Group

Editorial

Social neuroscience approaches to interpersonal sensitivity

Jean Decety
University of Chicago, Chicago Illinois, USA

C. Daniel Batson
University of Kansas, Lawrence, Kansas, USA

Interpersonal sensitivity refers to our ability to perceive and respond with care to the internal states (e.g., cognitive, affective, motivational) of another, understand the antecedents of those states, and predict the subsequent events that will result. This special issue brings together new research findings from empirical studies, including work with adults and children, genetics, functional neuroimaging, individual differences, and behavioral measures, which examine how we process and respond to information about our fellow individuals. By combining biological and psychological approaches, social neuroscience sheds new light on the complex and multi-faceted phenomenon of interpersonal sensitivity, including empathy. One should, however, be aware of the challenges and limits of such an approach.

Human beings are intrinsically social. Our survival critically depends on social interaction with others. Most of our actions are directed toward or are responses to others (Batson, 1990). No single factor can account for human social cognitive evolution (e.g., diet or climate), but the single most important factor is the increasing complexity of hominid social groups (Bjorklund & Bering, 2003).

For a long time, social sciences and neuroscience have developed independently. Social psychology is interested in the scientific study of how people's thoughts, feelings and behaviors are influenced by the actual, imagined or implied presence of other individuals. Neuroscience studies the nervous system and its constitutive unit, the neuron, and conceptualizes individuals as somewhat isolated units of analysis. Social neuroscience challenges these views by exploring the biological underpinnings of the processes traditionally examined by, but not limited to, social psychology (Decety & Keenan, 2006). It is an exciting endeavor. On the one hand, theories in social psychology can provide important guidelines for investigating the information-processing mechanisms and their neural instantiation. On the other hand, neuroscience may help disambiguate competing social theories, for instance the distinction between personal distress and empathic concern.

In this special issue, titled "Interpersonal Sensitivity: Entering Others' Worlds," we bring together new research findings from empirical studies that document how we process information about our fellow individuals. This title was chosen as a broad umbrella to accommodate a large array of approaches, including work with adults and children, genetics, neuroimaging, etc. By interpersonal sensitivity, we mean the ability to perceive and respond with care to the internal

Correspondence should be addressed to: Jean Decety, Departments of Psychology and Psychiatry, The University of Chicago, 5848 S. University Avenue, Chicago, IL 60637, USA. E-mail: decety@uchicago.edu

www.psypress.com/socialneuroscience DOI:10.1080/17470910701506060

states (e.g., cognitive, affective, motivational) of another, understand the antecedents of those states, and predict the subsequent events that will result. *Social Neuroscience* seeks to facilitate collaboration across a broad range of disciplines, including but not limited to neuroscience, philosophy, psychiatry, clinical, developmental, and social psychology. It also encourages a broad range of empirical measures, including but not limited to functional neuroimaging, psychophysiology, behavioral observation, comparison of patient and non-patient populations, and comparison of genetic alleles. In this special issue, we sought breadth in a third direction as well. We sought to include research focused on several different but related aspects of the complex and multi-faceted phenomenon that we have labeled interpersonal sensitivity.

ASPECTS OF INTERPERSONAL KNOWLEDGE AND INTERPERSONAL SENSITIVITY

A sensitive response to others requires at least some knowledge. Contributions to this special issue highlight the following different aspects of interpersonal knowledge and interpersonal sensitivity.

Interpersonal knowledge

Knowledge of the internal states of others includes at least three distinct but related phenomena. First is knowing *that* others are thinking and feeling (i.e., that they are cognizant and sentient). This is Theory of Mind (ToM) at the most basic level. Second is knowing *what* others are thinking and feeling (e.g., empathic accuracy). Third is *how* we know what others are thinking and feeling. This is more substantive ToM. Many possible ways of knowing are considered by contributors to this special issue: mimicry, mirroring, perception–action overlap, shared neural representations, simulation, theory-theory, narrative accounts, identification, similar prior experience, and perspective taking.

Interpersonal sensitivity

Interpersonal sensitivity also includes at least three distinct but related forms. One form is *feeling as* another person feels (i.e., "matching through catching" another's emotional state). This phenomenon has been variously called emotional contagion, resonance, or (parallel) empathy. It should be distinguished from both (a) a parallel response to the same environmental situation (e.g., use of another's emotional display as a cue to aspects of the situation that then evoke a parallel emotional response) and (b) association or conditioning (e.g., response to another's display as a conditioned stimulus due to prior pairing of one's own and others' response, evoking a parallel emotional response).

A second form of interpersonal sensitivity is *feeling for* another person (i.e., an other-oriented emotional response evoked by and congruent with the perceived welfare of another). This phenomenon has been called sympathy, compassion (reactive) empathy, or empathic concern. It should be distinguished from more self-oriented emotional reactions to another's plight, including personal distress, disgust, anxiety, or fear.

A third form if interpersonal sensitivity is *caring for* another person (i.e., love or attachment). We may care for or value another's welfare, making us vigilant or sensitive to the way he or she is affected by events. When a cared-for other is perceived to be in need, we are likely to feel for this person. Remove the need, and we no longer feel for—but we still care for—this person.

The contributions to this special issue address in a variety of ways each of these forms of interpersonal knowledge and interpersonal sensitivity. Interpersonal knowledge and sensitivity involve a complex mix of automatic and unconscious processing that can be described as bottom-up, and self-regulatory functions that tap into meta-cognitive resources and are best described as top-down. Several contributions focus on bottom-up processing whereas others investigate top-down processing. Both approaches are necessary to understand the whole complexity of social interaction. But exactly how these approaches relate is not yet clear. In this regard, the contributions raise a number of key theoretical issues.

CONTENT OF THE SPECIAL ISSUE

In the past decade an impressive body of work in various fields of research, including social psychology, developmental science, cognitive psychology and neuroscience has provided strong evidence for a direct link between perception and action (Sommerville & Decety, 2006). When we observe the actions of others, neural circuits related to performance of their actions are activated in us (mirror-neuron systems). Interestingly, this motor resonance seems specific to actions made by humans as compared to robots as shown by work in developmental psychology and neuroscience. Based on their previous research, which had shown that observing an action made by a human, but not by a robot, interferes with the execution of a different action, Kilner and Blakemore investigated what aspect of human movement, not present in robotic movement, causes this interference effect. In their study, participants made arm movements while observing a video of either a human making an arm movement or a ball moving across the screen. The executed and observed arm movements were either congruent (same direction) or incongruent (tangential direction) with each other. The results showed that observed movements are processed differently according to whether they are made by a human or a ball. For the ball videos, both minimum jerk and constant velocity incongruent movements significantly interfered with executed arm movements. In contrast, for the human videos, the velocity profile of the movement was the critical factor: only incongruent, minimum jerk human movements significantly interfered with executed arm movements.

Also using an interference technique, the study conducted by Oberman, Winkielman, and Ramachandran investigated the role of spontaneous facial mimicry in the recognition of facial emotions. Their results show that impairing one's ability to use facial muscles leads to a selective deficit in the recognition of emotion that involve those muscles.

To investigate the neural mechanisms underpinning facial emotional processing, Van der Gaag and his colleagues scanned participants during the observation, the discrimination and the imitation of dynamic facial expressions. Results demonstrate that even passive viewing of facial expressions activates a wide network of brain regions that are involved in the execution of similar expressions (aka mirror-neurons system), including the inferior frontal gyrus (IFG), the insula and the posterior parietal cortex. While the viewing of facial expressions recruited similar brain regions in all three experiments, adding an active task (discrimination and imitation) augmented the magnitude of these activations. However, no evidence was found for selectivity in hemodynamic response to specific emotion. The authors argue that understanding facial expressions of emotion involves a simulation process within motor, limbic and somatosensory systems.

This automatic mimicry mechanism may be modulated by social variables, such as whether the target person is a member of an in-group versus an out-group, as reported in a study conducted by Mondillon and colleagues. Their research investigated the automatic imitation of facial expressions of anger by in-group and out-group members, using a temporal estimation task. Individuals typically overestimate duration that is represented by emotional faces, and this bias appears to be due to increases in arousal. Overestimation is not observed, however, when imitation of the facial expressions is inhibited, suggesting that embodied simulation mediates the changes in arousal. This method thus provides an implicit measure of imitation and was used to test the hypothesis that individuals automatically imitate in-group, but not out-group members' facial expressions of emotion. Chinese and French-Caucasian female participants (all living in France), run in separate studies, were trained to categorize short (400 ms) and long (1600 ms) standard durations in a temporal bisection task. They then categorized standard and intermediate durations represented by angry and neutral faces. Half of the face stimuli were Chinese, and half Caucasian. Results revealed a bias in the temporal perception of emotion such that Caucasian perceivers imitated Caucasian facial expressions and not Chinese ones. Results suggested that Chinese individuals imitated faces of both in- and out-group members. The results of the Chinese participants are interpreted in terms of motivations to understand emotional expressions of members of a host culture.

Facial expressions of social emotions can also be modulated by personal dispositions. Burklund, Eisenberger, and Lieberman examined neural responses to a facial expression that primarily signifies a threat to social connection, namely a "disapproving" facial expression. They hypothesized that neural responses to disapproving facial expressions would be moderated by individual differences in rejection sensitivity. Study participants were scanned while they viewed brief video clips of facial expressions depicting disapproval, anger, and disgust. As expected, all three expressions yielded bilateral amygdala activation relative to a resting baseline. But even more interestingly, individuals who scored higher on a measure of rejection sensitivity exhibited greater dorsal anterior cingulate cortex activity in response to disapproving facial expressions, but not in response to anger or disgust facial expressions. Results suggest that, at the neural level, individuals high in rejection sensitivity may be more sensitive to facial expressions signaling potential rejection, but not to threatening facial expressions in general. Their results also suggest that disapproving facial expressions convey a distinct type of threat and should be considered in future studies of socially threatening facial expressions.

Human beings not only resonate with others as demonstrated in the previous studies, they also have the cognitive ability to intentionally adopt or imagine the psychological view point of others. This process, called perspective taking, is thought to play an important role in empathic concern and, possibly, moral reasoning. Preston and her collaborators examine perspective taking in an experiment that combined psychophysiological (i.e., heart rate, respiration rate, and skin conductance) and functional cerebral blood flow measurements with positron emission tomography. Study participants were requested to imagine personal experiences of fear or anger and similar emotional experiences from the point of view of another person. Their results suggest that substrates of cognitive empathy overlap with those of personal feeling states to the extent that one can relate to the state and situation of the other.

Most neuroimaging studies that have explored the overlap in brain response between the observation of behavior performed by others and the generation of the same behavior in self have relied on simple subtraction methods and generally highlight the commonalities between self and other processing, and ignore the differences. This is particularly true for the recent series of fMRI studies that have reported shared neural circuits for the first-hand experience of pain and the perception of pain in others (see Jackson, Rainville, & Decety, 2006, for an exception). It is, however, possible, as argued by Zaki and collaborators, that common activity in ACC and AI may reflect the operation of distinct but overlapping networks of regions that support perception of self or other-pain. To address this issue, they scanned participants while they received noxious thermal stimulation (self-pain condition) or watched short videos of other people sustaining painful injuries (other-pain condition). Analyses identified areas whose activity covaried with ACC and AI activity during self or other-pain either across time (intra-individual connectivity) or across participants (inter-individual connectivity). Both connectivity analyses identified clusters in the midbrain and periaqueductal gray with greater connectivity to the AI during self-pain as opposed to other-pain. The opposite pattern was found in the dorsal medial prefrontal cortex, which showed greater connectivity to the ACC and AI during other-pain than during self-pain using both types of analysis. Intra-individual connectivity analyses also revealed regions in the superior temporal sulcus, posterior cingulate, and precuneus that became more connected to ACC during other-pain as compared to self-pain. The results of this experiment document distinct neural networks associated with ACC and anterior insula in response to first-hand experience of pain and response to seeing other people in pain. These networks could not have been detected in prior work that examined overlap between self- and other-pain in terms of average activity, but not connectivity.

Lawrence and colleagues investigated affective and cognitive empathy with a series of self-report measures in individuals diagnosed with depersonalisation disorder (DPD), who report various disturbances related to "body image" such as feelings of "disembodiment" and "emotional numbing." In addition, they measured participants' tendency to rely on knowledge about themselves when attributing affective states to other people. As predicted, the DPD group showed intact performance on the task tapping cognitive empathy, alongside a disruption in affective empathy as evident from speech rate patterns. These data were also indicative of an over reliance by the DPD group on information relating to the self when attributing affective states to other people. The findings suggest that affective empathy is an embodied state that can be disrupted by changes in body image.

It is believed that disorganized attachment is an early predictor of development of psychopathology in childhood and adolescence. Gervai and colleagues previously had found an association between polymorphism of the *DRD4* gene and disorganized attachment, and in the present study, they investigated the interplay between genetic and care giving contributions to disorganized attachment. A total of 138 mother–infant dyads, 96 from a Hungarian low-social-risk sample and 42 from a US high-social-risk sample, were assessed for infant disorganized attachment behavior, for *DRD4* gene polymorphisms, and for disrupted forms of maternal affective communication with the infant. In accord with literature reports, the authors found a robust main effect of maternal AMBIANCE scores on infant disorganization. However, this relation held only for the majority of infants who carried the short form of the DRD4 allele. Among carriers of the 7-repeat DRD4 allele, there was no relation between quality of maternal communication and infant disorganization. This interaction effect was independent of degree of social risk and maternal *DRD4* genotype. These results suggest that the 7-repeat DRD4 allele reduces sensitivity to maternal affective communication.

Hobson, Lee, and Meyer elaborate a theoretical position and report an empirical study on a specific form of interpersonal engagement: the propensity to identify with the subjective orientation of another person. On the basis of a hypothesis that individuals with autism have a relative lack of this form of intersubjective connectedness, they predicted that children and adolescents with autism would contrast with matched participants without autism in specific aspects of communication when requested to "Get Pete to do this" after witnessing an adult demonstrating actions in Pete's absence. As predicted, on blind ratings of videotapes of participants' communication, those with autism achieved lower scores on four indices of identification that were selected a priori: emotional engagement, sharing experience in joint attention, communication of style, and shifting in communicative role. The two groups were almost completely separate on a composite measure of identification. Hobson et al. consider the implications of these findings for typical and atypical development.

One of the most complex and perplexing aspects of human interpersonal sensitivity is its links with moral reasoning. Recently cognitive neuroscience has begun to explore the neural underpinning of such complex behavior. Moll and his colleagues have argued that moral sensitivity depends on a sophisticated integration of cognitive, emotional, and motivational mechanisms that are modulated by individual experience in different cultural milieus. Different lines of investigation on agency and morality have pointed to overlapping neural systems. In the present research, Moll et al. used functional MRI to investigate the contribution of agency to brain activation by manipulating elicitors of specific moral emotions, such as interpersonal harm. Results showed that emotionally neutral agency recruited neural networks previously associated with agency, intentionality and moral cognition, encompassing ventral and subgenual sectors of the medial prefrontal cortex (PFC), insula, anterior temporal cortex and superior temporal sulcus (STS). Compared to emotionally neutral agency, different categories of moral emotions led to distinct activation patterns: (1) prosocial emotions (guilt, embarrassment, compassion) activated the anterior medial PFC and STS; with (2) empathic emotions (guilt and compassion) additionally recruiting the mesolimbic pathway; (3) other-critical emotions (disgust and indignation) were associated with activation of the amygdala—parahippocampal and fusiform areas. These findings indicate that agency related to norm-abiding social behaviors of emotionally neutral scripts share neural substrates both with the "default mode" of brain function and with the moral sensitivity network. Additional activation in specific components of this network is elicited by different classes of moral emotions, in agreement with recent integrative models of moral cognition and emotion.

Philosopher of mind Shawn Gallagher present critical arguments against both explicit and implicit versions of the simulation theory for intersubjective understanding. He contends that logical, developmental, and phenomenological evidence counts against explicit simulation if this is understood as the pervasive or default way that we understand others. Implicit (subpersonal) simulation, identified with neural resonance systems (mirror systems or shared representations), fails to provide the kind of simulation required by simulation theory, because it fails to explain how neuronal processes meet constraints that involve instrumentality and pretense. Implicit simulation theory also fails to explain how one can attribute a mental or emotion state that is different from one's own to another person. Building on his analysis, Gallagher sketches an alternative interpretation of neural resonance systems.

THEORETICAL ISSUES HIGHLIGHTED BY THE CONTRIBUTIONS

In addressing various aspects of interpersonal knowledge and interpersonal sensitivity, these contributions highlight a number of important theoretical issues. First, and perhaps most pressing, is the question of to what degree "shared" neural representations reflect automatic, bottom-up perceptual processes, and to what degree they reflect more constructive, top-down cognitive processes. For example, imagine seeing someone wince when receiving an electric shock. To what degree does this perception (a) directly activate regions of the brain associated with the affective experience of and response to pain, and then (b) this activation lead to cognitive understanding that the other is in pain? Alternatively, to what degree does this perception (a) lead to initial cognitive understanding that the other is in pain, which then (b) recruits regions associated with the affective experience of pain as part of a full understanding/appreciation of the other's pain? If the former occurs, can we feel another's pain without knowing the other is in pain? If the latter occurs, can the recruitment of regions associated with the affective experience of pain be regulated? Can we recognize but not fully understand/appreciate/care that another is in pain (e.g., someone we do not like)? Implicit in these questions is the more basic question of what "automatic" means when applied to psychological processes. Does it mean unconscious, inevitable, invariant, unmediated, unmoderated, effortless, self-moving, self-regulating, or something else?

Several of the contributions give attention to individual-difference or situational qualifiers of one or more aspects of interpersonal sensitivity. In the future, attention to such qualifiers will likely increase, raising more questions about how direct and automatic the various aspects of interpersonal sensitivity are.

Second, is there one basic or primary way we come to know how others think and feel? Or, when available, do we make use of multiple channels of information (including, perhaps, motor mimicry, facial mimicry, simulation, knowledge of the other's personality and values, knowledge of the other's situation, knowledge of what others have thought and felt in similar situations, knowledge of how we have thought and felt in similar situations, and perspective-taking/imagination)?

Third, how shared are "shared representations"? It is known that regions associated with the affective experience of pain (ACC and AI) are activated both during one's own experience of pain (self-pain) and when witnessing another in pain (other-pain). But does this common activation reflect the same experience of pain? Might one be pained for (distressed for) another in pain but directly pained (distressed) at one's own pain? If so, are the two experiences of pain embedded in distinct neural constellations or networks that lead to them being experienced as psychologically distinct emotions? Might the experience of pain in each case involve activation of regions associated with intention and desire, yet the intentions and desires be different in the two cases (relief of the other's suffering; relief of one's own suffering)? Does the level and character of other-pain change depending on one's personality, one's relation to the other, the relevance of the other's welfare to one's own, or prior experience in situations like the other's?

Finally, what is the role of self–other merging and self–other distinctiveness in interpersonal sensitivity? Do mimicry, mirroring, perception–action overlap, or some combination of these, merge the experience of the other with one's own experience? Or, even in these "shared" processes, does the awareness of whose experience is whose remain clear? That is, when a person reacts emotionally on witnessing someone else react emotionally, is there a loss of self–other distinctiveness, or is self–other distinctiveness maintained—and possibly even heightened?

We are just at the beginning of a long journey in search of answers to these and other questions about the nature of interpersonal sensitivity. The contributions to this special issue have helped get us off to a good start. They have also helped us to see some of the conceptual and methodological thickets that lie ahead. The journey will not be easy, but it promises to be exciting. Indeed, the contributions show how exciting it already is.

Interpersonal sensitivity, and social interaction more broadly, involve a complex interplay of personality and situational factors. One of the challenges for a social neuroscience approach to interpersonal sensitivity is the difficulty of taking into account situational variables. To provide interpretable data, neuroscience experiments require intra-individual comparisons and repeated-measures designs. To be financially feasible, they require small samples. These conditions limit opportunities to study the effects of potentially important situational variables.

This is but one example of the perennial challenge objective science faces in the attempt to understand human subjectivity in all its richness and complexity.

REFERENCES

Batson, C. D. (1990). How social an animal? *American Psychologist*, *45*, 336–346.

Bjorklund, D. F., & Bering, J. M. (2003). Big brains, slow development and social complexity: The development and evolutionary origins of social cognition. In M. Brune, H. Ribbert, & W. Schiedfenhovel (Eds.), *The social brain: Evolution and pathology* (pp. 113–151). Hoboken, NJ: Wiley.

Decety, J., & Keenan, J. (2006). Social neuroscience: A new journal. *Social Neuroscience*, *1*, 1–4.

Jackson, P. L., Rainville, P., & Decety, J. (2006). To what extent do we share the pain of others? Insight from the neural bases of pain empathy. *Pain*, *125*, 5–9.

Sommerville, J. A., & Decety, J. (2006). Weaving the fabric of social interaction: Articulating developmental psychology and cognitive neuroscience in the domain of motor cognition. *Psychonomic Bulletin & Review*, *13*(2), 179–200.

SOCIAL NEUROSCIENCE, 2007, 2 (3–4), 158–166

Psychology Press
Taylor & Francis Group

Interference effect of observed human movement on action is due to velocity profile of biological motion

James Kilner
University College London, London, UK

Antonia F. de C. Hamilton
Dartmouth College, Hanover, New Hampshire, USA

Sarah-Jayne Blakemore
University College London, London, UK

It has previously been shown that observing an action made by a human, but not by a robot, interferes with executed actions (Kilner, Paulignan, & Blakemore, 2003). Here, we investigated what aspect of human movement causes this interference effect. Subjects made arm movements while observing a video of either a human making an arm movement or a ball moving across the screen. Both human and ball videos contained either biological (minimum jerk) or non-biological (constant velocity) movements. The executed and observed arm movements were either congruent (same direction) or incongruent (tangential direction) with each other. The results showed that observed movements are processed differently according to whether they are made by a human or a ball. For the ball videos, both biological and non-biological incongruent movements interfered with executed arm movements. In contrast, for the human videos, the velocity profile of the movement was the critical factor: only incongruent, biological human movements interfered with executed arm movements. We propose that the interference effect could be due either to the information the brain has about different types of movement stimuli or to the impact of prior experience with different types of form and motion.

INTRODUCTION

The idea that actions are intrinsically linked to perception (James, 1890) has become increasingly accepted recently (Jeannerod, 1994; Prinz, 1997). This is in part due to the discovery of mirror neurons in monkey ventral premotor cortex. These neurons discharge both when the monkey executes specific hand movements and when it observes another monkey or human executing the same movements (Gallese, Fadiga, Fogassi, & Rizzolatti, 1996; Rizzolatti, Fadiga, Fogassi, & Gallese, 1996a; Rizzolatti, Fogassi, & Gallese, 2001). There is a large body of evidence that in humans several brain regions are activated both during action generation and during observation of others' actions (Decety et al., 1997; Hari et al., 1998; Rizzolatti et al., 1996b). In some brain regions the overlap between action observation and action execution is highly specific. Action observation activates premotor cortex in a somatotopic manner—watching mouth, hand,

Correspondence should be addressed to: Sarah-Jayne Blakemore, Institute of Cognitive Neuroscience, University College London, 17 Queen Square, London WC1N 3AR, UK. E-mail: E: s.blakemore@ucl.ac.uk

This research was supported by the Royal Society, UK, and the Wellcome Trust, UK.

We are grateful to Uta Frith and Daniel Wolpert for advice on experimental design. We are also grateful to Daniel Wolpert for the use of the high resolution video camera, and to James Ingram for help with programming.

www.psypress.com/socialneuroscience DOI:10.1080/17470910701428190

and foot movements activates the same functionally specific regions of premotor cortex as making those respective movements (Buccino et al., 2001).

Observing a movement also has measurable consequences on the peripheral motor system. During action observation there is a significant increase in the motor-evoked potentials from the hand muscles that would be used if making such a movement (Fadiga, Fogassi, Pavesi, & Rizzolatti, 1995). In addition, reaction times to initiate a finger movement are significantly slowed down after the visual presentation of photographs of different finger movements (Brass, Bekkering, & Prinz, 2001). This suggests that, during observation of action, the specific neural networks subserving that particular movement are already tuned for action (Jeannerod, 1994; Prinz, 1997).

Indeed, it has been shown that observing someone else make a movement has a measurable interference effect on qualitatively different simultaneously executed movements (Kilner, Paulignan, & Blakemore, 2003). In this experiment, subjects made sinusoidal movements with their right arm at the same time as observing movements that were either congruent (same direction) or incongruent (tangential direction) with their own movements. Variance in arm movement was used as an index of interference to the executed arm movement. Compared with the baseline condition in which subjects made arm movements without watching anything, there was significantly increased variance in subjects' arm movements when they observed another human making arm movements that were incongruent with their own. In contrast, variance was not increased when the observed incongruent movement was made by a robotic arm. Thus, the interference effect was not simply due to increased attentional demands or task complexity or any other artifactual effect of observing an incongruent movement. These results suggest that observed movements are processed differently according to whether they are made by a human or a robot.

However, it is not known what aspect of human movement, which is not present in robotic movement, causes the interference effect. There are several differences between humans and robots. First, human movements typically have a particular non-linear velocity profile. Point to point movements of the human arm tend to follow a minimum jerk (MJ) trajectory (Hogan, 1984), in which the movement starts slowly, accelerates smoothly to a peak velocity near the midpoint and then decelerates slowly. This results in a smooth, bell-shaped velocity profile (Abend, Bizzi, & Morasso, 1982), where mathematically the derivative of acceleration (jerk) is minimized over the movement. The bell-shaped velocity profile of human movements contrasts with the constant velocity profile of physical objects (such as the robot used in the above experiment) governed by Newton's laws. Second, humans do not move in perfectly straight lines, rather the trajectory of the hand is constrained by the dynamics of the arm and deviates in a systematic fashion from a linear path (Abend et al., 1982). Third, humans have features such as a face and limbs. The robot used in the Kilner et al. (2003) experiment was non-humanoid, and had no face-like structure, although in a very general sense it shared some resemblance to a human form since it comprised a main "trunk" with an attached "arm" that could move in three dimensions.

The aim of the current study was to investigate whether the interference effect of observed human action on executed action was specific to the observation of a biological motion trajectory, to the observation of a human form, or to some combination of these factors. The interference effect found in Kilner et al. (2003) is likely to be mediated by the human mirror neuron system (MNS), but the question of whether the human MNS responds to the behavior of non-human actors is currently controversial. Some studies (Tai, Scherfler, Brooks, Sawamoto, & Castiello, 2004) suggest that responses in premotor cortex to observed motion is specific to human action and this area does not respond to robotic movements. Other data suggest equivalent neural responses to both observed human action with a natural MJ trajectory and robotic action with a CV trajectory (Gazzola, Rizzolatti, Wicker, & Keysers, 2007). In the current study, we used a behavioral measure of motor interference to assess the impact of different types of observed action on the MNS.

Specifically, we investigated whether it was the form or movement type that caused the interference effect found in our previous study (Kilner et al., 2003). To distinguish between movement velocity profile and features of the moving effector, we created video stimuli and manipulated two aspects of the movement: (1) effector type, whether the stimulus was human or a ball; and (2) movement type, whether the stimulus moved with minimum jerk (MJ) or constant velocity (CV) profile. The velocity profile of the observed movement was manipulated by using a high-speed video camera to film a human arm movement and then digitally altering the video, frame-by-frame, to create a video of an arm movement with constant velocity. Therefore, there were two types of human movement video: keeping all other factors constant, one video type showed normal, biological (MJ) arm movement and the other showed the same human arm moving with non-biological (CV) motion. In addition, in another set of videos, we replaced the human with a ball, producing two further videos: a ball moving with biological motion (MJ), and a ball moving with CV. Subjects made arm movements while observing the above videos of a human or a ball moving either in the same direction (congruent) or tangential (incongruent) to the subject's own movements.

Based on our previous study, it was predicted that interference would occur to an executed movement when subjects observed another human making incongruent MJ movements. Since human CV movement is similar to the robot's movement in our previous study (Kilner et al., 2003), we did not expect to see interference in the condition in which subjects observed a human making incongruent CV movements. Predictions for the ball condition were more complex. If the interference effect is driven solely by the perception of incongruent MJ movements (regardless of form), we would also expect interference from the MJ ball but not from the CV ball. In contrast, if the interference effect is specific to the observation of a human form with MJ movement, we would not expect to see any interference from the ball. Finally, if the interference effect is modulated by the type of information present in the ball stimulus, or by prior experience of how balls tend to move in the world, interference could be found in both incongruent ball conditions. The results of this study will shed light on these possibilities.

METHODS

Subjects

Fifteen healthy, right-handed undergraduates and postgraduates from UCL took part in the study. Informed consent was obtained from each subject and the study was approved by the local ethics committee.

Experimental design

We generated video stimuli of humans and objects (a ball) with constant velocity (CV) and biological (MJ) motion by filming an actor performing several horizontal and vertical forearm movements with a high-speed digital video camera (Photron Europe, Buckinghamshire, UK, www.photron.com) at 250 frames and a resolution of 512 × 480 pixels. A single characteristic movement example of a horizontal movement and a vertical movement, each lasting 800 ms (200 frames) was selected and the location of the fingertip in each frame was recorded. To generate stimuli depicting the actor moving with a CV profile, 20 frames were selected from each set of 200, such that the fingertip maintained a CV over the frames. These 20 frames were repeated in reverse order to generate a 40 frame video clip of a complete horizontal or vertical movement cycle. To create stimuli depicting the actor moving with a MJ velocity profile, every tenth frame from the original high-speed movies was selected and assembled first forward and then in reverse order, maintaining the natural velocity profile of the actor's movement over the 40-frame clip.

Ball stimuli were created by generating a set of images with a blue background and a white disc at the location of the fingertip in each human movie. Thus each ball clip contained identical dynamic information to that in each human clip; the only difference between the human and ball videos was the form of the moving object. In all the clips, a complete cycle of movement took 1.6 seconds (40 frames).

There were nine video conditions, eight of which formed a 2 × 2 × 2 factorial design (see Figure 1). The factors were: (1) Observed movement velocity (MJ or CV); (2) Observed effector (dot or human); and (3) Congruency between observed and executed movement (same or tangential direction). In addition to these video conditions, there was a baseline condition in which a fixation cross on a blue screen was presented to the subject.

The videos were rear-projected onto a flat screen (width 100 cm × height 80 cm) positioned 50 cm in front of the subject who was sitting on a chair. While watching the videos each subject was

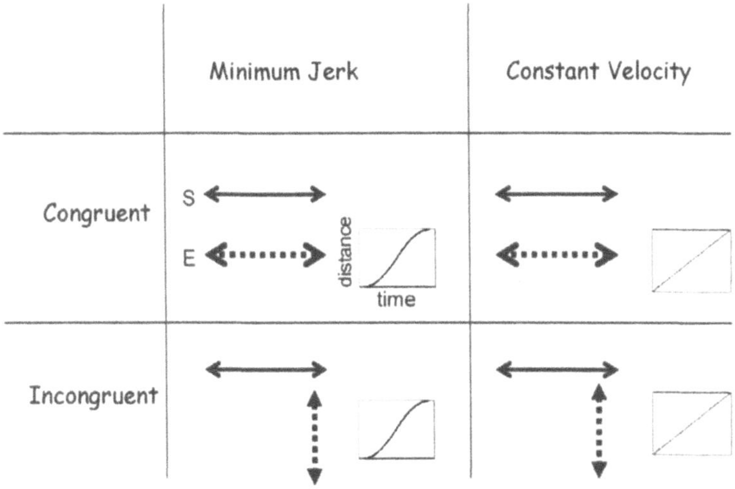

Figure 1. Experimental design. Subjects (S in left upper cell), who were seated, made sinusoidal movements with their right forearm at the same time as watching videos of movements of an effector (E in left upper cell). The observed effector was either a human or a ball. Observed movements were either minimum jerk (with a curved velocity profile as shown in the graphs in the left-hand column) or constant velocity (with a linear velocity profile as shown in the graphs in the right-hand column). Observed movements were either in the same direction (congruent) or tangential direction (incongruent) from the subject's executed movements. In addition, there were two baseline conditions in which the subject moved their forearm either horizontally or vertically while watching a fixation cross in the centre of the screen.

instructed to make movements (either horizontal or vertical) of their right forearm at a rate of 1.6 Hz. The subject was instructed to watch the index finger of the human, or the ball, or the fixation cross, in the videos. No other instruction was given. Before the experiment, subjects practised the movement until they were proficient at producing the desired forearm movements at the correct pace and amplitude. This took under one minute for each subject.

Each participant completed 5 trials for each of the nine conditions above while moving his or her arm in a horizontal direction, and 5 trials for each condition while moving his or her arm vertically. Trial order was fully randomized. Each trial began with a written instruction ("Move horizontally starting from the left" or "Move vertically starting from the top") on a black screen. The screen color then changed to blue and the computer gave a sequence of alternating low and high beeps at 800 ms intervals for three cycles. The participant was instructed to move his or her arm in time with the beeps, to reach the right, or bottom, position on the low beep and the left, or top, position on the high beeps. This ensured that when the video clip started after three cycles, the participant's movement was already synchronous with the visual image. After the sixth beep, the screen immediately displayed the video clip of a human or ball moving at 1.6 Hz, and the clip was repeated 10 times. Thus, participants performed 13 complete movement cycles on each trial, three cycles with beeps and no video and 10 with the video stimulus and no beeps. Breaks were provided between trials to prevent fatigue.

Data acquisition

The position of the participant's hand during every trial was recorded using a Flock of Birds (http://www.ascension-tech.com/products/flockofbirds.php) magnetic motion tracking system, which returns the position and orientation of markers in space at 100 Hz. A single marker was attached to the participant's fingertip, and the magnetic tracking unit was placed below the video screen. A customized C program was used to play the sounds and videos and to record the marker position in 3D space.

Data analysis

Data were low-pass filtered at 10 Hz using a 2nd order Butterworth filter. We analyzed only the final 12.8 seconds (8 cycles) of each trial for two reasons. First, we omitted the first two cycles to avoid the data being affected by the "surprise" of the visual stimulus appearing on the screen. Second, this allowed time for the subject's arm movement to align with the visual stimulus. Although subjects were not instructed to align their movement with the visual stimulus, they naturally did this and it took up to 2 cycles to achieve. Data were segmented offline into movements from right to left and those from left to right for horizontal conditions and for those from up to down and from down to up for the vertical movements (Figure 2). Therefore there was a maximum of 40 segmented movements per condition per subject. For each segmented movement the variance in the movement orthogonal to the dominant dimension of movement, and in the dominant dimension of the incongruent movement, was calculated. Thus if the subject made a movement from left to right, the X dimension was the dominant movement dimension and the variance was calculated for the movements in the Z dimension, and vice versa. The mean of these variances was calculated across all trials for each condition. The mean variance for each participant in each trial was analyzed using repeated measures ANOVAs and paired *t*-tests as described below.

RESULTS

An initial repeated measures $2 \times 2 \times 2 \times 2$ ANOVA was carried out on the factors performed Movement Direction (horizontal vs. vertical), Movement Congruency (congruent vs. incongruent), Observed Actor (human or ball) and Movement Type (MJ or CV). The results of this ANOVA revealed a significant main effect of movement direction, $F(1, 14) = 17.509$, $p = .001$, with a significantly greater variance for vertical than horizontal movements. There was also a significant main effect of movement congruency, $F(1, 14) = 12.827$, $p = .003$, with greater variance when observing incongruent movements than congruent movements. There was a significant interaction between movement direction and movement congruency, $F(1, 14) = 5.958$, $p = .029$ (see Figure 2A). There was a significant main effect of observed effector, $F(1, 14) = 9.205$, $p = .009$, and a significant interaction between movement direction and observed effector, $F(1, 14) = 6.266$, $p = .025$ (see Figure 2B). In addition, there was a significant interaction between observed effector and movement type, $F(1, 14) = 5.919$, $p = .029$ (see Figure 2C). No other main effects or interactions were significant.

As there was a significant interaction between the movement direction and movement congruency, subsequent analyses were performed to investigate effects for horizontal and vertical movements separately. Furthermore, because there was a significant interaction between observed actor and movement type, human and ball conditions were analyzed separately. Therefore, all subsequent analyses consisted of 2×2 ANOVAs where the factors were movement congruency and movement type for horizontal and vertical movements, and for observing a human or a ball. There were no significant main effects or interactions ($p > .05$) when the subjects were making horizontal movements. Therefore, the following analyses are for vertical executed movements only.

When subjects made vertical movements and observed a *human* there was a trend towards a main effect of congruency although this did not reach significance, $F(1, 14) = 3.945$, $p = .067$.

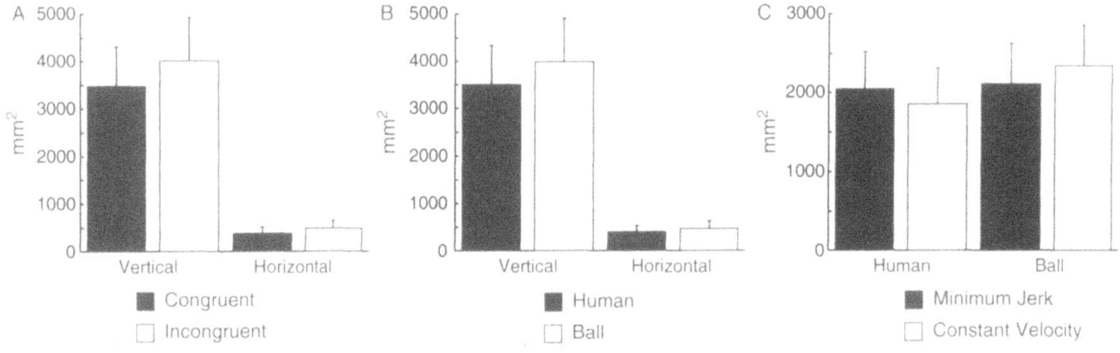

Figure 2. Interactions. The three panels show the mean variances in square millimeters tangential to the axis of movement. Panel A illustrates the interactions between movement direction and movement congruency. Panel B illustrates the interaction between movement direction and observed effect. Panel C illustrates the significant interaction between observed effector and movement type. In all panels the error bars represent the standard error.

There was a significant interaction between movement congruency and movement type, $F(1, 14) = 7.366$, $p = .017$ (see Figure 3). This interaction was driven by an increase in variance when observing an incongruent MJ relative to a congruent MJ human movement, $t(14) = 3.2551$, $p < .05$. Variance in the incongruent MJ condition was also significantly greater than variance in the baseline condition, $t(14) = 2.2960$, $p < .05$, paired t-test, two tailed.

When subjects made vertical movement observing a *ball* there was only a main effect of congruency, $F(1, 14) = 4.746$, $p = .047$. There was no significant interaction between congruency and movement type, $F(1, 14) = 0.203$, $p = .659$. There was significantly greater variance for both MJ and CV incongruent conditions compared with baseline (see Figure 3); $t(14) = 2.9601$, $p < .05$; $t(14) = 3.6152$, $p < .05$, respectively.

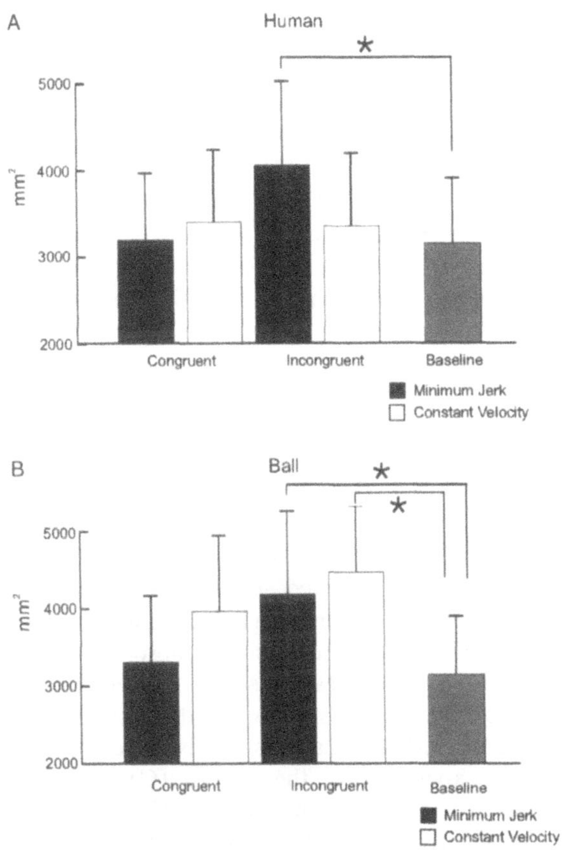

Figure 3. Vertical movements. The two panels show the mean variances in square millimeters tangential to the axis of movement for the vertical movements only. Panel A shows the mean variance for each of the conditions while observing a human. Panel B shows the mean variance for each of the conditions while observing a ball. The gray bar in both panels shows the mean variance during the baseline condition. In both panels the error bars represent the standard error.

DISCUSSION

The present results demonstrate that observing incongruent moving stimuli presented on a video display (rather than live) can have a measurable interference effect on simultaneously executed actions. The precise nature of this effect depends on type of movement present in the video stimuli, with interference found for observation of a human making biological (minimum jerk) movements but not for a human making non-biological (constant velocity) movements. In contrast, when the stimulus was more abstract (a moving ball), interference was found for both velocity profiles.

The results replicate our previous study demonstrating that observing arm movements made by a human has a significant interference effect on ongoing executed movements when the observed and executed movements are incongruent. As we found in our previous study, there was no significant interference to executed movements when the observed human movement was congruent with executed movements. However, these new data extend our previous study by showing that the interference effect is only present when the incongruent observed human movements are biological (MJ; Figure 3A). There was no significant interference to ongoing executed movements if the observed incongruent human movements were non-biological (CV). This condition is similar to the condition in our previous study in which subjects observed a robot (with a non-MJ velocity profile) making incongruent movements. In this condition there was no interference effect on executed movement. The finding that observing a human making incongruent CV movements had no significant effect on executed movement demonstrates that the interference effect found for human incongruent MJ movements is not simply due to increased attentional demands, task complexity or any other artifactual effect of observing an incongruent movement. Rather, it suggests that the brain processes MJ and CV movements differently when they are made by a human. Taken in isolation these results suggest that the interference effect when observing a human making incongruent movement is due to the biological nature of human movement, rather than to the form of the human body.

The question then remains, what happens when we change the form of the observed movement? Based on the results when observing a human we predicted that, for the ball, there would either be the same interference effect, i.e., the interference effect is form-invariant, or no interference effect, i.e., the presence of a human form is essential to illicit an interference effect. Neither of these predictions was realized. Instead, observing a ball interfered with executed movements if the observed ball movements were incongruent with executed movement direction, irrespective of whether the observed ball movements were biological or not. Thus, the finding that only MJ movements interfere with executed movement seems to be specific to observed human movement.

The reason for the difference in the interference effect between observing a ball and observing a human requires further investigation. Here we suggest two alternative explanations, based on knowledge of the different parts of the MNS that could mediate the interference effect. The first explanation focuses on the perception of biological motion stimuli in the superior temporal sulcus (STS), which is connected to both the premotor and inferior parietal regions of the MNS (Catani, Jones, & Ffytche, 2005). The second explanation focuses on the overlap of perceptual and motor processing of human actions in the parietal and premotor parts of the MNS (Gallese et al., 1996; Rizzolatti et al., 1996a, 2001).

There is extensive evidence that parts of the visual system discriminate between different types of motion stimuli, in particular between biological and non-biological motion. Three-month-old babies can discriminate between displays of moving dots that have biological motion and displays in which the same dots move randomly (Bertenthal, 1993). This ability has been associated with the STS, which responds selectively to biological motion in monkeys (Oram & Perrett, 1994) and in humans (Allison, Puce, & McCarthy, 2000; Frith & Frith, 1999). The STS may provide visual input to the MNS (Catani et al., 2005), so any modulation in neuronal activity in the STS could have a direct effect on the engagement of the MNS during action observation. The ventral bank of the posterior STS is activated by displays of moving dots showing a walking person (biological motion) compared with displays in which the same dots move at equal velocity in the same direction (Grossman et al., 2000) or have rigid object motion (Grèzes et al., 2001).

These data suggest that the detection of biological motion is hardwired in the human brain at an early age, which would lead to an automatic classification of observed motion as biological or non-biological. Our results for human videos and from our previous study (Kilner et al., 2003) are in line with the proposal that observed movements are classified as biological (MJ) or non-biological (CV) and treated differently accordingly. The next question is why such a distinction is only present for human movements, and not for ball movements, in the current study. One possibility is that there is not sufficient information in the ball stimulus for the biological and non-biological movements to activate differentially neurons in the STS. The ball stimuli in the current study are very different from those used in point-light displays. Point-light displays generally depict movements of whole bodies comprised of several points of light. Information about the nature of the movement can be extracted from the relative movements of the points of light. In contrast, the ball stimuli in the current study show just one dot moving across a screen and therefore there is no information about relative movements. Such a hypothesis predicts that there would not be differential STS activity when observing the ball with MJ or CV. In this case the CV ball would be processed in the same way as the MJ ball, which would explain why a similar interference effect was observed to both incongruent ball stimuli. Note that the relative spatial information is present when observing the human, which would result in a differential interference effect to incongruent CV and MJ human movements.

The second possible explanation is based on the proposal that the interference effect is mediated by the premotor or parietal regions of the MNS. The human MNS responds to both executed and observed actions (Rizzolatti et al., 2001), and therefore might mediate the interference between executed and observed actions found in the present study. Recent studies suggest that the response of the human MNS to observed actions is strongly modulated by prior experience of the actions (Calvo-Merino, Glaser, Grèzes, Passingham, & Haggard, 2005; Cross, Hamilton, & Grafton, 2006). In everyday life, we have extensive experience of seeing humans move with a MJ trajectory, but we never see humans move with a CV trajectory. Thus, the MNS is likely to respond, and interference be obtained, for human MJ but not for human CV stimuli. Humanoid figures that move with a MJ trajectory are likely to be familiar enough to be simulated and would cause interference, as found by Oztop, Franklin, Chaminade, and Cheng (2005). In contrast, the MNS is not likely to respond to observation of an industrial robot moving with an unfamiliar CV trajectory, as in Kilner et al. (2003), and interference was not seen in this condition.

This MNS explanation may also be able to account for the ball results of the current study. The premotor regions that form part of the human MNS and which respond to observed actions can also respond to the observation of sequenced, non-biological stimuli (Schubotz & von Cramon, 2004). Most people are familiar with balls or similar computer graphics moving with a wide variety of trajectories, including MJ and CV. It is possible that premotor regions are able to simulate, or predict, the movements of these familiar stimuli, and that this results in an interference effect for a ball regardless of the precise trajectory of the ball. In contrast, the MNS may not able to simulate the behavior of stimuli that are novel, for example some robots, and therefore such stimuli would not cause interference even when presented incongruent to the participant's own movement. A prediction of this theory is that participants with an unusual experience profile, for example, people who work with industrial robots, might also show interference effects from robots.

In the current study there were only significant effects when subjects made vertical, as opposed to horizontal, movements. In addition, the variance overall was significantly greater for vertical than horizontal effects. This is in contrast to the results of our previous study when horizontal movements were associated with more variance than vertical movements. These differences can be explained by a difference in the experimental design between the two studies. In our previous study, subjects were standing and movement in the vertical plane were made to the side of the body. In the current study subjects were seated. It is possible that, first, this stabilizes the movement in the horizontal direction resulting in lower sensitivity to the incongruency effects. Second, being seated might have prohibited a free vertical movement. Subjects typically made "J" shaped movements in all trials resulting in an increase in variance in the orthogonal dimension to that in which the intended movement should have been made. However, it should be noted that this can not explain the task-specific modulations in variance for the vertical movements. Thus, the interference effect can be seen in arm movements if subjects are seated or standing, but the direction differences should be borne in mind when designing future experiments.

Summary

In this study, we found that observed human movements interfere with incongruent executed arm movements only when they are biological (MJ). Observed non-biological (CV), incongruent human movements had no interference effect on executed movements. In contrast, observed ball movements interfere with incongruent executed arm movements whether they are MJ or CV. This effect could be due to the quantity of information the brain has to distinguish different types of motion stimuli, or the impact of prior experience with different types of form and motion. Further research is needed to discriminate between these possibilities.

REFERENCES

Abend, W., Bizzi, E., & Morasso, P. (1982). Human arm trajectory formation. *Brain, 105*, 331–348.

Allison, T., Puce, A., & McCarthy, G. (2000). Social perception from visual cues: Role of the STS region. *Trends in Cognitive Science, 4*(7), 267–278.

Bertenthal, B. I. (1993). Infants' perception of biomechanical motions: Intrinsic image and knowledge-based constraints. In C. Granrud (Ed.), *Visual perception and cognition in infancy* (pp. 175–214). Hillsdale, NJ: Lawrence Erlbaum Associates, Inc.

Brass, M., Bekkering, H., & Prinz, W. (2001). Movement observation affects movement execution in a simple response task. *Acta Psychologica (Amsterdam), 106*(1–2), 3–22.

Buccino, G., Binkofski, F., Fink, G. R., Fadiga, L., Fogassi, L., Gallese, V., et al. (2001). Action observation activates premotor and parietal areas in somatotopic manner: An fMRI study. *European Journal of Neuroscience, 13*, 400–404.

Calvo-Merino, B., Glaser, D. E., Grèzes, J., Passingham, R. E., & Haggard, P. (2005). Action observation and acquired motor skills: An fMRI study with expert dancers. *Cerebral Cortex, 15*(8), 1243–1249.

Catani, M., Jones, D. K., & Ffytche, D. H. (2005). Perisylvian language networks of the human brain. *Annals of Neurology, 57*(1), 8–16.

Cross, E. S., Hamilton, A. F., & Grafton, S. T. (2006). Building a motor simulation de novo: observation of dance by dancers. *NeuroImage, 31*(3), 1257–1267.

Decety, J., Grèzes, J., Costes, N., Perani, D., Jeannerod, M., Procyk, E., et al. (1997). Brain activity during observation of actions. Influence of action content and subject's strategy. *Brain, 120*, 1763–1777.

Fadiga, L., Fogassi, L., Pavesi, G., & Rizzolatti, G. (1995). Motor facilitation during action observation: A magnetic stimulation study. *Journal of Neurophysiology, 73*, 2608–2611.

Frith, C. D., & Frith, U. (1999). Interacting minds—A biological basis. *Science, 286*(5445), 1692–1695.

Gallese, V., Fadiga, L., Fogassi, L., & Rizzolatti, G. (1996). Action recognition in the premotor cortex. *Brain, 119*, 593–609.

Gazzola, V., Rizzolatti, G., Wicker, B., & Keysers, C. (2007). The anthropomorphic brain: the mirror neuron system responds to human and robotic actions. *NeuroImage, 31*, 1674–1684.

Grèzes, J., Fonlupt, P., Bertenthal, B., Delon-Martin, C., Segebarth, C., & Decety, J. (2001). Does perception of biological motion rely on specific brain regions? *NeuroImage, 13*, 775–785.

Grossman, E., Donnelly, M., Price, R., Pickens, D., Morgan, V., Neighbor, G., et al. (2000). Brain areas involved in perception of biological motion. *Journal of Cognitive Neuroscience, 12*, 711–720.

Hari, R., Fross, N., Avikainen, E., Kirveskari, E., Salenius, S., & Rizzolatti, G. (1998). Activation of human primary motor cortex during action observation: A neuromagnetic study. *Proceedings of the National Academy of Sciences of the USA, 95*, 15061–15065.

Hogan, N. (1984). An organizing principle for a class of voluntary movements. *Journal of Neuroscience, 4*, 2745–2754.

James, W. (1890). *Principles of psychology.* New York: Holt.

Jeannerod, M. (1994). The representing brain—Neural correlates of motor intention and imagery. *Behavioral and Brain Sciences, 17*, 187–202.

Kilner, J. M., Paulignan, Y., & Blakemore, S.-J. (2003). An interference effect of observed biological movement on action. *Current Biology, 13*(6), 522–525.

Oram, M. W., & Perrett, D. I. (1994). Responses of anterior superior temporal polysensory (STPa) neurons to biological motion stimuli. *Journal of Cognitive Neuroscience, 6*, 99–116.

Oztop, E., Franklin, D. W., Chaminade, T., & Cheng, G. (2005). Human–humanoid interaction: Is a humanoid robot perceived as a human? *International Journal of Humanoid Robotics, 2*(4), 537–559.

Prinz, W. (1997). Perception and action planning. *European Journal of Cognitive Psychology, 9*, 129–154.

Rizzolatti, G., Fadiga, L., Fogassi, L., & Gallese, V. (1996a). Premotor cortex and the recognition of motor actions. *Brain Research Cognitive Brain Research, 3*, 131–141.

Rizzolatti, G., Fadiga, L., Matelli, M., Bettinardi, V., Paulesu, E., Perani, D., et al. (1996b). Localization of grasp representations in humans by PET: 1. Observation versus execution. *Experimental Brain Research, 111*(2), 246–252.

Rizzolatti, G., Fogassi, L., & Gallese, V. (2001). Neurophysiological mechanisms underlying the understanding and imitation of action. *Nature Reviews Neuroscience, 2*(9), 661–670.

Schubotz, R. I, & von Cramon, D. Y. (2004). Sequences of abstract nonbiological stimuli share ventral premotor cortex with action observation and imagery. *Journal of Neuroscience, 24*(24), 5467–5474.

Tai, Y. F., Scherfler, C., Brooks, D. J., Sawamoto, N., & Castiello, U. (2004). The human premotor cortex is "mirror" only for biological actions. *Current Biology, 14*(2), 117–120.

SOCIAL NEUROSCIENCE, 2007, 2 (3–4), 167–178
Psychology Press
Taylor & Francis Group

Face to face: Blocking facial mimicry can selectively impair recognition of emotional expressions

Lindsay M. Oberman, Piotr Winkielman, and Vilayanur S. Ramachandran
University of California San Diego, La Jolla, California, USA

People spontaneously mimic a variety of behaviors, including emotional facial expressions. Embodied cognition theories suggest that mimicry reflects internal simulation of perceived emotion in order to facilitate its understanding. If so, blocking facial mimicry should impair recognition of expressions, especially of emotions that are simulated using facial musculature. The current research tested this hypothesis using four expressions (happy, disgust, fear, and sad) and two mimicry-interfering manipulations (1) biting on a pen and (2) chewing gum, as well as two control conditions. Experiment 1 used electromyography over cheek, mouth, and nose regions. The bite manipulation consistently activated assessed muscles, whereas the chew manipulation activated muscles only intermittently. Further, expressing happiness generated most facial action. Experiment 2 found that the bite manipulation interfered most with recognition of happiness. These findings suggest that facial mimicry differentially contributes to recognition of specific facial expressions, thus allowing for more refined predictions from embodied cognition theories.

In 1890, James observed that "Every representation of a movement awakens in some degree the actual movement which is its object." Since then, a large number of behavioral studies found that observers tend to overtly and covertly mimic behavior of those around them (Condon & Ogston, 1967; Kendon, 1970). Specifically, people tend to mimic others' gestures and body postures (Chartrand & Bargh, 1999), facial expressions (Dimberg, 1982; Dimberg, Thunberg, & Elmehed, 2000; Wallbott, 1991), tone of voice and pronunciation patterns (Goldinger, 1998; Neumann & Strack, 2000), and even breathing rates (McFarland, 2001; Paccalin & Jeannerod, 2000).

Early explanations of mimicry saw it as a simple by-product of previously established S–R links between perceiving and performing an action (Lipps, 1907). Contemporary research, however, provides evidence that in addition to action observation eliciting concurrent performance of that same action, the performance of an action influences the concurrent perception of that action. Specifically, concurrent performance of a compatible action tends to facilitate recognition whereas performance of incompatible actions tends to interfere with it (Reed & Farah, 1995; Tucker & Ellis, 1998). The findings that action execution and action observation reciprocally modulate each other suggest that mimicry occurs because action observation and action execution share a common representational code (Prinz & Hommel, 2002) and neural substrates (Grezes & Decety, 2001). More generally, the grounding of perceptual and conceptual understanding in the mechanisms underlying action in the world is the central tenet of embodied cognition theories (Niedenthal, Barsalou, Winkielman, Ric, & Krauth Gruber, 2005).

One of the most studied examples of the interaction between action observation and

Correspondence should be addressed to: Lindsay M. Oberman, 9500 Gilman Drive, La Jolla, CA 92093–0109, USA. E-mail: loberman@ucsd.edu

www.psypress.com/socialneuroscience DOI:10.1080/17470910701391943

action execution is mimicry of emotional facial expressions. Facial mimicry has been explained in two ways. Some researchers propose that mimicry reflects the process of emotional contagion (Hatfield, Cacioppo, & Rapson, 1994; Laird et al., 1994). That is, observation of other's emotional expressions first triggers the corresponding emotion in the observer which then elicits the same facial expression (Lundqvist & Dimberg, 1995). However, some researchers propose that facial mimicry reflects an internal simulation of the perceived facial expression in order to facilitate understanding of others' emotion (Atkinson & Adolphs, 2005; Niedenthal, Brauer, Halberstadt, & Innes-Ker, 2001; Wallbott, 1991). There are now several models explaining how "embodied simulation" could help in emotion perception and understanding (see Goldman & Sripada, 2005, for a review). The assumption underlying those models is that the internal re-enactment provides useful information either through generation of peripheral feedback (e.g., facial muscles) or engagement of an "as–if" loop in the somatosensory and motor cortices (e.g., face representation). In short, resonating with the topic in this special issue, mimicry allows us to increase our "interpersonal sensitivity" through "entering others' worlds."

Recently, Niedenthal and colleagues (2001) provided behavioral evidence for the role of facial mimicry in perception of emotional facial expressions. Participants were asked to identify the point at which a morphed face changed from happy to sad and vice versa. During this task, some participants could freely move their facial muscles, but other participants were prevented from doing so by a pen held sideways in their mouth. Compared to the free movement condition, participants in the pen condition detected the change in expression later in both directions, thus supporting the role of facial mimicry in recognition of facial expressions. However, this study leaves several questions open. It is unclear whether blocking mimicry impairs recognition of any facial expression or is restricted to select expressions (happy, sad, both?). This is interesting as concurrent performance of facial movements may be more critical for recognition of some expressions rather than others, as we discuss shortly. Further, it is unclear whether mimicry is only important for detecting a dynamic transition between two different expressions, or also is involved in perception of single facial expressions. Finally, we know little about what specific muscles are influenced by the various manipulations, and whether these muscles are involved in the actual generation of facial expressions. These questions, addressed in our current work, are important for a more precise understanding of the role of somatosensory and motor resources in emotion perception.

Neural substrates of recognition

Embodied theories of cognition and emotion date back to philosophers Lotze (1852) and James (1890) and have also been occasionally proposed in psychology (e.g., Zajonc & Markus, 1984). However, the recent revival of such theories has been inspired by several recent developments in neuroscience. One is the discovery of mirror neurons. While studying the premotor cortex in the macaque, Rizzolatti and colleagues came across a system of neurons in area F5 that responded not only when the monkey performed an action, but also when the monkey watched the researcher perform a similar action (Di Pellegrino, Fadiga, Fogassi, Gallese, & Rizzolatti, 1992). The team named this system of neurons the "mirror neuron system" (MNS) because it appeared that the observed action was reflected or internally simulated within the monkey's own motor system. Later studies by this same group further characterized this system as being selective for animate actions (Gallese, Fadiga, Fogassi, & Rizzolatti, 1996) and is somatotopically distributed based on the effectors used to perform the action (Buccino et al., 2001; Goldenberg & Karnath, 2006). In general, the MNS is assumed to form an important mechanism of shared representation (Decety & Sommerville, 2003; Oberman & Ramachandran, 2007). There is also some specific evidence for activation in the human equivalent of the mirror neuron area (Brodmann area 44, in the premotor cortex) when participants imitate other people's facial expressions (Carr, Iacoboni, Dubeau, Mazziotta, & Lenzi, 2003).

A related discovery contributing to the resurgence of interest in the embodied theories of cognition is the observation that motor and somatosensory areas are involved in what has been assumed to be purely perceptual tasks, such as stimulus recognition. In the area of recognition of emotional facial expression, the contribution of somatosensory cortices was examined in a pivotal study by Adolphs and colleagues (Adolphs, Damasio, Tranel, Cooper, & Damasio, 2000). In that study, 108 focal brain lesion patients and 30 normal control participants were asked to perform three visual emotion recognition tasks. In the first task, participants were asked to rate the intensity of basic emotional facial expressions. In the second task, participants were asked to match a facial expression with the name of the emotion it was meant to convey. The final task required participants to sort facial expressions into emotional categories. Though each task identified a slightly different group of regions, damage to right somatosensory-related cortices impaired performance in all of the three tasks. Interestingly, the study did not find any differential effects of right somatosensory damage across expression type, but this could be because the study assessed damage to fairly broad regions that possibly support a variety of simulation mechanisms, including re-enactment of facial, postural, and visceral aspects of emotional experience.

Current studies

Interestingly, to this date, only few psychological and neuroscientific studies have examined the specificity of the embodied systems in recognition of emotion and action. For example, the behavioral studies using manipulations that block mimicry did not assess which specific muscles were influenced and to what extent, and how this involvement related to recognition deficits for specific expressions (Niedenthal et al., 2001). Even the neuroscience studies investigating the somatotopic specificity of mirror neurons have worked on a fairly gross level (e.g., leg, hand, face in Buccino et al., 2001, or hand and finger in Goldenberg & Karnath, 2006).

However, if embodied simulation indeed plays a critical role in recognition of facial expressions, then one should be able to formulate more focused hypotheses. One prediction is that blocking the ability to engage in facial mimicry should differentially impair recognition of different emotions from the face. Note that, most generally, different emotions engage different peripheral and central mechanisms, with specific engagement varying across stimuli and contexts (Barrett, Mesquita, Ochsner, & Gross, 2007). More specifically, emotions differ in the degree to which they are expressed in the face (Ekman, 2004). For example, happiness is associated with distinct changes in many facial regions. As a result, its simulation might be more "external" and predominately draw on the perceiver's own facial musculature. In contrast, sadness is associated with more subtle and localized facial changes. As a result, its simulation might involve more "internal" experience. Based on these considerations, we predicted that interfering with a simulation using facial muscles by creating irrelevant activation (muscular noise) should be most detrimental to recognition of expressions that involve most facial action, such as happiness, but have little effect on recognition of emotions that involve less facial actions, such as sadness.

To examine these questions, our main study, reported in Experiment 2, investigated whether experimental manipulations that block mimicry differentially impair recognition of specific facial expressions. However, in order to answer our central question, we first needed to better understand the effects of our manipulations and stimuli on facial muscles. Thus, we conducted Experiment 1 using electromyography (EMG) and tested (1) what muscles and to what extent are influenced by mimicry-blocking manipulations, and (2) what muscles and to what extent are involved in production of specific facial expressions. The results of this study allowed us to make more focused predictions for Experiment 2.

EXPERIMENT 1

We employed two manipulations that required participants to activate facial muscles in a task-irrelevant fashion and thus should interfere with mimicry. Our primary manipulation involved constant activation from having to bite on a pen with the teeth while not allowing it to touch the lips. The inspiration for this manipulation came from Niedenthal et al. (2001). However, in order to ensure muscle activation, we required participants to exert a constant active pressure on the pen whereas Niedenthal's manipulation required no muscular activity to keep the pen in the mouth. In other words, our manipulation creates an irrelevant muscular "noise" whereas Niedenthal's manipulation prevents generation of muscular "signal."[1]

[1] Readers familiar with social psychology literature should note the difference between the above methods from manipulations aimed at temporarily inducing a specific emotional expression for the purpose of changing mood. Thus, Strack, Martin, and Stepper (1988) used a pen placed in a mouth like a writing instrument to induce a temporary increase in mood (when held in teeth and causing a light smile) and a decrease in mood (when held in lips so that it causes an expression of sadness). Note also that in our lip manipulation, the pen rests on the lips horizontally, not forming any particular expression.

Our second manipulation involved strong and cyclical activation of muscles around the mouth and jaw by a chewing action. This manipulation was adopted from a study that found that chewing gum while viewing neutral facial expressions might impair participants' performance at the later memory test for those expressions (Pietromonaco, Zajonc, & Bargh, 1981). Though this study dealt with neutral, rather than emotional facial expressions, and tested for a delayed memory for identity, rather than immediate discrimination of emotion type, we found it interesting to examine the gum manipulation as it potentially offers a powerful and widespread way of influencing mimicry (Zajonc & Markus, 1984).

We also included a control condition where participants held a pen horizontally with the lips while not allowing it to touch the teeth ("lip"). This action does not activate any mouth muscles and does not prevent mimicry and was included for the purpose of controlling for any possible effects due to performance of a concurrent task.

In addition to testing effects of mimicry-blocking manipulations on activity of specific facial muscles, we also tested which facial muscles are activated when performing facial expressions. Specifically, participants were asked to produce expressions of happiness, sad, fear, and disgust using either a model face presented on a computer screen, or on their own, in response to a verbal cue. Although many previous EMG studies examined the effects of making facial expressions (Tassinary & Cacioppo, 2000), we wanted to directly compare the degree to which each expression activated specific facial muscles, relative to the effects of external face manipulations.

Method

Participants

Participants were six individuals (4 male, 2 female) recruited through the psychology department at UCSD and ranged in age from 21 to 44 years ($M = 29$, $SD = 8.5$). Subjects volunteered their time and gave informed consent for the study.

Design and procedure

The study represented a within-subject design, with the presentation order of the conditions randomized. Following attachment of the electrodes, participants were asked to perform several tasks. The tasks fell into two groups designed to assess muscular effects of (1) external face manipulations and (2) expressions manipulations.

External manipulations. Participants were asked to (1) place a pen in their mouth horizontally and hold on to it using their teeth while not allowing their lips to touch the pen ("bite"); (2) chew gum ("gum"); and (3) place a pen in their mouth horizontally and hold onto it using their lips while not allowing their teeth to touch the pen ("lip").

Expression manipulations. Participants were asked to voluntarily produce facial expressions of happiness, sadness, fear, and disgust by (1) imitating the expressions of a model presented on a computer screen, and (2) in response to a verbal cue, without a model.

Finally, we had a neutral "rest" condition where participants were asked to sit and relax their face for 10 seconds. This condition served as a baseline control.

Stimulus presentation

During the external face manipulation conditions, participants were verbally instructed to perform the requested action (chew, bite, lip). During the conditions where the participant was asked to imitate the model, stimuli were presented on a 15-inch monitor located approximately 80 cm away from the participant. Pictures were approximately 10 × 10 cm in size and came from the "Pictures of Facial Affect" set (Ekman & Friesen, 1976). During the expression-generation condition, participants were verbally instructed to make a relevant face (happy, fear, sad, and disgust).

EMG measurement and equipment

EMG was used to assess activity of muscles involved in generating facial expressions of happiness, disgust, fear, and sadness. Specifically, we monitored activity in the cheek region by recording from the zygomaticus major (muscle that raises the lip corner) and buccinator (muscle that retracts the lip corner). Based on previous work, we expected zygomaticus to be particularly activated by the expressions of happiness (Tassinary & Cacioppo, 2000). Previous work has also associated buccinator with smiles, especially posed or "fake" smiles likely to be generated upon request to make an expression (Ekman & Friesen, 1982). We also monitored the nose region (levator) as it has been implicated in the expressions of disgust (Vrana, 1993). Finally, we monitored activity in the lip region by recording from orbicularis oris—a perioral muscle responsible for a variety of mouth movements, including the widening sometimes seen in fear (Ekman, 2004). We did not expect any specific muscle activation for sadness as this emotion is generally associated with overall reduction in muscle tone.

Muscle activity was measured with pairs of adjacent silver/silver-chloride electrodes placed on the left side of the participant's face. An additional ground electrode was placed in the upper portion of the forehead. The impedances of all electrodes were reduced to less than 10 kΩ. The location of the electrodes and recording technique conformed to the standards for EMG recording (Cacioppo, Tassinary, & Fridlund, 1990). The acquisition of EMG signals was controlled by MP 150 Amplifiers and Acknowledge software package by Biopac Corporation. The signals were amplified by 2000 and filtered on-line with a low pass of 500 Hz and a high pass of 10 Hz and sampled at 2000 times per second. First the signals were integrated, rectified, and screened for movement artifacts. Second, the data were logarithmically transformed, which reduces the impact of extreme values. Third, the data were standardized (i.e., expressed as *Z*-scores) within subjects and muscle sites, which attenuates the undue impact of highly reactive individuals on group scores and allows meaningful comparisons across sites. EMG values for each condition were derived by taking the average activity during the 5-second period following the first second after presentation of the instructions or the stimulus (the first second was excluded to avoid artifacts resulting from muscle action associated with orienting).

Results

The first goal of the data analysis was to determine what facial muscles are activated by the external face manipulations (bite, chew) as compared to control conditions (lip and rest). The second goal was to verify what facial muscles are activated when participants produce different expressions (happy, fear, disgust, sad) when voluntarily imitating a model or responding to verbal instructions to make a particular face.

External face manipulations

The data, presented in the left panel (A) of Figure 1, were analyzed using a Manipulation (4) by Muscle (4) within-subjects MANOVA. This analysis revealed an overall main effect of the face manipulations, $F(3, 45) = 16$, $p < .001$. Post hoc paired *t*-tests revealed the overall activity in the bite condition was significantly greater than the lip ($p < .001$) and rest ($p < .001$) conditions. The activity in the gum condition was greater than bite ($p < .01$), rest ($p < .001$) and lip ($p < .001$). Finally, the activity in the lip condition was slightly greater than the rest condition ($p < .05$).

In addition, the above MANOVA revealed a Manipulation by Muscle interaction, $F(9, 45) = 3.48$, $p < .01$. To understand this interaction, we

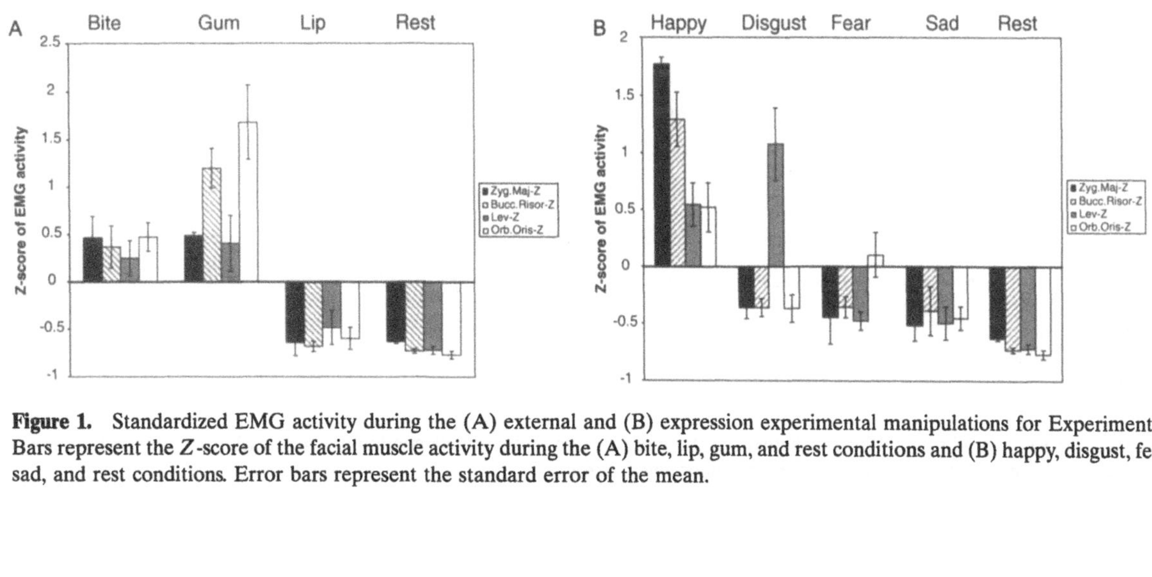

Figure 1. Standardized EMG activity during the (A) external and (B) expression experimental manipulations for Experiment 1. Bars represent the *Z*-score of the facial muscle activity during the (A) bite, lip, gum, and rest conditions and (B) happy, disgust, fear, sad, and rest conditions. Error bars represent the standard error of the mean.

compared relative activity of different muscles in each manipulation condition. In the bite condition, the increase in activity was distributed equally across all muscles, with all showing increased activity as compared to rest ($p < .05$) and no differences from each other. In the gum condition all muscles showed greater activity as compared to rest ($p < .05$). However, specific muscles showed greater activity than others. The orbicularis oris (lip muscle) and buccinator/risorius (cheek retractor) were enhanced when compared to zygomaticus major (cheek raiser) and levator (nose), $p < .01$. Finally, in the lip and rest conditions, there were no differences in activation across muscles.

Finally, using unfiltered data, we analyzed the temporal pattern of muscle activity caused by the gum and bite manipulations. Fast Fourier frequency spectrum analysis revealed that chewing gum generated low frequency activity, with average peak at 1.1 Hz ($SD = 0.33$). In the window between 0.5 and 1.5 Hz, the mean activity of the orbicularis oris muscle was significantly greater in the chew than bite condition ($p < .05$). The difference between the temporal patterns is illustrated in Figure 2.

In short, these data clearly suggest that facial muscles involved in producing emotional expressions are strongly activated by the bite and gum manipulations. The bite manipulation results in constantly elevated activity across muscles, whereas the gum manipulation causes most activation of muscles around the lips. Interestingly, this activity is not constant but occurs about every second. This finding is important because, as we discuss later, this pattern may allow for temporary engagement of muscles in expression-specific mimicry.

Expression manipulations

Preliminary analyses revealed no relevant effects of performing the expression to a model face compared to producing expression based on a verbal command. Therefore, subsequent analyses collapse across this manipulation.

The data, presented in the right panel (B) of Figure 1, were analyzed using Emotion (5) by Muscle (4) within-subjects MANOVA. This analysis revealed an overall main effect of condition on muscle activity, $F(4, 60) = 33.86$, $p < .001$. Post hoc paired *t*-tests revealed that overall level of activity in the happy condition was greater than disgust, sad, fear, and rest conditions (all $ps < .001$). Disgust condition resulted in overall more EMG activity compared to sad ($p < .01$) and rest ($p < .001$), but not fear. The fear condition was higher than rest ($p < .001$), but not sad. The sad condition was slightly higher than rest ($p < .01$).

In addition, there was an emotion by muscle interaction, $F(12, 60) = 11.42$, $p < .001$. The happy condition enhanced activation of the zygomaticus major when compared to all other muscles ($p < .001$). Additionally, buccinator/risorius was elevated when compared to the orbicularis oris ($p < .05$) and levator muscles ($p < .05$). The disgust condition selectively activated the levator, compared to other muscles ($p < .01$). The fear condition selectively activated the orbicularis oris compared to other muscles ($p < .05$). Again, the sad condition showed no significant differences in activation on any muscle, and there were no muscle differences at rest.

In short, as in earlier research, making a happy expression generated the greatest change in muscle activity, greater than any other emotion,

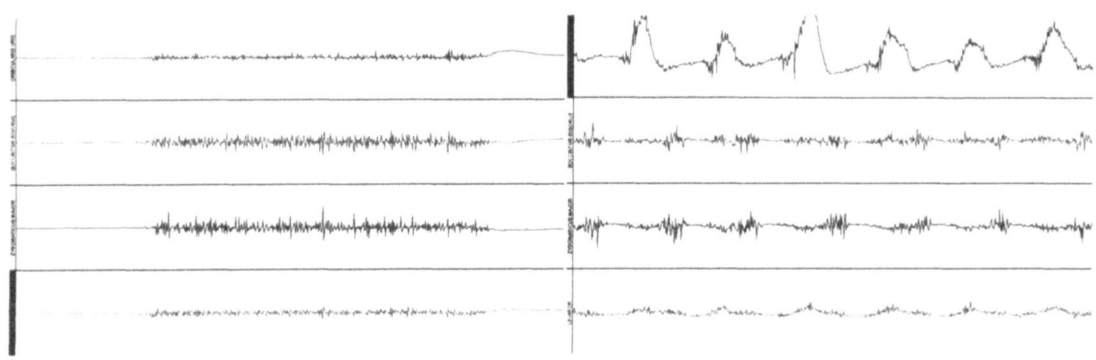

Figure 2. Examples of raw patterns of activity in the bite (left) and gum (right) condition. The muscle from top to bottom are orbicularis oris, buccinator/risorius, zygomaticus major, and levator.

with especially elevated activity of zygomaticus major and buccinator/risorius. The effects of disgust and fear expressions were limited to single muscles, levator and oris, respectively. Sadness showed no effects.

Effects of external versus expression manipulations

Finally, we compared overall levels of muscle activation generated by the external and expression manipulations. As can be seen in the left and right panels of Figure 1, the greatest level of activity was generated by the happy expression and the gum manipulation, followed by the bite manipulation. The activity generated by disgust, fear, and sadness expressions were lower than the bite, gum, and happy manipulations (all $ps < .001$)

Discussion

Experiment 1 had two goals. The first goal was to examine effects of "bite" and "gum" manipulations that are used to prevent mimicry on muscles involved in production of facial expressions. Results showed that the bite manipulation activates all measured facial muscles, with the pattern being constant over time. The gum manipulation also activates all measured muscles, with greatest activation in the buccinator/risorius and orbicularis oris, but the pattern over time is oscillating between activity and nonactivity. The second goal of Experiment 1 was to examine muscle involvements in facial expressions. Results revealed that the happiness expression was associated with most muscle activity, especially in the classic "cheek" region. Disgust and fear expressions also elevated the activity but the level was weaker than happiness and was restricted to single regions (nose for disgust, lip for fear). Sadness produced no elevated activity.

EXPERIMENT 2

Experiment 2 examined the effects of the mimicry-blocking manipulation on recognition of various emotional expressions. If, as previously suggested, facial mimicry is involved in emotion recognition, then non-specific activation of muscles involved in producing a specific expression should impair recognition of that expression (Atkinson & Adolphs, 2005; Niedenthal et al., 2001).

Based on the findings from Experiment 1, one can make some more specific predictions. Recognition should be particularly impaired by the bite manipulation as it generates continuous muscle activity. Further, the bite manipulation should particularly impair recognition of happy expressions as they are associated with the greatest muscle recruitment. Given that facial muscle activity is much weaker in other facial expressions, we should see weaker effects of the bite manipulation on the recognition of disgust, fear, and especially sad.

Regarding the gum manipulation, the predictions are less clear. This manipulation strongly activates multiple facial muscles, but that activation is intermittent and therefore could allow enough activity to engage in mimicry. Further, though Pietromonaco et al. (1981) report some impairments in delayed memory for neutral expressions studies while chewing gum, there are some interpretational ambiguities with that study that suggest that effects of the gum could be due to other factors (Graziano et al., 1996).

Overview

Participants saw a series of morphed photographs taken from the Ekman–Friesen database (Ekman & Friesen, 1976). Happy, sad, fearful and disgust facial expressions were morphed to create seven levels ranging from a low level to an extreme level of each emotion. On each trial, participants were shown one face, and then given a four-alternative, forced-choice recognition task. Recognition scores were obtained during four different blocked conditions: (1) baseline, no concurrent task; (2) bite; (3) lip; and (4) gum. The instructions of the manipulation were the same as in Experiment 1. The order of the conditions was counterbalanced for each subject.

Method

Participants

A total of 12 undergraduate students (6 females, 6 males) at the University of California, San Diego, participated in this study in exchange for class credit. As this was a within-subject design, each subject participated in each of four blocks, each consisting of 280 trials. Data from 2 participants (1 male, 1 female) were not included in the analyses because their recognition scores were outliers, falling two standard deviations below the mean for multiple blocks.

Materials

Emotional photographs.. Photographs of 10 actors/models (4 male, 6 female) displaying three levels of each of four emotions, happy, sad, fear and disgust were obtained from the standard set of Ekman and Friesen (1976) and morphed to create seven levels of each emotion. This resulted in a dataset of 280 stimuli. Pilot ratings revealed that the morphed photos were recognized correctly as conveying the intended emotion at levels ranging from just slightly above chance for the low levels of the emotion to 85% correct for the extreme faces. The original set of morphs had to be edited to make lower levels of happy faces, as these were recognized at a higher level than other facial expressions—a typical finding in the emotion literature.[2] After equalizing performance across facial expressions for the pilot subjects, that set was used for the experiment. Faces were presented on the screen for a period of 500 ms followed by a screen instructing the participants to rate the facial expression as conveying happiness, sadness, fear, or disgust. The forced-choice task was self-paced and we did not collect response times. The order of presentation of the faces was randomized for each subject.

Design and procedure

The experimental design was within-subject 4 (Observed Facial Expression) × 4 (Facial Manipulation), with the order of blocks randomized. After giving written consent, participants engaged in 20 practice trials followed by four counterbalanced blocks of a four-alternative, forced-choice recognition task. During each block the participant viewed the full set of 280 stimuli while they performed no task (baseline) or engaged in one of the three facial actions (bite, chew, lip).

[2] Very high recognition rates for happiness might account for variation in findings in some neuropsychological studies (e.g., Adolphs, Damasio, Tranel, & Damasio, 1996; Adolphs et al., 2000).

Results

As shown in Figure 3, despite our pretest aimed at adjusting all recognition rates to approximately the same level, the overall recognition for happy expressions was significantly lower (59%) than any of the other emotions ($p < .05$), which did not differ from each other (disgust at 71%, fear at 70%, and sad at 68%). However, note that the chance recognition for each expression was 25%, so that for each emotion there was enough room for detecting the inhibitory and facilitatory effects of external manipulations. Further, as we discuss shortly, statistically controlling for the overall recognition difficulty does not explain the impact of the mimicry manipulation.

Our primary analyses were conducted within each emotion condition and the results are shown in Figure 3. Specifically, we first conducted overall repeated measures MANOVAs with 4 levels of Manipulation Conditions (bite, gum, lip, rest), which we followed up with specific one-tailed paired-comparison tests between individual conditions. Within the happy condition there was a clear overall effect of manipulation type, $F(3, 27) = 3.52$, $p < .05$. Follow-up tests revealed that recognition of happy expressions was significantly worse during the bite condition as compared to the all other manipulation conditions combined and compared to each manipulation condition separately (all $ps < .05$). Importantly, the amount of bite-related impairment in recognition of happiness (bite minus rest) was unrelated to the participant's overall recognition performance and thus cannot be attributed to difficulty ($p > .25$).

Within disgust, there was no overall effect of manipulation type ($F < 1$). However, a paired-comparison test revealed that the bite

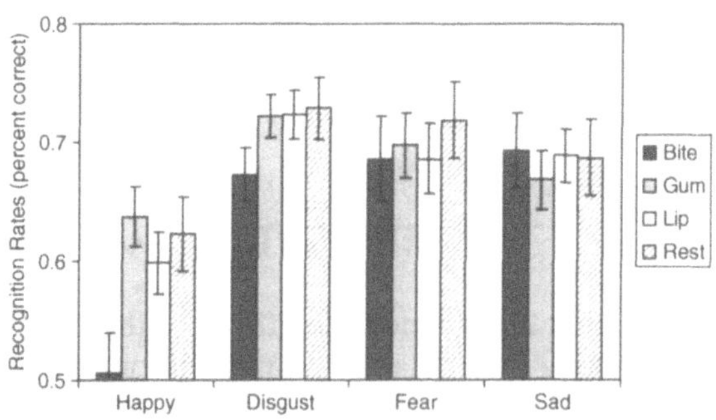

Figure 3. Recognition rates of emotional facial expressions. Bars represent the recognition rates for happy, disgust, fear, and sad facial expressions during each experimental manipulation. Error bars represent the standard error of the mean.

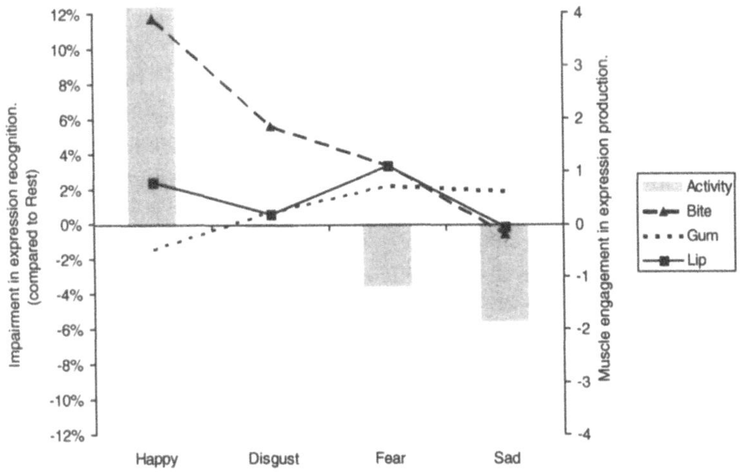

Figure 4. The relation between muscle engagement in producing an emotional facial expression and impairment in recognition of expression as a result of external face manipulation. On the right axis and shown as bars is the overall level of muscle engagement from Experiment 1. On the left axis, and shown as lines, is the relative impairment in recognition, as compared to rest, from Experiment 2.

manipulation lowered recognition of disgust when compared to the rest condition ($p < .05$), but not when compared to other conditions. There was no overall effect of manipulation type for fear ($F < 0.5$) or for sadness ($F\text{s} < 0.1$) and no specific difference between condition even approached significance.

Finally, it is interesting to relate the finding from Experiment 2 to Experiment 1 and compare the amount of impairment in expression recognition due to each manipulation (manipulation minus rest) to the overall level of muscle engagement in expression production (an average of activity across all measured muscles). Unfortunately, a proper statistical analysis is impossible as the different studies involved different subjects, thus allowing only aggregation across 4 emotion types. However, as illustrated in Figure 4, in the bite condition, the amount of recognition impairment appears positively related to the overall level of muscle engagement (i.e., most impairment for happy followed by disgust, fear, and sad). We will return to this observation in the discussion.

Discussion

Experiment 2 found that the bite manipulation specifically impairs the ability to recognize happy faces, and to some extent disgust. This finding is consistent with results from Experiment 1 that the biting manipulation (1) reliably activates several facial muscles and (2) making a happy expression generates substantial muscles activity. Experiment 2 found no reliable effects of blocking manipulations on recognition of fear and sadness, and only some indication for impairment in recognizing disgust. This finding is again consistent with Experiment 1 where these expressions generated much lower and more restricted facial activity. Reaction time (RT) data was not collected in this paradigm. Future studies including RT data would be useful to complement and clarify this finding.

One seemingly puzzling finding was the lack of interference from the gum manipulation. Apparently, college students can recognize emotions and chew gum at the same time! Jokes aside, this is puzzling as this manipulation caused strong activation of the facial muscles in Experiment 1. However, as mentioned, the chewing pattern is extremely stereotyped and occurs about every second with breaks that could allow for some muscle mimicry. The bite condition, on the other hand requires strong and continuous muscle activation, which prevents any differential muscle responses during perception.

GENERAL DISCUSSION

According to the embodiment theories, action recognition is supported by the same mental substrates as the actual performance of that same action (see Decety & Sommerville, 2003; Rizzolatti, Fadiga, Gallese, & Fogassi, 1996). Evidence for this account has been provided by studies that either facilitate or interfere with motor resources and find corresponding influences on perception (e.g., Reed & Farah, 1995; Tucker & Ellis, 1998). There is also emerging evidence for the role of simulation in perception of facial expression (Goldman & Sripada, 2005).

In the current studies, we examined a unique prediction of the embodiment theories that task-irrelevant activation of motor resources should selectively interfere with perception of stimuli which draw on these resources. We tested this idea with emotional facial expressions and predicted that blocking facial muscles involved in mimicry should selectively impair recognition of emotions that engage those muscles. Consistent with these predictions, in Experiment 2 "mimicry blocked" participants were selectively impaired in recognition of happy faces, which activated most motor resources in Experiment 1.

Our findings extend previous research showing that disruption of facial mimicry impairs detection of transition between happy and sad facial expression (Niedenthal et al., 2001). This previous study only used two expressions morphed into a continuum and therefore could not test for emotion specificity of mimicry blocking. Our studies included two additional expressions (fear and disgust) and presented the faces one expression at a time, thus allowing for independent assessment of motor engagement in each expression. Our findings suggest that facial mimicry might be particularly important for recognizing happiness, but is less important in sadness. In addition, Niedenthal et al.'s (2001) study only had two manipulation conditions: no task and holding pen passively in the mouth. This left unclear whether possible differences between conditions (effort, distraction, etc.) could account for the results, and whether similar effects could be obtained using a manipulation that actively engaged the muscles, rather then forced participants to keep the face still (i.e., cutting out any muscular signal). Our study used a pen-biting manipulation that required an active pressure, and included an additional condition that involve a similar amount of effort (chewing gum) as well a control for possible distraction (passively holding the pen between the lips). Given the overall pattern of results, it is unlikely that our and Niedenthal et al.'s (2001) results are driven by some irrelevant factors. However, it is possible, though unlikely, that this manipulation resulted in changes in emotional states. Future studies should control for this possibility.

The recognition of happiness was most impaired by the bite manipulation. Intriguingly, happiness also showed the strongest muscle activations during imitation. This finding has interesting implications. One is that simulation of happiness engages facial musculature more than other emotions, which are simulated internally. This could be due to the fact that expression of happiness is less restricted by display rules, thus creating a particularly strong pairing between perception and action (Ekman, 2004). Importantly, the fact that recognition of disgust, fear and sadness were less affected by muscular blocking does not imply lack of simulation, but rather that their simulation may draw on different types of somatosensory resources. For example, recognition of disgust might involve simulation of interoceptive states (e.g., feeling nauseous) and recruit somatic maps in the insula (Wicker, Keysers, Plailly, Royet, Gallese, et al., 2003) whereas recognition of sadness might involve simulation of more postural components and draw more on physical body schema (Reed, 2002). The relative role of simulation of specific external and internal aspects of emotion in recognition of different expressions represents an exciting direction for future research.

It is also possible that facial mimicry depends on a system that is somatotopically specific, perhaps down to the level of utilized muscles. Although we tried to measure and manipulate the most relevant facial muscles, it is possible that we were particularly successful in interfering with muscles related to happiness. This possibility is intriguing in the light of recent research showing that the same neurons in the premotor cortex respond to both the observed and the executed action of the same somatotopic area (Buccino et al., 2001). Future studies may address this question with more focused manipulation of individual facial muscles or selective somatosensory resources.

More generally, our findings are consistent with the proposal that people's ability to understand emotions in others involves simulating their states and might be supported by the somatosensory system working in conjunction with the MNS (Atkinson & Adolphs, 2005; Gallese & Goldman, 1998). Thus, if the relevant neurons are already busy with the execution of the action, as in this study, or they are immobilized due to natural or artificial lesions, they are unable to be modulated by the observation of the same action. These suggestions are consistent with studies showing activation in the premotor cortex when participants watch people's facial expressions (Carr et al., 2003). It is also supported by research in autism. It is well known that individuals with autism spectrum disorders have difficulty understanding other people's emotional state. This impairment could be related to the fact that when viewing emotional expressions these individuals show no spontaneous activation of the mirror neuron system (Dapretto et al., 2006) and do not engage in spontaneous facial mimicry, as measured by EMG (McIntosh, Reichman-Decker, Winkielman, & Wilbarger, 2006).

In conclusion, the current results converge with previous research highlighting the contribution of embodied simulation to emotion recognition (Goldman & Sripada, 2005) and to emotion processing in general (Winkielman, Niedenthal, & Oberman, in press). More specifically, we show that impairing one's ability to use facial muscles leads to a selective deficit in the recognition of the emotions that engage those muscles. This novel finding highlights potential specificity in the simulation system underlying facial mimicry. Future studies will hopefully clarify the role of specific peripheral contributions, such as individual muscles, as well as central contribution, such as the MNS, to the observed effects.

REFERENCES

Adolphs, R., Damasio, H., Tranel, D., Cooper, G., & Damasio, A. (2000). A role for somatosensory cortices in the visual recognition of emotion as revealed by three-dimensional lesion mapping. *Journal of Neuroscience*, *20*, 2683–2690.

Adolphs, R., Damasio, H., Tranel, D., & Damasio, A. (1996). Cortical systems for the recognition of emotion in facial expressions. *The Journal of Neuroscience*, *16*, 7678–7687.

Atkinson, A., & Adolphs, R. (2005). Visual emotion perception: Mechanisms and processes. In L. Feldman-Barret, P. M. Niedenthal, & P. Winkielman (Eds.), *Emotion and consciousness* (pp. 150–184). New York: Guilford Press.

Barrett, L. F., Mesquita, B., Ochsner, K. N., & Gross, J. J. (2007). The experience of emotion. *Annual Review of Psychology*, *58*, 387–403.

Buccino, G., Binkofski, F., Fink, G. R., Fadiga, L., Fogassi, L., Gallese, V., et al. (2001). Action observation activates premotor and parietal areas in a somatotopic manner: An fMRI study. *European Journal of Neuroscience*, *13*, 400–404.

Cacioppo, J. T., Tassinary, L. G., & Fridlund, A. J. (1990). The skeletomotor system. In J. T. Cacioppo & L. G. Tassinary (Eds.), *Principles of psychophysiology* (pp. 325–384). Cambridge, UK: Cambridge University Press.

Carr, L., Iacoboni, M., Dubeau, M. C., Mazziotta, J. C., & Lenzi, G. L. (2003). Neural mechanisms of empathy in humans: A relay from neural systems for imitation to limbic areas. *Proceedings of the National Academy of Sciences*, *100*(9), 5497–5502.

Chartrand, T. L., & Bargh, J. A. (1999). The chameleon effect: The perception–behavior link and social interaction. *Journal of Personality and Social Psychology*, *76*(6), 893–910.

Condon, W. S., & Ogston, W. D. (1967). A segmentation of behavior. *Journal of Psychiatric Research*, *5*, 221–235.

Dapretto, M., Davies, M. S., Pfeifer, J. H., Scott, A. A., Sigman, M., Bookheimer, S. Y., et al. (2006). Understanding emotions in others: Mirror neuron dysfunction in children with autism spectrum disorders. *Nature Neuroscience*, *9*(1), 28–30.

Decety, J., & Sommerville, J. A. (2003). Shared representations between self and other: A social cognitive neuroscience view. *Trends in Cognitive Sciences*, *7*(12), 527–533.

Di Pellegrino, G., Fadiga, L., Fogassi, L., Gallese, V., & Rizzolatti, G. (1992). Understanding motor events: A neurophysiological study. *Experimental Brain Research*, *91*, 176–180.

Dimberg, U. (1982). Facial reactions to facial expressions. *Psychophysiology*, *19*(6), 643–647.

Dimberg, U., Thunberg, M., & Elmehed, K. (2000). Unconscious facial reactions to emotional facial expressions. *Psychological Science*, *11*(1), 86–89.

Ekman, P. (2004). *Emotions revealed: Recognizing faces and feelings to improve communication and emotional life*. New York: Owl Books.

Ekman, P., & Friesen, W. V. (1976). *Pictures of facial affect*. Palo Alto, CA: Consulting Psychologists.

Ekman, P., & Friesen, W. V. (1982). Felt, false and miserable smiles. *Journal of Nonverbal Behavior*, *6*, 238–252.

Gallese, V., Fadiga, L., Fogassi, L., & Rizzolatti, G. (1996). Action recognition in the premotor cortex. *Brain*, *119*, 593–609.

Gallese, V., & Goldman, A. (1998). Mirror neurons and the simulation theory of mind-reading. *Trends in Cognitive Science*, *3*, 493–501.

Goldenberg, G., & Karnath, H. O. (2006). The neural basis of imitation is body part specific. *The Journal of Neuroscience*, *26*(23), 6282–6287.

Goldinger, S. D. (1998). Echoes of echoes? An episodic theory of lexical access. *Psychological Review*, *105*, 251–279.

Goldman, A. I., & Sripada, C. S. (2005). Simulationist models of face-based emotion recognition. *Cognition*, *94*, 193–213.

Graziano, W. G., Smith, S. M., Tassinary, L. G. Sun, C. R. & Pilkington, C. (1996). Does imitation enhance memory for faces? Four converging studies. *Journal of Personality and Social Psychology*, *71*, 874–887.

Grezes, J., & Decety, J. (2001). Functional anatomy of execution, mental simulation, observation, and verb generation of actions: A meta-analysis. *Human Brain Mapping*, *12*, 1–19.

Hatfield, E., Cacioppo, J. T., & Rapson, R. L. (1994). *Emotional contagion.* Paris: Cambridge University Press.

James, W. (1890). *The principles of psychology.* New York: Holt.

Kendon, A. (1970). Movement coordination in social interaction: Some examples described. *Acta Psychologica, 32,* 101–125.

Laird, J. D., Alibozak, T., Davainis, D., Deignan, K., Fontanella, K., Hong, J., et al. (1994). Individual differences in the effects of spontaneous mimicry on emotional contagion. *Motivation and Emotion, 18,* 231–247.

Lipps, T. (1907). Das Wissen von fremden Ichen. In T. Lipps (Ed.), *Psychologische untersuchungen* (Band 1, pp. 694–722). Leipzig, Germany: Engelmann.

Lotze, H. (1852). *Medicinische Psychologie oder Physiologie der Seele.* Leipzig, Germany: Weidmannsche Buchandlung.

Lundqvist, L. O., & Dimberg, U. (1995). Facial expressions are contagious. *International Journal of Psychophysiology, 9,* 203–211.

McFarland, D. H. (2001). Respiratory markers of conversational interaction. *Journal of Speech, Language, and Hearing Research, 44,* 128–143.

McIntosh, D. N., Reichmann-Decker, A., Winkielman, P., & Wilbarger, J. L. (2006). When the social mirror breaks: Deficits in automatic, but not voluntary mimicry of emotional facial expressions in autism. *Developmental Science, 9,* 295–302.

Neumann, R., & Strack, F. (2000). "Mood contagion": The automatic transfer of mood between persons. *Journal of Personality & Social Psychology, 79,* 211–223.

Niedenthal, P. M., Barsalou, L., Winkielman, P., Ric, F., & Krauth-Gruber, S. (2005). Embodiment in attitudes, social perception, and emotion. *Personality and Social Psychology Review, 9,* 184–211.

Niedenthal, P. M., Brauer, M., Halberstadt, J. B., & Innes-Ker, A. H. (2001). When did her smile drop? *Cognition and Emotion, 15,* 853–864.

Oberman, L. M., & Ramachandran, V. S. (2007). The simulating social mind: The role of simulation in the social and communicative deficits of autism spectrum disorders. *Psychological Bulletin, 133,* 310–327.

Paccalin, C., & Jeannerod, M. (2000). Changes in breathing during observation of effortful actions. *Brain Research, 862*(1–2), 194–200.

Pietromonaco, P. R., Zajonc, R. B., & Bargh, J. A. (1981). *The role of motor cues in recognition for faces.* Paper presented at the meeting of the American Psychological Association, Los Angeles.

Prinz, W., & Hommel, B. (Eds.). (2002). Common mechanisms in perception and action: *Attention and performance XIV.* Oxford, UK: Oxford University Press.

Reed, C. L. (2002). What is the body schema? In W. Printz & A. Meltzoff (Eds.), *The imitative mind: Development, evolution, and brain bases* (pp. 233–243). Cambridge, UK: Cambridge University Press.

Reed, C. L., & Farah, M. J. (1995). The psychological reality of the body schema: A test with normal participants. *Journal of Experimental Psychology: Human Perception & Performance, 21,* 334–343.

Rizzolatti, G., Fadiga, L., Gallese, V., & Fogassi, L. (1996). Premotor cortex and the recognition of motor actions. Brain Research. *Cognitive Brain Research, 3*(2), 131–141.

Strack, F., Martin, L. L., & Stepper, S. (1988). Inhibiting and facilitating conditions of the human smile: A nonobtrusive test of the facial feedback hypothesis. *Journal of Personality and Social Psychology, 54,* 768–777.

Tassinary, L. G., & Cacioppo, J. T. (2000). The skeletomotor system: Surface electromyography. In J. T. Cacioppo, L. G. Tassinary, & G. G. Berntson (Eds.), *Handbook of psychophysiology* (2nd ed., pp. 163–199). New York: Cambridge University Press.

Tucker, M., & Ellis, R. (1998). On the relations between seen objects and components of potential actions. *Journal of Experimental Psychology: Human Perception and Performance, 24,* 830–846.

Vrana, S. R. (1993). The psychophysiology of disgust: Differentiating negative emotional context with facial EMG. *Psychophysiology, 30,* 279–286.

Wallbott, H. G. (1991). Recognition of emotion from facial expression via imitation? Some indirect evidence for an old theory. *British Journal of Social Psychology, 30,* 207–219.

Wicker, B., Keysers, C., Plailly, J., Royet, J.-P., Gallese, V., & Rizzolatti, G. (2003). Both of us disgusted in my insula: The common neural basis of seeing and feeling disgust. *Neuron, 40,* 655–664.

Winkielman, P., Niedenthal, P., & Oberman, L. (in press). The embodied emotional mind. In G. R. Semin & E. R. Smith (Eds.), *Embodied grounding: Social, cognitive, affective, and neuroscientific approaches.* New York: Cambridge University Press.

Zajonc, R. B., & Markus, H. (1984). Affect and cognition: The hard interface. In C. Izard, J. Kagan, & R. B. Zajonc (Eds.), *Emotions, cognition and behavior* (pp. 73–102). Cambridge, UK: Cambridge University Press.

Psychology Press
Taylor & Francis Group

Facial expressions: What the mirror neuron system can and cannot tell us

Christiaan van der Gaag, Ruud B. Minderaa, and Christian Keysers
University of Groningen, Groningen, The Netherlands

Facial expressions contain both motor and emotional components. The inferior frontal gyrus (IFG) and posterior parietal cortex have been considered to compose a mirror neuron system (MNS) for the motor components of facial expressions, while the amygdala and insula may represent an "additional" MNS for emotional states. Together, these systems may contribute to our understanding of facial expressions. Here we further examine this possibility. In three separate event-related fMRI experiment, subjects had to (1) observe (2) discriminate and (3) imitate facial expressions. Stimuli were dynamic neutral, happy, fearful and disgusted facial expressions, and in Experiments 1 and 2, an additional pattern motion condition. Importantly, during each experiment, subjects were unaware of the nature of the next experiments. Results demonstrate that even passive viewing of facial expressions activates a wide network of brain regions that were also involved in the execution of similar expressions, including the IFG/insula and the posterior parietal cortex. Only a subset of these regions responded more during the observation of facial than pattern motion (bilateral ventral IFG, bilateral STS/MTG, bilateral amygdala, SMA). While the viewing of facial expressions recruited similar brain regions in all three experiments, adding an active task (discrimination, imitation) augmented the magnitude of these activations. Brain activations reflected differences in observed facial expressions, with emotional expressions activating relatively more the insula/frontal operculum, and neutral ones (blowing up the cheeks) the somatosensory cortices (SII). Using movies, fear activated the amygdala and disgust the insula, but other emotions activated these structures to a similar degree.

INTRODUCTION

Understanding the emotional facial expressions of others is important for daily social functioning. A number of different approaches have been used to investigate the way we understand the emotions of other people. Lesion studies looked at the consequences of damage to specific brain sites on the capacity to read the emotions of others (see Adolphs, Damasio, Tranel, Cooper, & Damasio, 2000, for the most extensive study). They found that damage to the right somatosensory cortex impaired subjects in recognizing basic emotions from visually presented facial expressions. According to Adolphs et al. (2000), this finding supports the idea that emotion recognition is mediated by a simulation mechanism within the observer, which generates somatosensory representations as if the observer feels similar emotional states to the ones being observed.

Correspondence should be addressed to: Christian Keysers, School of Behavioral and Cognitive Neurosciences NeuroImaging Center, University Medical Center Groningen, University of Groningen, NL-9713 AW Groningen, The Netherlands. E-mail: c.m.keysers@rug.nl

The research was supported by the University Medical Centre Groningen and an NWO VIDI grant and a Marie Curie Excellence Grant to CK.

We thank Remco Renken for his valuable help in data analysis and Anita Kuiper for data acquisition and performing the independent movement observation during the imitation task. Mbemba Jabbi is thanked for his help in creating the film stimuli and Valeria Gazzola for her assistance in programming the presentation software and data analyses.

www.psypress.com/socialneuroscience DOI:10.1080/17470910701376878

More recently, the understanding of other people through facial expressions has been studied through the mirror neuron system (MNS). The critical property of this system is that the observation of a particular action activates regions involved in the execution of that particular action. Both in monkeys and humans, a vast literature exists describing the classical MNS to be composed of the posterior parietal cortex and the premotor cortex (BA 6/BA 44; e.g., Rizzolatti & Craighero, 2004; Rizzolatti, Fogassi, & Gallese, 2001). The superior temporal sulcus (STS) is closely related to the classical mirror neuron system in terms of function and connections but lacking clear motor properties (Gazzola, Aziz-Zadeh, & Keysers, 2006; Keysers & Gazzola, 2006).

Classically, the mirror-neuron matching system has been investigated through transitive, goal-directed actions. Facial expressions are a specific class of actions, not directed at a particular object, although facial expressions can be viewed as instrumental devices to influence other people (e.g., Blair, 2003; Russell, Bachorowski, & Fernandez-Dols, 2003), which in a sense is goal directed. Interestingly, neurons in the monkey's ventral premotor cortex (F5) have been described as responding to the observation of communicative mouth actions (Ferrari, Gallese, & Rizzolatti, 2003).

In humans, non-invasive neuroimaging techniques like fMRI have been applied to study the involvement of the MNS in the understanding of facial expressions. One fMRI approach required subjects in alternating blocks to observe facial expressions and to imitate them (Carr, Iacoboni, Dubeau, Mazziotta, & Lenzi, 2003; Dapretto et al., 2006). The blocks of facial expression observation consisted of a mix of static displays of several basic emotions. Results showed that premotor and parietal cortex were both involved in facial expression observation *and* execution, supporting the idea of the involvement of the MNS in understanding facial expressions. In addition, Carr et al. (2003) showed that two limbic regions, the amygdala and insula, were involved in both the observation and execution of facial expressions. Next, Leslie, Johnson-Frey, and Grafton (2004) used a paradigm in which subjects had to observe and imitate hand and face actions (smile and frown condition) using film clips instead of static displays. The right ventral premotor cortex was commonly activated during observation and imitation of facial expressions. A study by Hennenlotter et al. (2005) specifically looked into the MNS involvement during the production and observation of pleasant affect (smiling). Right premotor cortex and inferior frontal cortex were involved in both conditions, in addition to right parietal operculum (SII) and left anterior insula. Considering only these studies, it is still an open question whether activated regions are confounded by the upcoming execution part of the study, as subjects knew they had to either imitate (Carr et al., 2003; Dapretto et al., 2006; Leslie et al., 2004) or to execute similar movements (Hennenlotter et al., 2005) subsequently. Reported MNS activations may thus reflect a preparatory state for future execution, and may therefore differentiate from naturalistic situations.

Another approach to studying the involvement of the MNS during facial expression observation is to use emotion *experiencing* and emotion *observation*. When subjects observed and experienced the emotion "disgust", Wicker et al. (2003) found the anterior insula and the anterior cingulate gyrus to be involved in both cases. These regions are located outside the classical MNS. Some of the overlapping anterior insula regions extended into the inferior frontal regions, although not in BA 44/BA 45. This finding supports the idea that humans have both a "classical" MNS for goal-directed actions (parietal and premotor regions) and an "extended" MNS for emotional states (insula, anterior cingulate cortex and, potentially, other brain regions). For the latter, the term "mirror neuron system" is used in an extended context, where neurons do not require visual and motor properties, but visual properties and matching properties in another domain—emotions or sensations (Keysers & Gazzola, 2006; Gallese, Keysers, & Rizzolatti, 2004). In order to understand further the respective contribution of these two mirror neuron systems involved in the processing of facial expressions, the observation of facial expressions can thus be combined with either an emotional experiencing paradigm to place the emphasis on the extended MNS or with a facial expression production paradigm to emphasize the classical MNS. This distinction bears a relation to the distinction found in psychological theories of emotion processing. Theories positing that understanding other people's emotions requires the induction of a similar *emotional* state in the observer would predict an important role for the extended MNS while theories positing that the production of a facial expression similar to the

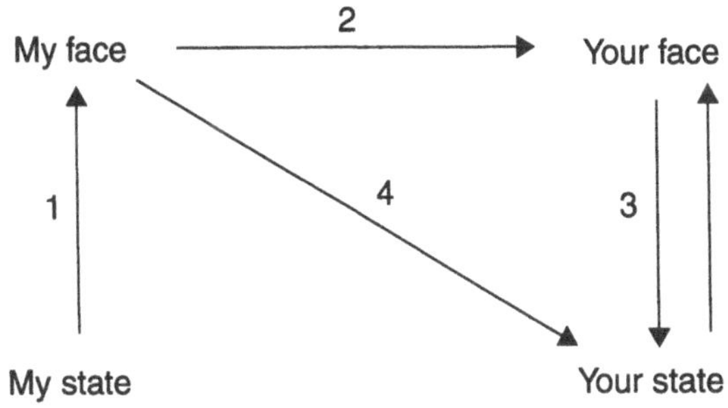

Figure 1. Empathy through simulation: possible processes that explain the sharing of feelings. Hypothesized processes are reflected by numbered arrows, which reflect the direction of influence: (1) Both your and my feeling state is reflected by a congruent facial expression; (2) The mimicry process by which your facial musculature reflects my facial expression; (3) Emotional contagion through a facial feedback mechanism causes the configuration of your facial musculature to initiate a congruent feeling state; (4) Direct contagion of my emotional state when you observe my facial expression: an alternative mechanism for empathy through mimicry and facial feedback mechanisms. Notice that after the direct contagion of an emotional state your face can also reflect this state (1). Psychological theories of understanding other people's emotions differ in the causal relationship they see between sharing the emotional state and sharing the facial expression of others. Lipps (1907) posited that the observer automatically imitates the observed facial expression—facial mimicry (McIntosh, 1996), and that this mimicry is essential for understanding the emotion. Through facial feedback, the mimicked facial expression then leads to the experience of the emotion in the observer (indirect emotional contagion). Hess, Philippot, and Blairy (1998) on the other hand posited that observing the facial expression leads to the induction of a similar emotional state in the observer without the need for mimicry. Mimicry *can* then occur at a later point, as a reflection of the shared emotional state, but social context greatly modulates this process. Most theories lie somewhere between these two standpoints. Although some theories make clear predictions about what should come first—sharing the facial expression or the emotional state—the temporal resolution of fMRI is unsuited for resolving this difference reliably.

What is the relationship of these psychological models and the MNS? Lipps' model is in tight relationship with motor mirror neurons, as they could provide a neural basis for the facial mimicry. Hess's model on the other hand attributes a marginal role to motor mirror neurons. Her model predicts the existence of mirror neurons for emotional states, that would be involved both in emotion experience and emotion observation (see Keysers & Perrett, 2004). Such "emotional" mirror neurons have never been recorded in monkeys, but fMRI experiments suggests that they might exist, as voxels in the human brain show this property (Singer et al., 2004; Wicker et al., 2003). For a model that also adds reflective brain regions involved in Theory of Mind processes to the understanding of facial expressions see Keysers and Gazzola (2007).

observed is an essential component for emotion understanding ("facial feedback mechanism") would predict an important role for the classical MNS (see Figure 1).

A reason for caution with a "facial expression-production" approach is that one might assume that with the mere production of a particular emotional expression one can find brain regions involved in both the motor part of a facial expression *and* the emotional state (e.g., by means of a facial feedback mechanism), and as a consequence find emotional-related brain structures. A more careful viewpoint is that through the execution of a facial expression one can only state something about the motor part of facial expressions and not the emotional state. This is in fact what previous studies using an imitation paradigm argue, that action representation in premotor cortex during facial expression observation mediates emotional recognition (Carr et al., 2003; Hennenlotter et al., 2005; Leslie et al., 2004).

The emotion specificity of the MNS has received limited attention. While Carr et al. (2003) explicitly excluded emotion specificity by mixing all emotions in imitation blocks, Hennenlotter et al. (2005) looked only at one emotion (smiling). The only study that investigated multiple emotions (disgust and pleasure) was performed by Wicker et al. (2003), although they only studied one positive and one negative emotion. One region commonly activated in all of these studies was the anterior insula, which has strong associations with disgust processing (Calder, Keane, Manes, Antoun, & Young, 2000). This leaves us wondering whether the anterior insula activation found by Carr et al. (2003) is specifically related to disgust processing or is commonly shared by several emotions within their paradigm as suggested by Jabbi, Swart, and Keysers (2007). Another limbic region showing mirror neuron properties in the Carr study was the amygdala. Data from animal studies and lesion patients have shown the importance of this structure in fear processing (Adolphs et al., 1999; Calder, Lawrence, & Young, 2001; LeDoux, 2000). The amygdala activations in the Carr study could therefore be exclusively related to fear processing. Alternatively, these activations may reflect the processing of multiple emotions as suggested by a number of previous studies (e.g., Fitzgerald, Angstadt, Jelsone, Nathan, & Phan, 2005; Van der Gaag, Minderaa, & Keysers, 2007; Winston, O'Doherty, & Dolan, 2003; Yang et al., 2002).

In this experiment, we will study the MNS with a "production" paradigm using a similar imitation task to that performed by Carr et al. (2003), therefore being able to compare results. In an attempt to contribute to the understanding of facial expressions through mirroring, adjustments were made to the stimuli and paradigm to increase ecological validity and to be able to answer remaining questions: (a) Does the MNS become spontaneously active when observing facial expressions, even when subjects do not need to execute facial expressions? (b) Do different task instructions during the viewing of facial expressions change the pattern of activations in the MNS? (c) Is there biological motion specificity within the mirror neuron circuit? (d) Are the individual emotional facial expressions treated differently and/or specifically within the classical MNS? (e) Are the involvement of the insula and the amygdala during facial expression observation and production emotion specific? To answer these questions, subjects first had to passively observe in an event-related design *dynamic* neutral, disgusted, fearful and happy facial expressions. Prototypical examples of the mentioned basic emotions were used, as they are most likely to be associated with specific brain areas. An extra control condition comprised of pattern motion was included to be able to investigate biological motion specificity in the MNS. Next, subjects had to perform a delayed-match-to-sample task on the same movies. Finally, they were asked to imitate the four facial expressions. Important in this fixed-order experiment, subjects were unaware of the tasks to come.

If the observation and production of facial expressions in our experiment triggered an internal feeling of specific emotions, based on results of previous observation and/or experiencing studies on specific emotions, we expected to find categorical mirror activity for disgust in the anterior insula/frontal operculum (IFO), anterior cingulate and/or basal ganglia (Calder et al., 2000; Goldman & Sripada, 2005; Hennenlotter et al., 2004; Wicker et al., 2003) and for fear in the amygdala (Calder et al., 2000; Goldman & Sripada, 2005; Phan, Wager, Taylor, & Liberzon, 2002). Two meta-analyses did not find brain specificity for happiness (Murphy, Nimmo-Smith, & Lawrence, 2003; Phan et al., 2002), although two other studies found some evidence for specificity for happiness in the supplementary motor area (Fried, Wilson, MacDonald, & Behnke, 1998; Krolak-Salmon et al., 2006).

If, on the other hand, emotional states are not triggered by facial expressions we predicted that ventral premotor regions would be activated equally by the different facial expressions.

MATERIALS AND METHODS

Subjects

Seventeen healthy young adults (age range 19–27 years of age; mean age 23.3 years; 9 women, 8 men) were recruited from the University of Groningen community. All subjects were right-handed (assessed with the Edinburgh Handedness Questionnaire; Oldfield, 1971) and were screened for neurological and psychiatric diseases. Informed consent was obtained from each subject in accordance with the human subjects research protocol approved by the Medical Ethical Committee of the University Medical Center Groningen.

Stimuli

Visual stimuli consisted of 3-second movie clips depicting the emotions happiness, disgust, fear or neutral. Unlike previous studies using dynamic displays (Kilts, Egan, Gideon, Ely, & Hoffman, 2003; LaBar, Crupain, Voyvodic, & McCarthy, 2003; Sato, Kochiyama, Yoshikawa, Naito, & Matsumura, 2004; Wicker et al., 2003), we had a range of facial expressions and more naturalistic specimens (no unecological morph technique was used). The happy condition consisted of spontaneous laughs by our actors, triggered by jokes. Fearful and disgusted expressions resulted from instructions to the actors to display prototypical, strong expressions of these emotions. The neutral condition consisted of an actor blowing up his/her cheeks. Actors, all Caucasian, were filmed from the shoulders up and were asked to express clearly the different emotions while limiting rigid head movements as much as possible. The movies were recorded digitally with a Sony DSR-PDX10P digital camcorder. Adobe Premier Pro Software was used to cut 3 s movie-clips starting with a neutral expression (slightly friendly) lasting for 0.5 s followed by the unfolding of the facial expression and ending on a strong facial expression. An extra control condition was used, displaying abstract pattern motions, composed of 0.5 s of a static oval shape patterned with vertical or horizontal stripes. Part of this pattern then started to swirl for 2.5 s. Twenty different actors

(10 males, 10 females) were filmed for all facial conditions. Each film was then rated on the content of basic emotions on a 7-point intensity scale (range 1–7) by an independent group of 15 subjects (according to methods used in Adolphs, Tranel, Damasio, & Damasio, 1994). On a separate 3-point scale (range 1–3), subjects rated how genuine the emotions looked. A score of 1 corresponded with a facial expression that looked "*fake*", a score of 2 meant "*doubtful genuineness*" while a score of 3 corresponded with a "*genuine looking*" facial expression. Some of the actors were rated as displaying some emotions more intensely than other emotions, while other actors displayed their emotions more homogeneously. Out of the 20 actors, we selected five males and five females that showed the least differences between the intensity ratings of their target emotions. For this subset of actors, there were no significant differences between the intensity of the displayed target emotions (see Figure 2). On average, the chosen movie sets for the different conditions scored above 2 on the 3-point genuineness scale.

Procedure

The experiment was composed of three smaller experiments, each with a different instruction to the subjects. The observation and the discrimination tasks were conducted on the first scanning day, the imitation task on a second scanning day. Stimuli were presented using Presentation software (Neurobehavioral Systems, Inc., Albany, CA, USA).

Observation task. Subjects were instructed to pay careful attention to the different movie clips, without further explicit tasks. Stimuli were shown in a randomized event-related design with movies lasting 3 s and being separated by a random interval (average 6 s, range 4–8 s, hereafter written 6 ± 2 s) during which a white fixation cross was shown on a black background. The experiment was split into two functional runs lasting ~7.5 minutes. In each run, 5 out of the 10 movies of each condition were presented twice. After the functional runs, during the acquisition of an anatomical image, subjects had to perform a surprise memory task: They were shown 40 movies of single facial expressions, half of which had been shown to them during the functional runs. The 20 new movies were either movies of the same actors performing different emotions than those used in this experiment or from different actors showing the same or different emotions as used in this experiment. Subjects had to indicate by means of a button press in a two-alternative forced-choice task if they had previously seen the movie.

Discrimination task. After the observation task, subjects received instructions for this second task. Subjects performed four functional runs (each lasting ~10 minutes) of a delayed match-to-sample task on emotions: movies were shown in pairs, subjects had to report by means of a button press whether the emotion of the first and second movie was the same or different. Only brain activity during presentation of the first stimulus in the pair was analyzed here, because it represents the brain activity during the deliberate extraction of the emotion from a facial expression, without the motor planning involved during the presentation of the second stimulus. Movie pairs were shown pseudo-randomly, 50% of the trials displayed the same emotion. Every movie was shown twice at the first position of movie pairs to enable comparisons between all tasks. Random intervals separated the two movies of a trial (4 ± 2 s) and two consecutive trials (6 ± 2 s). A red fixation cross was shown within movie pairs and between movie pairs the fixation cross was white. After this task, subjects were informed about the imitation task to follow a week later. Subjects were instructed to generate three personal scenarios that could help them to induce the tested emotions (e.g., sudden encounter with a spider, a dirty toilet, a good joke).

Imitation task (second scanning day). Participants had to watch a facial expression. After a random pause (4 ± 2 s) subjects had to (a) imitate the movements of the first movie during a 3-second period and (b) generate the corresponding emotion using the scenarios they had prepared ahead of time. In particular, subjects were instructed not simply to generate facial movements that are generic to the demonstrator's emotion, but to imitate the particularities of that exemplar of the displayed emotion. Only the brain activity during the presentation of the movie was analyzed, to avoid contamination by motor execution. Data from this session thus represents brain activity while subjects explicitly pay attention to the motor aspects of a facial expression. Subjects were cued to start imitation by means of a change

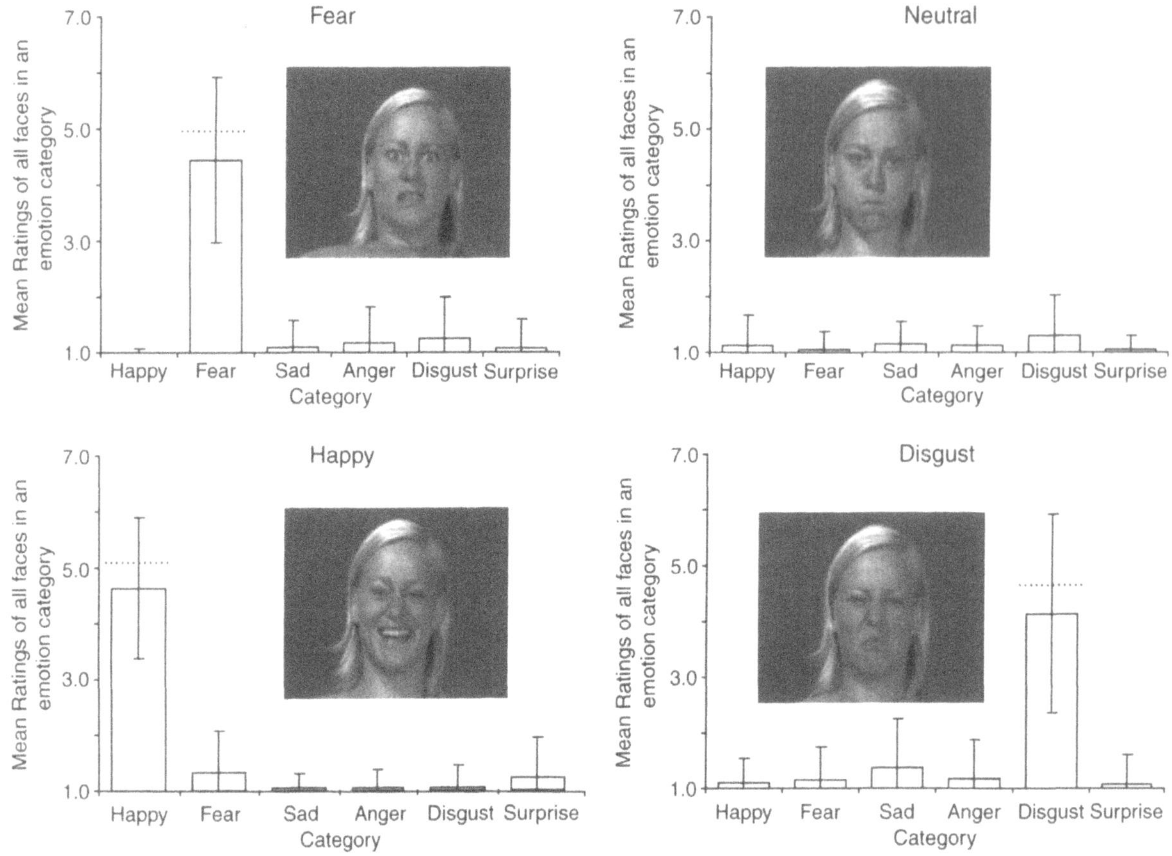

Figure 2. Ratings of emotional facial expressions on emotion words. Data are from an independent sample of 15 normal subjects. The photos are snapshots from one of the stimulus movies. The words in capitals above each graph refer to the stimulus category, those under each bar to the scale along which subjects had to rate the stimulus. The *y*-axis refers to reported intensity of the emotion, with 1 referring to "*none*" and 7 "*very intense*", and error bars represent the standard-error of the mean (*SEM*) across 15 subjects. The second bar from the left on the top left plot thus refers, for example, to how much fear subjects saw in the fear stimuli. Dotted lines are mean ratings for the Ekman faces of the same category as adapted from Figure 2a in Adolphs et al. (1994). Note how these dotted lines fall within the error bars of the ratings of our stimuli.

in color of the fixation cross from red to green. Turning white of the same fixation cross was the signal to stop the imitation. Every movie-clip was shown twice.

For the purpose of analyses, the same number of trials of each condition were analyzed in all three tasks, namely 20 trials of each emotion. The discrimination task, due to the delayed matching to sample paradigm contained twice as many movies, but only the sample movies were analyzed. This allowed direct comparisons between the three sub-experiments.

Importantly, to avoid biasing mental processes during free viewing, subjects were held naive about the discrimination and imitation tasks ahead. Similarly, during the discrimination task, subjects did not yet know whether they had to perform an imitation task later. Tasks had to be conducted in the fixed order: observation, discrimination, imitation for all subjects.

MRI data acquisition

Imaging data were acquired with a 3T Philips Intera MRI scanner (Philips, Best, The Netherlands). The standard 6-channel SENSE head coil was used to acquire whole brain echo-planar functional images (EPIs). Thirty-nine axial slices were acquired with the following parameters: TR 2000 ms; TE 30 ms; flip angle 90°; SENSE factor 2; field of view 224 mm; matrix 64×64; slice thickness 3.5 mm with no slice gap, yielding isotropic voxels of $3.5 \times 3.5 \times 3.5$ mm in size. In addition, two anatomical images were acquired: one 3D-FFE to co-register and normalize functional data (TR = 25 ms, TE = 4.6 ms, flip angle = 30°,

FOV =256 mm, matrix 256 ×256 mm, slice thickness 1.0 mm) and a high contrast 3D-T1TFE to trace the amygdalae (TR =8.2 ms, TE =3.7 ms, flip =8.0°, FOV =256 mm, matrix 256 ×256 mm, slice thickness 1.0 mm).

General data processing

A voxel-based analysis implemented in SPM2 (www.fil.ion.ucl.ac.uk/spm) was used first to analyze the fMRI data. Functional images were temporally adjusted for interleaved slice acquisition and then realigned to the first functional image of the first run. High quality T1 images were co-registered to the mean EPI image and segmented. Low-frequency signal drift was corrected by applying a high-pass temporal filter with a temporal cut-off of 250 s. Data preprocessing ended here for the analysis of signals in anatomically defined regions of interest (ROIs), here the left and right amygdala. For the whole-brain analyses, the co-registered grey matter segment was normalized onto the MNI grey matter template and the resulting normalization parameters applied to all EPI images. The functional data were spatially smoothed with a 6 mm isotropic Gaussian Kernel before the statistical analysis.

General linear model (GLM)

In all analyses below, a similar general linear model (GLM) random effect analysis was performed. For each subject, a GLM considering the time course of each condition, convoluted with the hemodynamic response was used to derive an average parameter estimate for each condition and subject. Parameter estimates could then be subtracted from each other to calculate contrast values. The contrast values from the 17 subjects were then compared against zero using one-sample *t*-tests to implement a random effect analysis (RFX). Resulting *p*-values are reported uncorrected for multiple comparisons. Input for this analysis could be the mean activity extracted from a region of interest (Brett, Anton, Valabregue, & Poline, 2002; http://marsbar.sourceforge.net) or the voxel by voxel activity of the entire brain using SPM2 (see below). Head movements were modeled as co-variates according to Friston, Williams, Howard, Frackowiak, and Turner (1996) for the imitation runs only.

Analyses of motor data alone. The data of the imitation task were modeled using separate predictors for the viewing of the movies, the retention interval and the production phase. Analysis relative to the viewing phase are reported below. The retention phase was not analyzed further in this manuscript. Data relative to the motor production phase were modeled with four predictors of interest—neutral (N), disgusted (D), fearful (F) and happy (H) imitation—and 24 motion covariates (see above) to remove confounding effects of head motion despite the fact that head motion never exceeded one voxel in this study.

In order to get a general view of activations during imitation, the sum of the parameter estimates for all four predictors was compared against zero at the second level and thresholded for display purposes at $p < .005$ (unless specified otherwise). To differentiate sub-regions more involved in producing emotional compared to neutral facial expressions, a contrast D+F+H−3N was thresholded at $p < .005$ and inclusively masked with D+F+H+N at $p < .005$.

To investigate responses specific for particular emotions, two approaches were used. First the production of that emotion was contrasted against the production of the neutral emotion. To determine areas more involved in the production of one emotion compared to the other emotions, pairwise contrasts between the target emotion and the three remaining facial conditions were calculated. We required that all of these contrast needed to be significant at $p < .1$. In addition, the parameter estimate of the target emotion needed to be above zero at $p < .05$. Requiring that these four tests be significant results in a false positive rate under a global null hypothesis of $p_{global} < .1^3 \times .05 = .00005$ (see Friston, Penny, & Glaser, 2005, for a discussion of related issues). Unlike a traditional conjunction, this analysis has the advantage that each individual contrast needs to be above a certain threshold, making it more controllable. The logic behind requiring that the production of the particular emotion be above zero is to exclude regions from this analysis that do not respond to the target emotion but show only BOLD decreases to the other emotions.

Analyses of viewing data alone (observation, discrimination and viewing before imitation). In these analyses a GLM was generated in which 4 or 5 predictors of interest were identified per task (four facial conditions and an additional pattern condition for the observation and discrimination task). For the observation task, no other predictors were modeled. For the discrimination task, 10 additional predictors were modeled, which were not analyzed here (the retention intervals and the second stimuli for all five conditions). For the imitation task, only the four predictors relative to the viewing of the stimuli were analyzed here (with the motor execution predictors analyzed as discussed in the motor section before). To restrict ourselves to mirror activations, results of the viewing phases were masked inclusively by the motor production maps obtained from the analyses described above.

Analyses of mirror activities combining all facial conditions. To investigate mirror activations, we combined data from viewing a certain stimulus with data obtained while subjects produced similar facial expressions. The two sets of data were combined using logical "AND" analysis, identical to inclusive masking (Keysers et al., 2004; Wicker et al., 2003). For viewing, we analyzed either the "*viewing* of all facial expressions – rest" to study the general MNS, the "*viewing* of all faces – pattern" to search for biological motion specificity in the MNS or the "*viewing* of emotional faces – neutral faces" to identify the emotional MNS. For motor production, we compared "*production* of all facial expressions – rest" to study the general and biological motion specific MNS or the "*production* of emotional faces – neutral faces" to study the emotional MNS. Moderate thresholds of both (viewing and production) group *t*-maps ($p < .005$) were combined, resulting in an overall false positive under a global null hypothesis of $p < .000025$. Results were then displayed using a spatial threshold of 20 voxels ($k = 20$).

To study task effects in the MNS, the abovementioned analyses were performed for each vision task individually (observation, discrimination and viewing before imitation) and consecutively compared as a paired *t*-test to the parameter estimates under the different instructions. The emotional MNS was studied by combining the viewing of facial expressions during both the observation and discrimination tasks, therefore increasing statistical power.

Analyses of mirror activities of individual emotions. Mirror activations for specific emotions were studied in two ways. The first used a classical approach by searching for overlap between the group *t*-map of viewing of that emotion minus the neutral condition (e.g., fear–neutral) and the group *t*-map of imitation of that facial emotion minus neutral (e.g., fear–neutral). Thresholds were used as described above ($p < .005$, $k = 20$).

For mirror activations for specific emotions we used a similar approach as the one described above to analyze activations related to the *production* of specific emotions, differing only by increasing the number of statistical tests that were combined: here we required that comparing the target emotion (including neutral) against the other facial conditions needed to be significant in each case both during viewing and during production, and that both during viewing and production the target emotion by itself exceeded zero, i.e., for fear, we tested F–N, F–H, F–D for viewing and for execution at $p < .1$ and F by itself for execution and viewing at .05. Combining these 8 tests leads to a false positive rate under the global null of $p < .025 \times 10^{-8}$.

ROI analysis of the amygdala. A previous study by Carr et al. (2003) found activation overlap within the right amygdala during both emotional facial expression imitation and observation. In a previous study, the role of the amygdalae during viewing was studied using a voxel-wise analysis and an ROI approach with MarsBar and individually defined amygdalae (see Van der Gaag et al., 2007; Appendix 1). During viewing, we could not find a difference between any emotional facial and neutral facial motion observation applying the two mentioned methods. Therefore, to further clarify the role of the amygdalae during imitation, the ROI approach was applied here as well.

RESULTS

We will first discuss the results of the motor execution phase of the imitation task in order to identify voxels in the brain involved in the production of facial expressions (section a). Resulting maps will be used as masks to locate mirror circuits during the different viewing conditions (section b). The rationale behind that procedure is that "mirror-voxels" have to be active both during the production and the observation of facial expressions. In section c, we will test the specificity for biological motion within the MNS. Section d will investigate whether certain voxels within the MNS might be more involved in particular emotions, including the involvement of the amygdala and insula.

Behavioral data. Behavioral results of three subjects could not be analyzed due to malfunctioning of the computer that recorded key-presses during scanning sessions. Behavioral data presented are therefore based on the analysis of the remaining 14 subjects. Data from all 17 subjects were kept for the fMRI analysis.

Surprise memory task. Subjects showed a good performance on this task: on average 93% of the new stimuli were correctly identified while 84% of the familiar movie clips were identified as such. All subjects scored >70%, calculated as (correct rejection+hits)/number of trials, on the memory task, suggesting that subjects paid attention to our stimuli during the observation task.

Discrimination task. On average, 92% of the "same emotion" trials, and 97% of the "different emotion" trials were correctly identified, respectively. All subjects scored >80% on this task, justifying the conclusion that all our subjects paid attention during the delayed match-to-sample-task.

No performance data were acquired during the imitation task, although gross motor movements of the subjects were monitored by an independent observer with an infra-red camera installed in the MRI environment. Facial movements during the appropriate moment of the imitation task were confirmed for all our subjects.

(a) The neural circuit involved in facial imitation

First, combining all motor execution conditions (emotional and neutral facial expressions) we found a circuit very comparable to that reported in previous studies (Carr et al., 2003; Dapretto et al., 2006; Hennenlotter et al., 2005; Leslie et al., 2004) composed of primary motor, pre/supplementary motor, somatosensory, posterior parietal, and visual areas, in particular STS/MTG (Appendices 2 and 3). The premotor activations extended into BA 45. In addition, we found sectors of the insula, hypothalamus, basal ganglia and cerebellum to be involved. Comparing the execution of the emotional with that of the neutral facial expressions revealed that in the vast majority of the areas involved in the production of facial expressions, sub-areas were more active during the production of the emotional facial expressions. (A detailed description of the areas is reported in Appendices 4 and 5.)

While the activations found in frontal and parietal areas in our data can be attributed to the motor execution and somatosensory consequences of these movements, activations in visual areas may have been related to either the generated emotional scripts, the working memory trace of the facial stimuli subjects had to imitate or to the generation of a visual image of what the facial expression they are currently producing would look like in order to match the target facial expression (Iacoboni et al., 2001; Keysers & Perrett, 2004).

These motor execution maps will be used in the remainder of the manuscript to restrict the analysis of visual responses to areas also involved in motor production in order to identify mirror regions. The potential influence of the visual stimuli inherent to our imitation task need to be kept in mind when interpreting the results obtained from masking visual activations with the motor production maps, particularly in areas known to have prominent visual activations (e.g., MTG/STS).

(b) The MNS of facial expressions and its task effects

The overlap between viewing during the observation of dynamic facial expressions (without an explicit task) and facial expression production included areas classically considered to be the MNS, bilateral inferior frontal gyrus and rostral sectors of the posterior parietal cortex (Figure 3A; Appendix 6). In addition, overlaps were found in SMA, SI and SII, the insula, amygdala, hippocampus, cerebellum, and a number of visual areas (Area 17/18, fusiform gyrus, MTG). In essence, the circuit we found is similar to that of Carr et al. (2003).

Activations in SI and SII have been reported before in action observation (Avikainen, Forss, & Hari, 2002) and, in the case of facial perception, might be important for emotion understanding (Adolphs et al., 2000). The study by Carr et al. (2003) only showed SI activation during imitation and not during observation, however, their stimuli were static in comparison with our dynamic movies. Of the reported limbic regions, insula

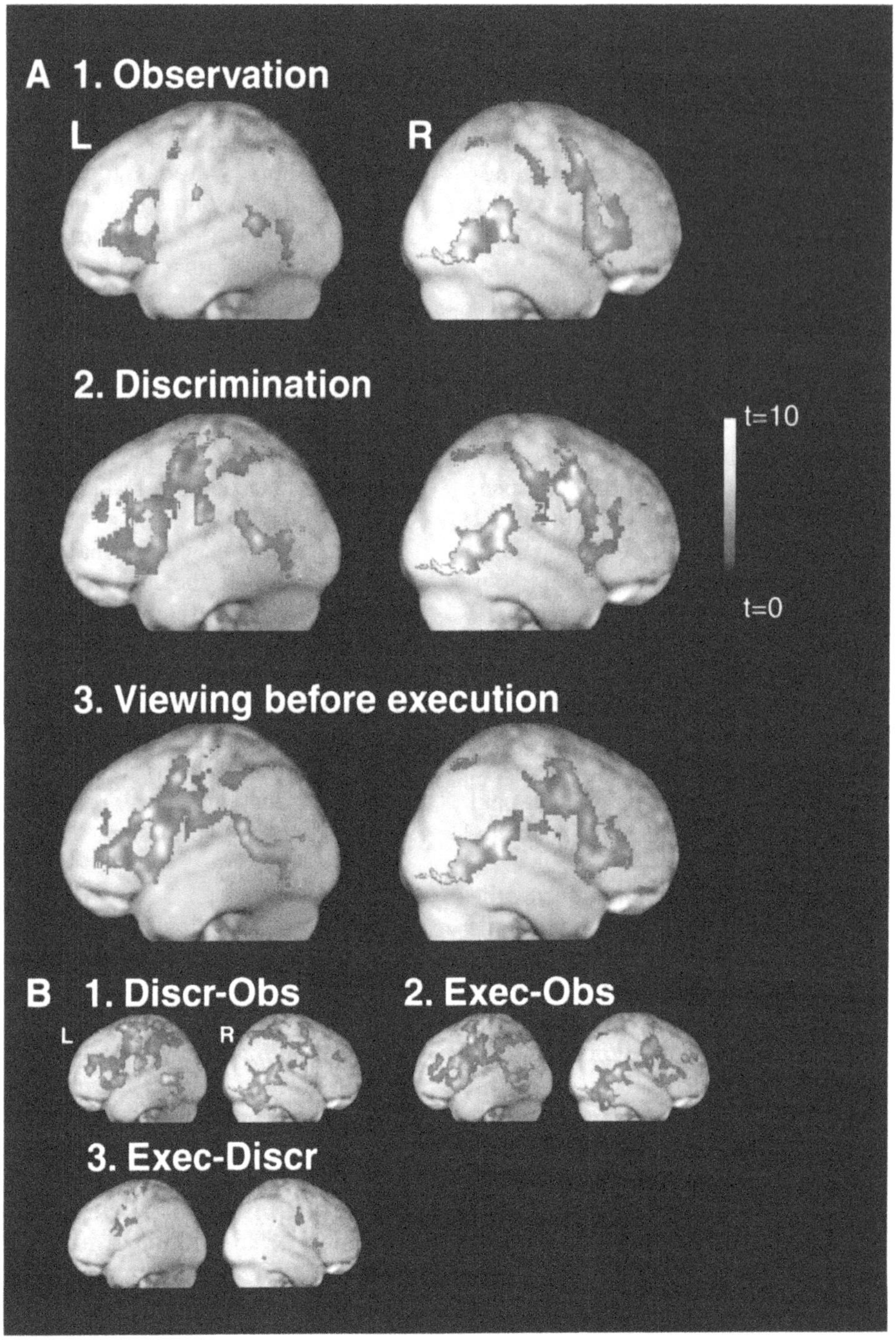

Figure 3. MNS analyses during the different task conditions. (A) shows those brain regions that are active during facial expression viewing, all four facial conditions combined – rest, $t(16) > 2.92$, $p < .005$, $k = 20$, inclusively masked with facial expression execution. The same analysis was performed with passive observation (1), discrimination (2) and viewing-before-imitation (3) as viewing conditions. (B) shows brain regions that significantly differentiate between facial viewing tasks, matched pair t-test, $t(16) > 2.92$, $p < .005$, $k = 20$, inclusively masked with facial expression production, all four facial conditions – rest, $t(16) > 2.92$, $p < .005$. The following pairwise comparisons were performed: discrimination contrasted against observation (1), viewing-before-imitation contrasted against observation (2), and viewing-before-imitation contrasted against discrimination (3). Inclusive masking at $p < .005$ of a contrast thresholded at $p < .005$ resulted in an overall false positive rate of $p < .005^2 \geq p < .000025$. Results are overlaid on an averaged T1-weigthed 3D render of the 17 subjects. (See Appendices 6 & 7 for coordinates of local maxima.)

involvement in the processing of facial emotions using a mirror-paradigm has also been observed before (Carr et al., 2003; Hennenlotter et al., 2005; Wicker et al., 2003), but only the study by Carr et al. (2003) reported amygdala involvement. The amygdala involvement during facial expression imitation and its mirror neuron capacities will be discussed further in section d. Overlap in visual areas is unlikely to be caused by the presence of mirror neurons with true motor properties in these regions. As discussed above, overlaps are more likely to be due to visual components of the imitation task.

With this analysis we have shown that inferior frontal regions become active when subjects view unfolding facial expressions, even if they are unaware of an upcoming imitation task. This is important, as it shows that premotor results of previous studies (Carr et al., 2003; Hennenlotter et al., 2005; Leslie et al., 2004) can not entirely be explained by anticipated motor requirements of the task.

In general, our two active tasks (discrimination and imitation) lead to stronger and more extensive brain activations. From the discrimination task (delayed match-to-sample), only the sample facial expression was analyzed. Defining the mirror neuron circuit based on the viewing phase of the discrimination task instead of the observation task, reveals a number of additional brain regions: the bilateral middle frontal gyrus, putamen, inferior parietal cortex, middle cingulate gyrus and the left primary motor area (Figure 3A; Appendix 6). From the imitation task, only the vision of facial expression movies were analyzed here, the imitation episode was temporally separated and analyzed in section a. Results based on the viewing part of imitation resembled those during the discrimination task.

Comparing the different tasks directly (Figure 3B; Appendix 7), we found that the viewing phase of discrimination and imitation resulted in significantly stronger brain activations than the more passive observation task demonstrating task influence on the mirror system. Increased brain activation during the active tasks could be related to increased visual attention or to working memory (Druzgal & D'Esposito, 2003; Haxby, Petit, Ungerleider, & Courtney, 2000). Premotor and parietal augmentations may additionally be linked to motor requirements of the response requested from the subjects (button press/facial expression production; Grezes, Costes, & Decety, 1999; Grezes & Decety, 2001). Only the anterior right superior frontal gyrus was more involved in observation compared to imitation. No brain area was more active during the observation task than during the discrimination task. While our data thus suggest that the activity of the MNS can be modulated by attentional factors, the origin of this modulation is likely to lie outside of the MNS (e.g., see Corbetta & Shulman, 2002).

Comparing the viewing phase during the discrimination and imitation tasks directly revealed differences in areas related to the motor system, premotor, primary motor, cerebellum, putamen and globus pallidus (Figure 3B; Appendix 7). Interestingly, the primary motor differences were compatible with the differences in task requirement: the discrimination task requiring the pressing of a button with the right hand determined stronger activations in a dorsal sector of the contralateral M1, while the imitation task requiring facial movements determined larger activations in a ventral sector of bilateral M1 known to relate to facial movements (Carr et al., 2003; Dapretto et al., 2006; Hennenlotter et al., 2005; Leslie et al., 2004). This finding suggests that a readiness to provide motor output can bias activation in paradigms investigating mirror activity, and advocates the use of paradigms not requiring motor output in order to get a "pure" estimate of mirror activity. Despite these task differences, the overall pattern of areas defined as "mirror" in all three tasks (Figure 3A) is reassuringly similar.

In summary, the classical MNS and some affective regions like the insula and amygdala resonate during the observation of facial expressions. This circuit becomes spontaneously active when we watch facial expressions, even when subjects are unaware of future imitation tasks. The addition of cognitive tasks to the observation process that have to result in a motor output increases not only the amplitude of activated brain regions, but also the extent of the activated neural circuit. Paying attention to the muscular movements itself during the viewing episode of the imitation task or adding meaning to these muscular movements during the discrimination task influences the circuit similarly.

(c) Specificity in the MNS for biological motion

Figure 4 shows the results of the brain activation during the viewing of faces compared with that

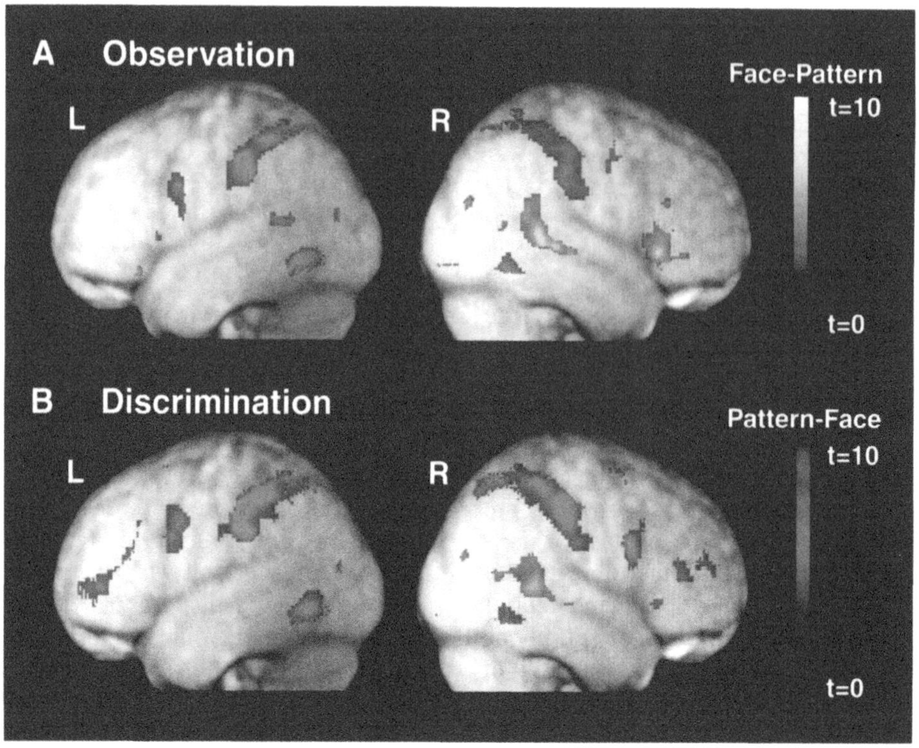

Figure 4. Signal increases in the MNS for the contrast dynamic faces – dynamic patterns during the passive observation (A) and the discrimination (B) task. Warm colors show brain regions that prefer the observation of dynamic faces over dynamic patterns, all four facial conditions combined-patterns, matched pair t-test, $t(16) > 2.92$, $p < .005$, $k = 20$, while cold colors represent brain regions that prefer the observation of dynamic patterns over facial expressions, matched pair t-test, $t(16) > 2.92$, $p < .005$, $k = 20$. All regions shown are also inclusively masked with facial expression production, all four facial conditions – rest, $t(16) > 2.92$, $p < .005$. The same false positive rate as explained in Figure 3 applies. (See Appendix 8 for coordinates of local maxima.)

during the observation of patterns, within the MNS described in section b. The bilateral ventral inferior frontal gyrus, bilateral STS/MTG, bilateral amygdala, SMA, the right lateral fusiform gyrus and the bilateral cerebellum (Appendix 8) all responded significantly more during the viewing of facial movements compared to pattern movements (in addition to being involved in the production of facial movements). This is in agreement with studies that compared static faces and patterns (Gorno-Tempini et al., 2001; Kesler-West et al., 2001). This biologically specific MNS is substantially more restricted than the MNS defined under section b. Interestingly, the parietal nodes of the classical MNS are activated more by moving patterns than faces while parts of the MTG and inferior frontal gyrus are more selective for faces. This is in agreement with a series of fMRI studies by Schubotz (Schubotz & von Cramon, 2001, 2002, 2004), showing activation of the ventral premotor cortex and posterior parietal regions in abstract pattern sequence processing. In a study containing biological motion and pattern sequence observation, only the pattern condition activated the parietal regions (Schubotz & von Cramon, 2004). Besides a role in action observation and execution (Fogassi & Luppino, 2005), the posterior parietal area in monkeys has been shown to be involved in visuomotor transformations and space representations (Fogassi et al., 2005). Recent studies suggest a role of parietal areas in object recognition (Fogassi et al., 2005; Janssen, Vogels, Liu, & Orban, 2001), and widespread connections with inferior temporal regions (Luppino, Murata, Belmalih, Calzavara, & Rozzi, 2004) could provide the detailed visual input necessary for this function. Indeed, we also find higher inferior temporal activations during our pattern condition than during our face conditions. Alternatively, the identification of unfamiliar pattern movements might simply require more attentional resources, and this attentional effect might overshadow the significant parietal activations found during facial observation located in section b. Pattern processing might require more parietal involvement because pattern processing is less holistic than face processing and may thus require more explicit processing of the spatial relationship between the elements composing the moving pattern. During facial processing, more activation was observed in areas known to specialize in the processing of faces, the fusiform gyrus and MTG/STS (Kanwisher, McDermott, & Chun, 1997; Perrett, Rolls, & Caan, 1982).

Premotor mirror regions identified in section b were much reduced if contrasted with pattern motion. Of Broca's area and its right-sided homologue, only some ventral and anterior parts survived a direct contrast. This is in accordance with data by Schubotz and von Cramon (2004), showing comparable frontal activations during observation of biological hand movements compared with the observation of pattern sequences. Regions of premotor cortex commonly activated during facial expression and pattern observation might reflect general sequence processing qualities (Sakreida, Schubotz, Wolfensteller, & von Cramon, 2005). Because our ventral inferior frontal gyrus activations are bilateral with stronger activations on the right side, it is unlikely that they are due to silent naming of, for example, the facial expressions (Grezes & Decety, 2001).

From the present data, BA 44/45 could thus contain neurons that would selectively map the sight of facial movements onto motor programs involved in facial expressions. Due to the limitations of fMRI, this conclusion can only be tentative. Single-cell recordings in primates, however, have demonstrated the existence of neurons responding to facial expressions more than to other stimuli (O Scalaidhe, Wilson, & Goldman-Rakic, 1997) in an area of the inferior frontal cortex that could correspond to BA 44/45 (Petrides, Cadoret, & Mackey, 2005). The same area has been shown to contain neurons selectively involved in orofacial actions. Unfortunately, whether the same neurons in this area respond both to the observation and execution of facial expressions has not yet been tested.

As the only form of biological movement in this experiment was of facial nature, the stronger responses in BA 44/45 could reflect specificity for *facial* movements or specificity for *biological* movements. A number of studies have found ventral premotor areas to respond more strongly to face compared to hand actions (Buccino et al., 2001; Sakreida et al., 2005; Wheaton, Thompson, Syngeniotis, Abbott, & Puce, 2004) supporting the idea that BA 44/45, although not exclusively devoted to the face, may indeed be particularly involved with the execution and observation of facial actions.

A similar pattern of results was observed when facial and pattern processing was analyzed during the discrimination task, although premotor activations to faces were less pronounced during discrimination. Habituation due to the fact that the same stimuli were used in both tasks and that discrimination always occurred after observation may explain this difference. In addition, the requirement for motor output during discrimination even for the patterns may have influenced these results.

In sum, this analysis shows that only limited premotor regions show biological motion specificity. Anterior portions of the ventral premotor cortex might be the site for the direct matching of observed facial expressions onto the motor representation of facial expressions in the observer. More dorsal and posterior premotor regions might be involved in sequence processing, be it of biological motion or not. Pre-SMA and the cerebellum showed areas of biological selectivity.

(d) Emotion specificity in the MNS

First, we searched for an overlap in the observation and execution of facial expressions. Therefore, a similar contrast comparing the emotional facial conditions with the neutral condition was performed both during viewing and production tasks. The bilateral insula/frontal operculum (IFO; including bilateral BA 44 and right BA 45), the right anterior STS and the pre-SMA (Figure 5; Appendix 9) were involved in both tasks. All these regions are biologically specific as well, as they are also activated in the contrast described under section c. Previous studies found increased ventral premotor activations due to observation (e.g., Gorno-Tempini et al., 2001; Kesler-West et al., 2001; Sprengelmeyer, Rausch, Eysel, & Przuntek, 1998), volitional and spontaneous production (Wild et al., 2006) or observation and imitation (e.g., Carr et al., 2003) of facial expressions. The insula was reported to be active in both facial expression observation and execution in two studies, one that combined several emotions in their analysis (Carr et al., 2003) and one that only studied smiling facial expressions (Hennenlotter et al., 2005). In both studies the neutral facial condition did not contain implicit or explicit movements. We therefore show for the first time that emotional facial expression

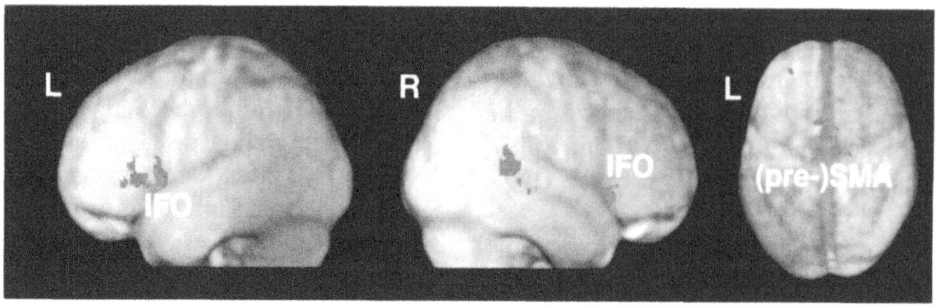

Figure 5. Signal increases for the viewing of emotional facial expressions minus neutral facial expressions, after combining the observation and discrimination task, average of three emotional facial conditions – neutral facial condition, $t(16) > 2.92$, $p < .005$, $k = 20$. Blue regions are areas that do not differentiate between the execution of emotional and neutral facial expressions, average of three emotional facial conditions combined – neutral facial condition, $t(16) > 2.92$, $p < .005$, $k = 20$, despite the inherent difference during viewing, while orange areas differentiate between the execution of emotional and neutral facial expressions, average of three emotional facial conditions combined – neutral facial condition, $t(16) > 2.92$, $p < .005$, $k = 20$, in addition to the viewing difference (congruent mapping). The same false positive rate as explained in Figure 3 applies. (See Appendix 9 for coordinates of local maxima.) L = left hemisphere; R = right hemisphere; IFO = insula/frontal operculum; STS = superior temporal sulcus; SMA = supplementary motor area.

observation and execution overlaps in bilateral IFO, when both the neutral and emotional facial expressions contained facial movements. Motor programs for emotional facial expressions located in the ventral inferior frontal cortex of the observer might resonate when emotional facial expressions are witnessed. This effect might simply be a reflection of the increased complexity of the emotional facial expressions, as the movements in the neutral condition were only limited to the lower regions of the face. The increased complexity entails at the same time an (emotional) intention (see Figure 2), which has recently been shown to increase the activation within inferior frontal cortex (Iacoboni et al., 2005).

Looking at the reverse contrast, neutral–emotional, we could locate brain regions that are significantly more active during both the viewing and execution of the neutral condition over emotional facial expressions. Bilateral SI and right SII (Appendix 9) are significantly more active during the viewing and execution of neutral facial expressions compared with emotional facial expressions. Although we cannot exclude the possibility that subjects overtly blew up their cheeks during our viewing condition, located brain regions are very similar as found by a study by Blakemore, Bristow, Bird, and Frith (2005) that analyzed the overlaps between the viewing and experiencing of touching the face.

In conclusion, there seems to exist a dissociation between the viewing and experiencing of neutral facial movements like blowing up the cheeks and the viewing and experiencing of emotional facial movements: blowing up the cheeks activates relatively more somatosensory regions (SI and SII), while emotional facial movements activate the IFO relatively more. This makes sense, as emotional facial expressions are more related to interoceptive representations of emotions supported by the insula (Craig, 2002; Damasio et al., 2000), while blowing up the cheeks may activate relatively more extero- and proprioceptive pathways, which eventually terminate in parietal somatosensory cortex like SI and SII (Craig, 2002).

Second, we tested whether we could find evidence of mirror neuron systems for specific facial conditions. We will focus our analyses on the IFO and amygdala as previous studies have shown that the insula is important for disgust recognition and experiencing and the amygdala might be important for fear recognition and experiencing (Calder et al., 2001; Keysers & Gazzola, 2006; however see Adolphs & Spezio, 2006; Van der Gaag et al., 2007). Moreover, these two regions were found to be active during both the observation and production of a combined analysis of six basic facial expressions (Carr et al., 2003). We showed above that only the (bilateral) IFO is involved in a direct matching analysis of emotional facial expressions, in which we combined all emotions to compare it with neutral. The amygdala was not activated in this general analysis because, unlike Carr et al. (2003), we used a more stringent analysis, contrasting emotional facial expressions not against a fixation condition but against a neutral facial expression (Van der Gaag et al., 2007). We also performed analyses on the somatosensory cortex as we located this region bilaterally in our direct matching analysis contrasting neutral facial expressions with emotional facial expressions. The somatosensory cortex has been attributed to have an important role in the recognition of facial expressions (Adolphs, 2002; Adolphs et al., 2000; Damasio et al., 2000).

To look into the question of facial condition specificity, we extracted the signal from the regions found in the general analysis (Figure 5; Appendix 9) for each condition individually (this was not possible for the amygdala as this structure was not activated in the general analysis) and tested the regions on condition specificity. A two-way analysis of variance (ANOVA) was performed with Task (viewing, execution) and Facial Condition (neutral, disgust, fear, happy) as predictor variables. We label areas as "specific" for a target facial condition if the target condition is significantly more active than all the other facial conditions in this analysis.

In the right IFO (Figure 6A), there was a significant main effect of task, $F(1, 16) = 11.1$, $p < .004$, with higher parameter estimates during execution. There was also a significant main effect of emotion, $F(3, 48) = 15.0$, $p < .001$. There was no significant interaction on parameter estimates, $F(3, 48) = 2.2$, $p > .102$, between task and condition. Simple contrasts comparing disgust with the remaining facial conditions (*a priori* hypothesis) showed that the parameter estimates during neutral, $F(1, 16) = 39.4$, $p < .001$, differed significantly from disgust, however, fear and happy did not significantly differ from disgust ($p > .25$), in accordance with the finding of Jabbi et al. (2007).

In the left IFO (Appendix 10A), the same pattern was true, with significant main effects of task, $F(1, 16) = 4.94$, $p < .04$, and emotion, $F(1.63, 26) = 15.8$, $p < .001$ (Greenhouse–Geisser corrected). There was no significant interaction on parameter estimates, $F(3, 48) = 1.54$, $p > .22$. Simple contrasts comparing disgust with the remaining facial conditions showed again that parameter estimates during neutral, $F(1, 16) = 47.3$, $p < .001$, differed significantly from disgust, however, fear and happy did not significantly differ from disgust ($p > .15$). The SMA and STS (as defined in Appendix 9) also failed to show specificity for any of the emotions used (Appendix 10A).

Fear specificity in the amygdala was tested by means of extracting the signal from *anatomically* defined amygdala (see materials and methods). During the different viewing conditions, the amygdala was not specific for fear or any of the other facial conditions, as reported elsewhere (Van der Gaag et al., 2007). During facial expression production, we performed a two-way analysis of variance (ANOVA) with Hemisphere (left, right) and Facial Condition (neutral, disgust, fear, happy) as predictor variables. There was a significant main effect of facial condition, $F(1, 16) = 7.7$, $p < .011$, however no significant main effect of hemisphere, $F(3, 48) = 5.9$, $p > .45$, and no significant interaction, $F(3, 48) = 15.0$, $p > .45$. The left amygdala was less active during fear compared to all other emotions (Figure 6C), however this difference was significant only compared to happiness ($p < .05$) explored by Newman–Keuls post hoc analyses. In the right amygdala, the same trend was true, fear significantly lower than happiness ($p < .0005$) with significantly less activation during fear against neutral ($p < .005$). In conclusion, also during the different imitation conditions, the amygdala was not specific for fear or any of the other facial conditions. A voxel-wise analysis of the IFO and amygdala led to similar results (Appendix 9: no IFO in the disgust specificity analysis and no amygdala in the fear specificity analysis).

In the right SI (Appendix 10B), there was a significant main effect of task, $F(1, 16) = 6.9$, $p < .02$, with higher parameter estimates during execution. There was also a significant main effect of facial condition, $F(3, 48) = 9.8$, $p < .001$. There was no significant interaction, $F(3, 48) = 0.01$, $p > .99$, between task and condition. Simple contrasts comparing neutral with the remaining facial conditions showed that neutral differed significantly from all the other facial conditions ($p < .01$).

In the left SI, the same pattern was true (Appendix 10B), with a significant main effects of task, $F(1, 16) = 10.9$, $p < .005$, and facial condition, $F(3, 48) = 16.7$, $p < .001$. There was no significant interaction, $F(3, 48) = 1.0$, $p > .4$. Simple contrasts comparing neutral with the remaining facial conditions showed again that the neutral condition differed significantly from the emotional conditions ($p < .005$).

In the right SII (Figure 6B), there was a significant main effect of task, $F(1, 16) = 19.2$, $p < .001$, and facial condition, $F(3, 48) = 15.4$, $p < .001$, without a significant interaction, $F(3, 48) = 1.6$, $p > .1$, between task and condition. Simple contrasts comparing neutral with the remaining facial conditions showed that the

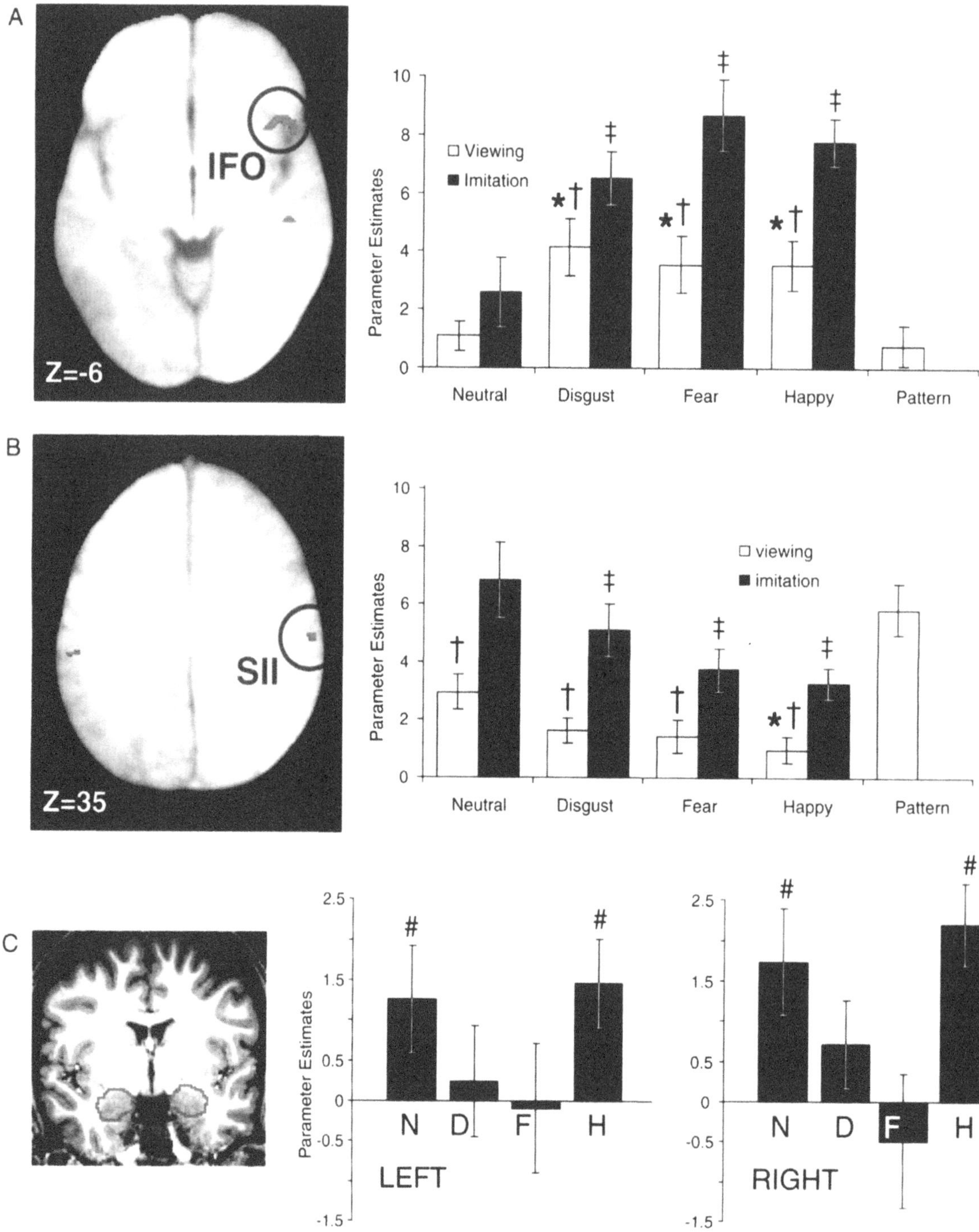

Figure 6. (See facing page for caption.)

neutral condition differed significantly from all facial emotional conditions (p <.005).

In summary, the IFO seems to be a region that is more involved in the observation and production of emotional facial expressions compared with neutral facial expressions, which is in agreement with previous findings (e.g., Carr et al., 2003). The insula might be a region with emotional mirror neuron properties, although we did not find evidence for a disgust specificity in the insula (like Damasio et al., 2000; Morris et al., 1998; Phan et al., 2002; Schienle et al., 2002). As has been suggested previously (Carr et al., 2003), the insula might be a relay station between the classical MNS and limbic regions like the amygdala important for emotional processing. We found the amygdala to be activated during facial expression observation and production (see section b), however, activations during viewing and production of *emotional* facial expressions were not significantly higher than during *neutral* facial expressions when both neutral and emotional facial expressions contain facial movements. Specificity of the amygdala in fear recognition and production could not be established. We therefore did not find evidence for the existence of an *emotional* MNS in the amygdala. This fits with more recent data showing that the amygdala is not an essential site for fear recognition (Adolphs et al., 2005; Adolphs & Spezio, 2006) and not essential for the volitional production of facial expressions including fear (Anderson & Phelps, 2000).

In contrast, bilateral SI and right SII show higher activations for blowing up the cheeks (our neutral condition) than for our emotional facial expressions.

Finally, we performed a general and preliminary search for regions specifically involved in the observation and production of particular facial movements (Appendices 9, 11 & 12). For the neutral facial condition, specific overlaps between observation and execution were found in right SI, bilateral SII, and premotor cortex, which fits with our previous analysis contrasting neutral with emotional facial expressions during both observation and execution (see above). For disgust, the left putamen, the left lingual gyrus, the right fusiform gyrus and occipital regions showed mirror properties. Overlap in observation and production of fearful facial expressions were found in bilateral inferior frontal cortex, right middle frontal gyrus, right premotor cortex and left middle temporal gyrus. Only two regions were happy specific, which were the right superior medial gyrus and right superior temporal gyrus.

Summarizing the previous sections, only minor differences in the MNS were found between facial expressions. Comparing emotional facial expressions with blowing up the cheeks there is, however, a clear difference. Emotional facial expressions selectively activate the IFO region more in both viewing and execution, supporting the occurrence of motor and emotional simulation during the viewing of emotional facial expressions. During the viewing of our neutral expression we found evidence for increased somatosensory simulation compared with our emotional conditions. We hypothesize that in combining simulation at motor, somatosensory and limbic levels one really gains insight into what is happening in other people.

Several limitations apply to this part of the study: the displayed actors on the movies were unfamiliar to the subjects, the context in which the actors showed the facial expressions was absent in addition to the absence of a real interaction between subjects and actors when using film clips. As a result, the viewing conditions of the emotions displayed were of limited self-relevance to the subjects being scanned. Activations of our viewing tasks might therefore be more related to physical aspects of the stimuli used than to emotional processing. Second, the cognitive generation of an emotion is likely to be different from a spontaneous emotional experience in daily life. Moreover, a spontaneous emotional facial expression might be processed through a partly different neural circuit than the volitional production of one

Figure 6 (opposite). Contrast values of the different viewing and execution conditions from several ROIs. Contrast values from the different ROIs were extracted from individual subjects using the MarsBar toolbox (see material and methods). (A) Left: Right IFO ROI resulting from the congruent mapping analysis of the emotional facial expression minus neutral facial expression contrast, as depicted in Figure 5 and listed in Appendix 9; Right: Contrast values for the different stimuli during both viewing, combining observation and discrimination (white bars) and execution (black bars) from our IFO ROI. Error bars represent the *SEM*. The significant main effect of condition (see results) was further explored using Newman–Keuls post hoc comparisons: emotional facial conditions that significantly differed from the neutral condition during *viewing* ($p < .05$) are marked with an asterisk (*), facial conditions that significantly differed from the pattern condition during *viewing* ($p < .05$) are marked with a dagger (†), emotional facial conditions that significantly differed from the neutral condition during *execution* ($p < .05$) are marked with a double dagger (‡). Comparisons of the disgust condition with the remaining facial conditions within the IFO (*a priori* hypothesis) is discussed further in the text. (B) Left: Right SII ROI resulting from the congruent mapping analysis of the neutral facial expression minus emotional facial expression contrast as described in Appendix 9; Right: contrast values taken from the SII ROI during the different conditions. The same explanation for the graph and the symbols apply as described under (A). (C) Results from the anatomically defined amygdala analysis. Left: example tracing of left and right amygdala shown on one slice; Right: Contrast values for the different *execution* conditions for the left and right amygdalae. Conditions marked with a hash (#) produced activations that significantly differed from zero according to a one-tailed t-test ($p < .05$). Emotional facial conditions were never significantly larger than the neutral condition. The fear specificity analysis (*a priori* hypothesis) is described further in the text. IFO = insula/frontal operculum; SII = secondary somatosensory cortex; N = neutral facial condition; D = disgust; F = fear; H = happy.

(Hopf, Muller-Forell, & Hopf, 1992; Iwase et al., 2002; Wild et al., 2006). These drawbacks might explain the limited limbic system involvement in our emotional mirror system analyses. Our analysis has therefore to be regarded as preliminary, located "mirror areas" have to be replicated for the different emotions but can be useful in generating hypotheses.

GENERAL DISCUSSION

Mirroring of facial expressions has recently generated a lot of attention (Adolphs, 2006; Carr et al., 2003; Dapretto et al., 2006; Goldman & Sripada, 2005), not least because the MNS is hypothesized to be the neurological basis of fundamental human capacities such as empathy (Gallese et al., 2004). Encouraging evidence has been reported that the MNS is indeed involved in the observation of facial expressions in humans, although the exact nature of this MNS involvement remains rather sketchy. Moreover, important evidence from comparative single-cell studies in monkeys on the mirroring of facial expressions is still lacking.

Using dynamic displays of both emotional and neutral facial expressions, we found the bilateral inferior frontal gyrus (including sections of BA 44 and BA 45) and inferior parietal cortex to be involved in both the observation and execution of facial expressions. In addition to these "classical" MNS sites, we found a number of other regions to be involved in the observation and execution of facial expressions: STS–MTG, insula, amygdala (pre-)SMA and somatosensory cortex (SI and SII), hereafter called "extended" MNS. These findings replicate previous studies on this topic (Carr et al., 2003; Dapretto et al., 2006; Leslie et al., 2004). We studied the extent to which activations in the "classical" and "extended" MNS could depend on specific task instructions given to the subjects during the viewing of the facial conditions. Task effects within the "classical" and "extended" MNS could be established, although the different instructions essentially activated the same circuit. A passive observation task spontaneously activated the "classical" and "extended" MNS, even when subjects were unaware of an upcoming execution task. This finding is important, as several previous studies that reported activations within the MNS during facial expression observation all used paradigms in which observation and execution were intermixed conditions (Carr et al., 2003; Dapretto et al., 2006). Similarly, in other studies subjects knew during the observation of facial expressions that an execution task composed the next step of the experiment (Hennenlotter et al., 2005; Leslie et al., 2004). During mere observation, brain activations within premotor and parietal regions in these experiments could therefore be biased by the (implicit) attention of the subjects towards their own motor programs. The other viewing tasks that were used (paying attention to the emotion of facial expressions in a delayed match-to-sample task and paying attention to facial motor programs during an imitation task) increased the amplitude and the extent of the activations within the "classical" and "extended" MNS. This finding is at odds with a study by Iacoboni et al. (2005), which did not report differences in "classical" MNS activations due to differential task instructions during the observation of actions, so further studies will be necessary to resolve this discrepancy. The motor response requirement in our discrimination and imitation task but not our observation task may have been a factor in explaining the higher activations during the discrimination and imitation tasks. It should be noted, however, that our results suggest on the one hand that the MNS is recruited even in the absence of an explicit task but on the other hand that it can be modulated by instructions. The MNS may thus be "automatic" only in the sense of being recruited without explicit instructions or efforts, but not in the strong psychophysical sense of being immune to attention modulation.

Interestingly, only a very limited number of studies have investigated the issue of biological motion specificity within the MNS (Schubotz & von Cramon, 2004). We have shown biological specificity in parts of the MNS, as specific sites within inferior frontal cortex showed more involvement during the observation of moving faces than moving patterns. In addition, the pre-SMA, STS and amygdala also showed more activation during the observation of moving faces than moving patterns. Since we only compared pattern motion with a single category of biological motion, it is impossible to conclude from our data whether brain regions located through this analysis show true specificity for facial actions or specificity for more general biological motion. Interestingly, parietal regions showed greater activity during pattern motion observation than during facial movement observation. This finding is in apparent contrast with monkey single-cell data, which showed the presence of mirror neurons in the inferior parietal lobule that responded both to the execution of a particular action and the observation of a similar action (Fogassi & Luppino, 2005). Moreover, several neuroimaging studies in humans have shown congruent mapping of particular motor actions during both execution and observation in the parietal cortex (see Rizzolatti & Craighero, 2004, for a review). A possible explanation for the higher parietal activation during pattern viewing compared with face viewing could be that since the moving patterns were less familiar to our subjects, pattern processing had to draw more heavily on attentional resources like the parietal cortex. Face processing on the other hand could occur in more specialized regions like the Fusiform Face Area (FFA; Kanwisher et al., 1997).

One of the fundamental pillars of the mirror-matching hypothesis of emotion attribution from faces is that a specific observed expression is being transformed into a motor and/or affective and/or somatosensory representation similar enough to this observed expression to understand the emotion. So far, only one study has examined the MNS using different emotions (Wicker et al., 2003). A major aim of our study was therefore to take a careful look whether fMRI can provide evidence of this selective mapping property in the brain. In order to accomplish this, we used several different facial expressions including a neutral facial condition that also contained facial dynamics. The inclusion of facial dynamics in our neutral condition is crucial in limiting the motion confound contrasting moving emotional expressions with immobile neutral expressions as present in previous studies (e.g., Hennenlotter et al., 2005).

Prime candidates for finding facial expression specificity in the MNS are sub-regions of the premotor cortex that map the different expressions, and limbic brain regions more intimately linked with specific emotions, e.g., amygdala and insula (Calder et al., 2001; Goldman & Sripada, 2005; Keysers & Gazzola, 2006). The "classical" mirror neurons in the premotor cortex could map the motor aspects of witnessed facial expressions (Leslie et al., 2004), while limbic regions could contain mirror neurons resonating the emotion behind the facial expressions (Carr et al., 2003; Wicker et al., 2003; see also Figure 1). In this study, we found emotional facial expressions compared with a neutral dynamic expression to activate premotor, insula and pre-SMA regions during both observation and execution. On the other hand, SI and SII were more strongly activated to neutral facial expressions compared to emotional facial expressions during both observation and execution. Stronger somatosensory activations during the observation and execution of blowing up the cheeks might be a reflection of the accumulation of tension around the mouth region while the cheeks are being inflated. This type of sensory information is processed through extero- and proprioceptive pathways, which eventually terminate in parietal somatosensory cortex such as SI and SII (Craig, 2002). Our finding of differential processing of neutral and emotional facial expressions in the brain is important, as it shows that the brain discriminates between neutral and emotional facial movements within specific brain regions in a congruent manner. Understanding other people from facial expressions might thus be a combined endeavor of simulation processes within different modalities: somatosensory, motor and limbic systems are all important. Simulation processes within all of these individual systems have been described previously (Gallese, Fadiga, Fogassi, & Rizzolatti, 1996; Keysers et al., 2004; Wicker et al., 2003). At the neuronal level, premotor mirror neurons might resonate the facial movement and its implied intention (Carr et al., 2003; Iacoboni et al., 2005), insula mirror neurons might process the emotional content (Wicker et al., 2003), and somatosensory neurons might resonate proprioceptive information contained in the observed facial movement (Keysers et al., 2004). This process fits with the theories of facial expression understanding (Adolphs, 2006; Carr et al., 2003; Leslie et al., 2004) as it proposes that several brain systems corroborate during the reading of facial expressions. Premotor cortex, insula and somatosensory regions can resonate while observing facial movements in other people, the amount and pattern of activation within each system depending on what expression is being witnessed.

However, the picture of the understanding of facial expressions by the MNS remains incomplete, as differences between individual emotional facial expressions within the MNS were minor. Disgust specificity in the insula and fear specificity in the amygdala could not be established. Studies have found specificity for individual emotions and sensations at the neuronal level in amygdala (Paton, Belova, Morrison, &

Salzman, 2006) and IFO (Scott & Plata-Salaman, 1999). However, neurons that code for opposite emotional valences were found side by side within the spatial resolution that would correspond with an fMRI voxel. In that case, the fMRI BOLD response will be unable to differentiate between neural activations due to different emotional valences. fMRI might therefore not be a very suitable technique to solve the specificity problem, new paradigms or other techniques are necessary to look more careful into this issue. Another limitation of our study is the limited self-relevance of the stimuli we used for the subjects, although we believe that the range of dynamic stimuli we adopted is still more suitable for studying the MNS than the still pictures and the restricted number of dynamic stimuli implemented in previous studies. In addition, although our execution part of the study might be reasonable in studying the motor aspects of the MNS, the study of emotional aspects is questionable as we asked subjects to generate emotional feelings by a cognitive induction method. A more natural induction of emotions during scanning is a difficult but essential next step in establishing emotional resonance within the brain (Wicker et al., 2003).

Strikingly, the pre-SMA showed up in all our analyses: it is spontaneously active during facial expression observation even in the absence of an explicit task, it responded more to the vision of facial movements than pattern motion and it responded more strongly to the observation and execution of emotional facial expressions than neutral facial expressions. Recent research has reported an anatomical and functional distinction between the pre-SMA and the SMA proper. While the more rostral pre-SMA might be more involved in the preparation and selection of the future execution of movements, the more caudal SMA might be more closely tied to the actual execution of movements (Cunnington, Windischberger, & Moser, 2005). A similar distinction in the function of pre-SMA and SMA-proper can be applied to the domain of language as well (Alario, Chainay, Lehericy, & Cohen, 2006), where pre-SMA is involved in the selection of words and SMA-proper in their production. It is interesting that surgical lesions to the (pre-)SMA have been reported to lead to impairments in the production of emotional facial expressions while other facial movements were less affected (Laplane, Talairach, Meininger, Bancaud, & Orgogozo, 1977). Some discussion, however, exists on whether the (pre-) SMA is crucial for the execution of emotional facial expressions as it has been hypothesized that a more ventral region in the anterior cingulate cortex might be responsible for difficulties in expression production (Morecraft, Stilwell-Morecraft, & Rossing, 2004). Recently, it has also been speculated that the pre-SMA is involved in attention to timing aspects of stimuli (Coull, Vidal, Nazarian, & Macar, 2004) and the representation of intention (Lau, Rogers, Haggard, & Passingham, 2004). These functions do not contradict the possibility that the pre-SMA contains mirror properties, as is suggested by the results of our study. In monkeys, the homologue of area pre-SMA (Area 6αβ or F6) shows complex neural responses that suggest that it may modulate ventral premotor cortex depending on whether an action is appropriate within a certain context or not (Rizzolatti et al., 1990). However, single-cell studies in this structure during action observation have, to our knowledge, not yet been performed. It may thus be that pre-SMA exerts a modulatory function on the ventral premotor MNS, for instance by inhibiting inappropriate imitation, in analogy to its modulatory function during action execution, or contain mirror neurons itself. By using depth electrodes, Krolak-Salmon et al. (2006) showed that the pre-SMA in humans is selectivity activated during both the observation and production of happy facial expressions. We could not replicate this promising finding of selectivity in the pre-SMA for happy facial expressions, differences in spatial resolution between the depth electrode technique and fMRI voxels could account for this discrepancy. As single-cell studies of the pre-SMA in monkeys and humans are lacking, conclusions about the mirror properties of this structure remain tentative.

In conclusion, findings from this study are encouraging, despite the limitations outlined before. In short, the "classical" mirror neuron system (composed of premotor and parietal areas), limbic regions and the somatosensory system become spontaneously active during the monitoring of facial expressions and the production of similar facial expressions. Restricted regions of this extensive network are selective for biological motion. Although we did not find evidence on the selectivity for specific emotions in the "classical" and "extended" MNS, we report a congruent discrimination between blowing up the cheeks (our neutral condition) and emotional facial expressions. Premotor and insula regions are more involved in the observation and execution of emotional facial expressions, while somatosensory regions are more active during the witnessing and production of blowing up cheeks. Based on our findings, the understanding of facial expressions of other people may be best conceived as involving a combined effort of simulation processes within motor, limbic, and somatosensory systems. This process might reflect the translation of the motor program, emotions and somatosensory consequences of facial expressions, respectively (Keysers & Gazzola, 2006). Through these processes we might translate the bodily and emotional states of others into our own. Understanding the simulated state of the other then boils down to understanding our own states, a process that probably requires additional neural circuits (Keysers & Gazzola, 2007).

REFERENCES

Adolphs, R. (2002). Neural systems for recognizing emotion. *Current Opinion in Neurobiology*, *12*, 169–177.

Adolphs, R. (2006). How do we know the minds of others? Domain-specificity, simulation, and enactive social cognition. *Brain Research*, *1079*, 25–35.

Adolphs, R., Damasio, H., Tranel, D., Cooper, G., & Damasio, A. R. (2000). A role for somatosensory cortices in the visual recognition of emotion as revealed by three-dimensional lesion mapping. *Journal of Neuroscience*, *20*, 2683–2690.

Adolphs, R., Gosselin, F., Buchanan, T. W., Tranel, D., Schyns, P., & Damasio, A. R. (2005). A mechanism for impaired fear recognition after amygdala damage. *Nature*, *433*, 68–72.

Adolphs, R., & Spezio, M. (2006). Role of the amygdala in processing visual social stimuli. *Progress in Brain Research*, *156*, 363–378.

Adolphs, R., Tranel, D., Damasio, H., & Damasio, A. (1994). Impaired recognition of emotion in facial expressions following bilateral damage to the human amygdala. *Nature*, *372*, 669–672.

Adolphs, R., Tranel, D., Hamann, S., Young, A. W., Calder, A. J., Phelps, E. A., et al. (1999). Recognition of facial emotion in nine individuals with bilateral amygdala damage. *Neuropsychologia*, *37*, 1111–1117.

Alario, F. X., Chainay, H., Lehericy, S., & Cohen, L. (2006). The role of the supplementary motor area (SMA) in word production. *Brain Research*, *1076*, 129–143.

Anderson, A. K., & Phelps, E. A. (2000). Expression without recognition: Contributions of the human amygdala to emotional communication. *Psychological Science*, *11*, 106–111.

Avikainen, S., Forss, N., & Hari, R. (2002). Modulated activation of the human SI and SII cortices during observation of hand actions. *NeuroImage*, *15*, 640–646.

Blair, R. J. (2003). Facial expressions, their communicatory functions and neuro-cognitive substrates. *Philosophical Transactions of the Royal Society of London, Series B: Biological Sciences*, *358*, 561–572.

Blakemore, S. J., Bristow, D., Bird, G., Frith, C., & Ward, J. (2005). Somatosensory activations during the observation of touch and a case of vision–touch synaesthesia. *Brain*, *128*, 1571–1583.

Brett, M., Anton, J. L., Valabregue, R., & Poline, J. B. (2002). Region of interest analysis using an SPM toolbox. *NeuroImage, 16*, abstr. 497.

Buccino, G., Binkofski, F., Fink, G. R., Fadiga, L., Fogassi, L., Gallese, V., et al. (2001). Action observation activates premotor and parietal areas in a somatotopic manner: An fMRI study. *European Journal of Neuroscience*, *13*, 400–404.

Calder, A. J., Keane, J., Manes, F., Antoun, N., & Young, A. W. (2000). Impaired recognition and experience of disgust following brain injury. *Nature Neuroscience*, *3*, 1077–1078.

Calder, A. J., Lawrence, A. D., & Young, A. W. (2001). Neuropsychology of fear and loathing. *Nature Reviews Neuroscience*, *2*, 352–363.

Carr, L., Iacoboni, M., Dubeau, M. C., Mazziotta, J. C., & Lenzi, G. L. (2003). Neural mechanisms of empathy in humans: A relay from neural systems for imitation to limbic areas. *Proceedings of the National Academy of Sciences of the USA*, *100*, 5497–5502.

Corbetta, M., & Shulman, G. L. (2002). Control of goal-directed and stimulus-driven attention in the brain. *Nature Reviews Neuroscience*, *3*, 201–215.

Coull, J. T., Vidal, F., Nazarian, B., & Macar, F. (2004). Functional anatomy of the attentional modulation of time estimation. *Science*, *303*, 1506–1508.

Craig, A. D. (2002). How do you feel? Interoception: the sense of the physiological condition of the body. *Nature Reviews Neuroscience*, *3*, 655–666.

Cunnington, R., Windischberger, C., & Moser, E. (2005). Premovement activity of the pre-supplementary motor area and the readiness for action: studies of time-resolved event-related functional MRI. *Human Movement Science*, *24*, 644–656.

Damasio, A. R., Grabowski, T. J., Bechara, A., Damasio, H., Ponto, L. L., Parvizi, J., et al. (2000). Subcortical and cortical brain activity during the feeling of self-generated emotions. *Nature Neuroscience*, *3*, 1049–1056.

Dapretto, M., Davies, M. S., Pfeifer, J. H., Scott, A. A., Sigman, M., Bookheimer, S. Y., et al. (2006). Understanding emotions in others: mirror neuron dysfunction in children with autism spectrum disorders. *Nature Neuroscience*, *9*, 28–30.

Druzgal, T. J., & D'Esposito, M. (2003). Dissecting contributions of prefrontal cortex and fusiform face area to face working memory. *Journal of Cognitive Neuroscience*, *15*, 771–784.

Ferrari, P. F., Gallese, V., Rizzolatti, G., & Fogassi, L. (2003). Mirror neurons responding to the observation of ingestive and communicative mouth actions in the monkey ventral premotor cortex. *European Journal of Neuroscience*, *17*, 1703–1714.

Fitzgerald, D. A., Angstadt, M., Jelsone, L. M., Nathan, P. J., & Phan, K. L. (2005). Beyond threat: Amygdala reactivity across multiple expressions of facial affect. *NeuroImage*, *30*, 1441–1448.

Fogassi, L., Ferrari, P. F., Gesierich, B., Rozzi, S., Chersi, F., & Rizzolatti, G. (2005). Parietal lobe: from action organization to intention understanding. *Science*, *308*, 662–667.

Fogassi, L., & Luppino, G. (2005). Motor functions of the parietal lobe. *Current Opinion in Neurobiology*, *15*, 626–631.

Fried, I., Wilson, C. L., MacDonald, K. A., & Behnke, E. J. (1998). Electric current stimulates laughter. *Nature*, *391*, 650.

Friston, K. J., Penny, W. D., & Glaser, D. E. (2005). Conjunction revisited. *NeuroImage*, *25*, 661–667.

Friston, K. J., Williams, S., Howard, R., Frackowiak, R. S., & Turner, R. (1996). Movement-related effects in fMRI time-series. *Magnetic Resonance in Medicine*, *35*, 346–355.

Gallese, V., Fadiga, L., Fogassi, L., & Rizzolatti, G. (1996). Action recognition in the premotor cortex. *Brain*, *119*(2), 593–609.

Gallese, V., Keysers, C., & Rizzolatti, G. (2004). A unifying view of the basis of social cognition. *Trends in Cognitive Sciences*, *8*, 396–403.

Gazzola, V., Aziz-Zadeh, L., & Keysers, C. (2006). Empathy and the somatotopic auditory mirror system in humans. *Current Biology*, *16*, 1824–1829.

Goldman, A. I., & Sripada, C. S. (2005). Simulationist models of face-based emotion recognition. *Cognition*, *94*, 193–213.

Gorno-Tempini, M. L., Pradelli, S., Serafini, M., Pagnoni, G., Baraldi, P., Porro, C., et al. (2001). Explicit and incidental facial expression processing: An fMRI study. *NeuroImage*, *14*, 465–473.

Grezes, J., Costes, N., & Decety, J. (1999). The effects of learning and intention on the neural network involved in the perception of meaningless actions. *Brain*, *122*(10), 1875–1887.

Grezes, J., & Decety, J. (2001). Functional anatomy of execution, mental simulation, observation, and verb generation of actions: A meta-analysis. *Human Brain Mapping*, *12*, 1–19.

Haxby, J. V., Petit, L., Ungerleider, L. G., & Courtney, S. M. (2000). Distinguishing the functional roles of multiple regions in distributed neural systems for visual working memory. *NeuroImage*, *11*, 380–391.

Hennenlotter, A., Schroeder, U., Erhard, P., Castrop, F., Haslinger, B., Stoecker, D., et al. (2005). A common neural basis for receptive and expressive communication of pleasant facial affect. *NeuroImage*, *26*, 581–591.

Hennenlotter, A., Schroeder, U., Erhard, P., Haslinger, B., Stahl, R., Weindl, A., et al. (2004). Neural correlates associated with impaired disgust processing in pre-symptomatic Huntington's disease. *Brain*, *127*, 1446–1453.

Hess, U., Philippot, P., & Blairy, S. (1998). Facial reactions to emotional facial expressions: Affect or cognition? *Cognition and Emotion*, *12*, 509–531.

Hopf, H. C., Muller-Forell, W., & Hopf, N. J. (1992). Localization of emotional and volitional facial paresis. *Neurology*, *42*, 1918–1923.

Iacoboni, M., Koski, L. M., Brass, M., Bekkering, H., Woods, R. P., Dubeau, M. C., et al. (2001). Reafferent copies of imitated actions in the right superior temporal cortex. *Proceedings of the National Academy of Sciences of the USA*, *98*, 13995–13999.

Iacoboni, M., Molnar-Szakacs, I., Gallese, V., Buccino, G., Mazziotta, J. C., & Rizzolatti, G. (2005). Grasping the intentions of others with one's own mirror neuron system. *PLoS Biology*, *3*, e79.

Iwase, M., Ouchi, Y., Okada, H., Yokoyama, C., Nobezawa, S., Yoshikawa, E., et al. (2002). Neural substrates of human facial expression of pleasant emotion induced by comic films: a PET Study. *NeuroImage*, *17*, 758–768.

Jabbi, M., Swart, M., & Keysers, C. (2007). Empathy for positive and negative emotions in the gustatory cortex. *NeuroImage*, *34*, 1744–1753.

Janssen, P., Vogels, R., Liu, Y., & Orban, G. A. (2001). Macaque inferior temporal neurons are selective for three-dimensional boundaries and surfaces. *Journal of Neuroscience*, *21*, 9419–9429.

Kanwisher, N., McDermott, J., & Chun, M. M. (1997). The fusiform face area: a module in human extrastriate cortex specialized for face perception. *Journal of Neuroscience*, *17*, 4302–4311.

Kesler-West, M. L., Andersen, A. H., Smith, C. D., Avison, M. J., Davis, C. E., Kryscio, R. J., et al. (2001). Neural substrates of facial emotion processing using fMRI. *Brain Research. Cognitive Brain Research*, *11*, 213–226.

Keysers, C., & Gazzola, V. (2006). Towards a unifying neural theory of social cognition. *Progress in Brain Research*, *156*, 379–401.

Keysers, C., & Gazzola, V. (2007). Integrating simulation and theory of mind: from self to social cognition. *Trends in Cognitive Science, 11,* 194–196.

Keysers, C., & Perrett, D. I. (2004). Demystifying social cognition: a Hebbian perspective. *Trends in Cognitive Science*, *8*, 501–507.

Keysers, C., Wicker, B., Gazzola, V., Anton, J. L., Fogassi, L., & Gallese, V. (2004). A touching sight: SII/PV activation during the observation and experience of touch. *Neuron*, *42*, 335–346.

Kilts, C. D., Egan, G., Gideon, D. A., Ely, T. D., & Hoffman, J. M. (2003). Dissociable neural pathways are involved in the recognition of emotion in static and dynamic facial expressions. *NeuroImage*, *18*, 156–168.

Krolak-Salmon, P., Henaff, M. A., Vighetto, A., Bauchet, F., Bertrand, O., Mauguiere, F., et al. (2006). Experiencing and detecting happiness in humans: the role of the supplementary motor area. *Annals of Neurology*, *59*, 196–199.

LaBar, K. S., Crupain, M. J., Voyvodic, J. T., & McCarthy, G. (2003). Dynamic perception of facial affect and identity in the human brain. *Cerebral Cortex*, *13*, 1023–1033.

Laplane, D., Talairach, J., Meininger, V., Bancaud, J., & Orgogozo, J. M. (1977). Clinical consequences of corticectomies involving the supplementary motor

area in man. *Journal of Neurological Science*, *34*, 301–314.
Lau, H. C., Rogers, R. D., Haggard, P., & Passingham, R. E. (2004). Attention to intention. *Science*, *303*, 1208–1210.
LeDoux, J. E. (2000). Emotion circuits in the brain. *Annual Review of Neuroscience*, *23*, 155–184.
Leslie, K. R., Johnson-Frey, S. H., & Grafton, S. T. (2004). Functional imaging of face and hand imitation: Towards a motor theory of empathy. *NeuroImage*, *21*, 601–607.
Lipps, T. (1907). Das wissen von fremden Ichen. In T. Lipps (Ed.), (Ed.), *Psychologische untersuchungen* (Vol. 1). Leipzig, Germany: Engelmann.
Luppino, G., Murata, A., Belmalih, A., Calzavara, R., & Rozzi, S. (2004). Ventral visual stream information to the AIP-F5 circuit for grasping: A tracing study in the macaque monkey [abstract]. *Society for Neuroscience Program*, no. 655.5.
McIntosh, D. N. (1996). Facial feedback hypotheses: evidence, implications, and directions. *Motivation and Emotion*, *20*, 121–147.
Morecraft, R. J., Stilwell-Morecraft, K. S., & Rossing, W. R. (2004). The motor cortex and facial expression: new insights from neuroscience. *Neurologist*, *10*, 235–249.
Morris, J. S., Friston, K. J., Buchel, C., Frith, C. D., Young, A. W., Calder, A. J., et al. (1998). A neuromodulatory role for the human amygdala in processing emotional facial expressions. *Brain*, *121*(1), 47–57.
Murphy, F. C., Nimmo-Smith, I., & Lawrence, A. D. (2003). Functional neuroanatomy of emotions: A meta-analysis. *Cognitive, Affective and Behavioral Neuroscience*, *3*, 207–233.
O Scalaidhe, S. P., Wilson, F. A., & Goldman-Rakic, P. S. (1997). Areal segregation of face-processing neurons in prefrontal cortex. *Science*, *278*, 1135–1138.
Oldfield, R. C. (1971). The assessment and analysis of handedness: the Edinburgh inventory. *Neuropsychologia*, *9*, 97–113.
Paton, J. J., Belova, M. A., Morrison, S. E., & Salzman, C. D. (2006). The primate amygdala represents the positive and negative value of visual stimuli during learning. *Nature*, *439*, 865–870.
Perrett, D. I., Rolls, E. T., & Caan, W. (1982). Visual neurones responsive to faces in the monkey temporal cortex. *Experimental Brain Research*, *47*, 329–342.
Petrides, M., Cadoret, G., & Mackey, S. (2005). Orofacial somatomotor responses in the macaque monkey homologue of Broca's area. *Nature*, *435*, 1235–1238.
Phan, K. L., Wager, T., Taylor, S. F., & Liberzon, I. (2002). Functional neuroanatomy of emotion: A meta-analysis of emotion activation studies in PET and fMRI. *NeuroImage*, *16*, 331–348.
Rizzolatti, G., & Craighero, L. (2004). The mirror-neuron system. *Annual Review of Neuroscience*, *27*, 169–192.
Rizzolatti, G., Fogassi, L., & Gallese, V. (2001). Neurophysiological mechanisms underlying the understanding and imitation of action. *Nature Reviews Neuroscience*, *2*, 661–670.
Rizzolatti, G., Gentilucci, M., Camarda, R. M., Gallese, V., Luppino, G., Matelli, M., et al. (1990). Neurons related to reaching-grasping arm movements in the rostral part of area 6 (area 6a beta). *Experimental Brain Research*, *82*, 337–350.
Russell, J. A., Bachorowski, J. A., & Fernandez-Dols, J. M. (2003). Facial and vocal expressions of emotion. *Annual Review of Psychology*, *54*, 329–349.
Sakreida, K., Schubotz, R. I., Wolfensteller, U, & von Cramon, D. Y. (2005). Motion class dependency in observers' motor areas revealed by functional magnetic resonance imaging. *Journal of Neuroscience*, *25*, 1335–1342.
Sato, W., Kochiyama, T., Yoshikawa, S., Naito, E., & Matsumura, M. (2004). Enhanced neural activity in response to dynamic facial expressions of emotion: An fMRI study. *Brain Research. Cognitive Brain Research*, *20*, 81–91.
Schienle, A., Stark, R., Walter, B., Blecker, C., Ott, U., Kirsch, P., et al. (2002). The insula is not specifically involved in disgust processing: An fMRI study. *Neuroreport*, *13*, 2023–2026.
Schubotz, R. I., & von Cramon, D. Y. (2001). Functional organization of the lateral premotor cortex: fMRI reveals different regions activated by anticipation of object properties, location and speed. *Brain Research. Cognitive Brain Research*, *11*, 97–112.
Schubotz, R. I., & von Cramon, D. Y. (2002). Dynamic patterns make the premotor cortex interested in objects: influence of stimulus and task revealed by fMRI. *Brain Research. Cognitive Brain Research*, *14*, 357–369.
Schubotz, R. I., & von Cramon, D. Y. (2004). Sequences of abstract nonbiological stimuli share ventral premotor cortex with action observation and imagery. *Journal of Neuroscience*, *24*, 5467–5474.
Scott, T. R., & Plata-Salaman, C. R. (1999). Taste in the monkey cortex. *Physiology and Behavior*, *67*, 489–511.
Singer, T., Seymour, B., O'Doherty, J., Kaube, H., Dolan, R. J., & Frith, C. D. (2004). Empathy for pain involves the affective but not sensory components of pain. *Science*, *303*, 1157–1162.
Sprengelmeyer, R., Rausch, M., Eysel, U. T., & Przuntek, H. (1998). Neural structures associated with recognition of facial expressions of basic emotions. *Proceeding of the Royal Society, Series B: Biology Science*, *265*, 1927–1931.
Van der, Gaag, C., Minderaa, R. B., & Keysers, C. (2007). The BOLD signal in the amygdala does not differentiate between dynamic facial expressions. *Social, Cognitive and Affective Neuroscience*, *2*, 93–103.
Wheaton, K. J., Thompson, J. C., Syngeniotis, A., Abbott, D. F., & Puce, A. (2004). Viewing the motion of human body parts activates different regions of premotor, temporal, and parietal cortex. *NeuroImage*, *22*, 277–288.
Wicker, B., Keysers, C., Plailly, J., Royet, J. P., Gallese, V., & Rizzolatti, G. (2003). Both of us disgusted in

my insula: The common neural basis of seeing and feeling disgust. *Neuron*, *40*, 655–664.
Wild, B., Rodden, F. A., Rapp, A., Erb, M., Grodd, W., & Ruch, W. (2006). Humor and smiling: Cortical regions selective for cognitive, affective, and volitional components. *Neurology*, *66*, 887–893.
Winston, J. S., O'Doherty, J., & Dolan, R. J. (2003). Common and distinct neural responses during direct and incidental processing of multiple facial emotions. *NeuroImage*, *20*, 84–97.
Yang, T. T., Menon, V., Eliez, S., Blasey, C., White, C. D., Reid, A. J., et al. (2002). Amygdalar activation associated with positive and negative facial expressions. *Neuroreport*, *13*, 1737–1741.

A BRIEF GUIDE TO THE APPENDICES

Appendix 1. Illustration of the tracing of the amygdalae in the 17 subjects.

1. Facial Execution

Appendix 2. Areas of increased signal during the execution of facial expressions.

Appendix 3. Results of facial expression imitation (combining all facial conditions – rest) as shown by a random effects analysis of 17 subjects.

Appendix 4. Comparisons of emotional and neutral execution (combining all emotional facial conditions – neutral facial condition) as shown by a random effects analysis of 17 subjects.

Appendix 5. Areas selectively involved in the production of one facial condition compared to the remaining facial conditions (execution specificity).

2. Task influences

Appendix 6. MNS analyses during the different task conditions.

Appendix 7. Brain regions that significantly differentiate between facial viewing tasks inclusively masked with facial expression production.

3. Biological specificity

Appendix 8. Brain regions that significantly differentiate between dynamic faces and dynamic patterns during the passive Observation (1) and the Discrimination (2) task, inclusively masked with facial expression production.

4. Facial condition specificity

Appendix 9. Emotional and neutral facial expression processing in the MNS.

Appendix 10. Contrast values of the different viewing and execution conditions from remaining ROIs which were not shown in Figure 6.

Appendix 11. Signal increases for the contrast Emotional facial condition minus Neutral facial condition during the viewing condition (combining observation and discrimination) for consecutively, the disgusted (A), fearful (B) and happy (C) expressions.

Appendix 12. Specificity for individual facial conditions within the MNS.

APPENDIX 1

(See p. 204 for caption.)

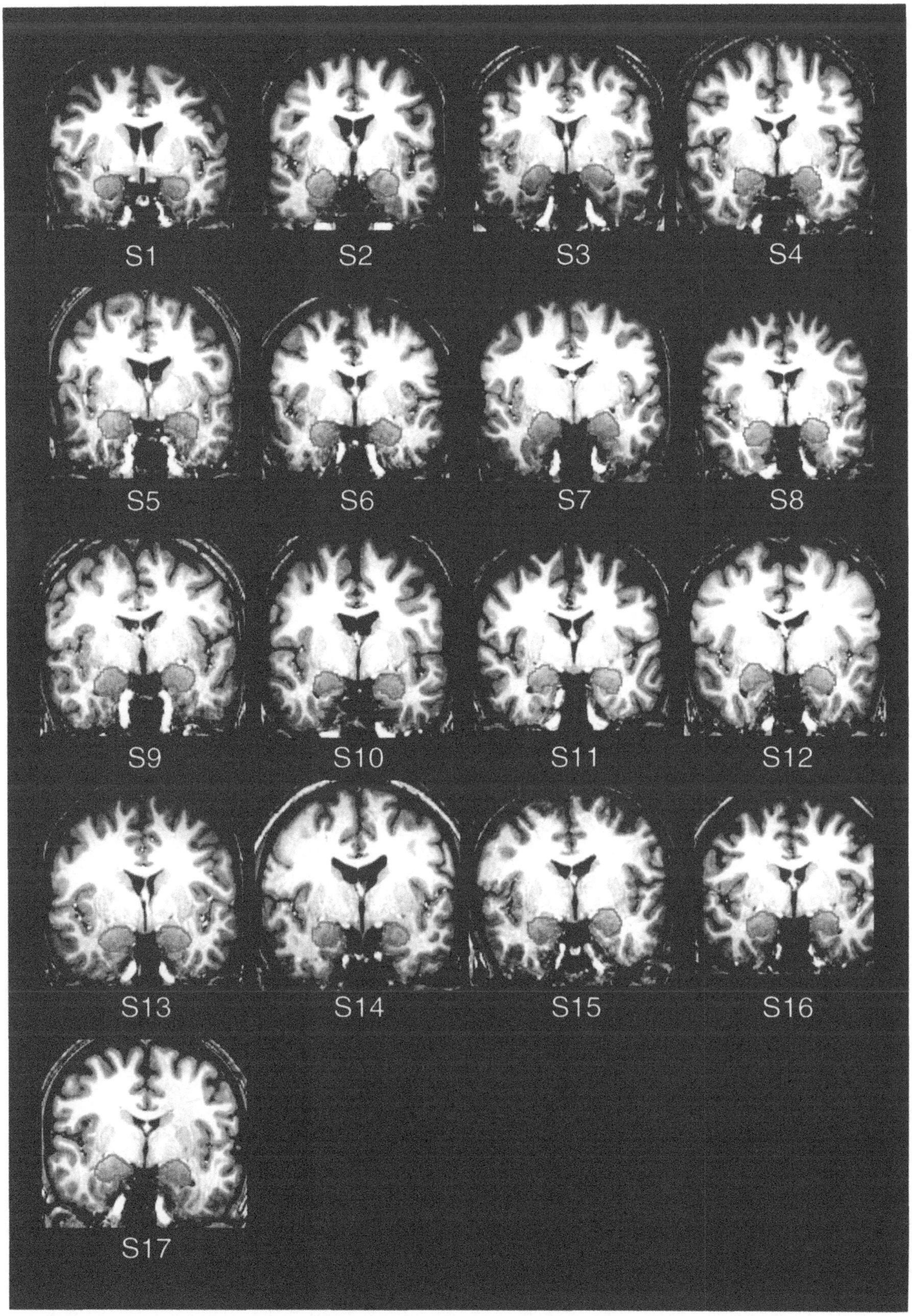

APPENDIX 1 (previous page)

Illustration of the tracing of the amygdalae in the 17 subjects (S1 . . . S17). A coronal slice running approximately through the middle of the amygdalae is shown for each subject. The red contour delineates the outside border of the traced region, being the first voxel outside of the traced amygdalae.

APPENDIX 2 (opposite)

Areas of increased signal during the execution of facial expressions.

(A) Shows those brain regions that are active during facial expression execution [all four facial conditions combined − rest, $t(16) > 2.92$, $p < .005$, $k = 20$]. Maps are shown with the same threshold as used for the inclusive masking for our MNS analyses.

(B) The same map as displayed in (A), only separated in areas that are more involved in emotional expression execution than neutral expression execution [three emotional facial conditions combined − rest, $t(16) > 2.92$, $p < .005$, $k = 20$] (in orange) and areas which do not differentiate between emotional and neutral expression execution [three emotional facial conditions combined − rest, $t(16) < 2.92$, $p < .005$, $k = 20$] (in blue).

(C) Areas selectively involved in the production of one facial condition compared to the remaining facial conditions (execution specificity). Pair-wise contrasts between the target emotion and the three remaining facial conditions were calculated ($p < .1$). In addition, the parameter estimate of the target emotion needed to be above zero at $p < .05$. Requiring that these four tests be significant results in a false positive rate under a global null hypothesis of $p_{global} < .1^3 \times .05 = .00005$ (see also the Materials and methods section). See Appendices 3, 4, and 5 for coordinates of local maxima.

APPENDIX 2

(See opposite for caption.)

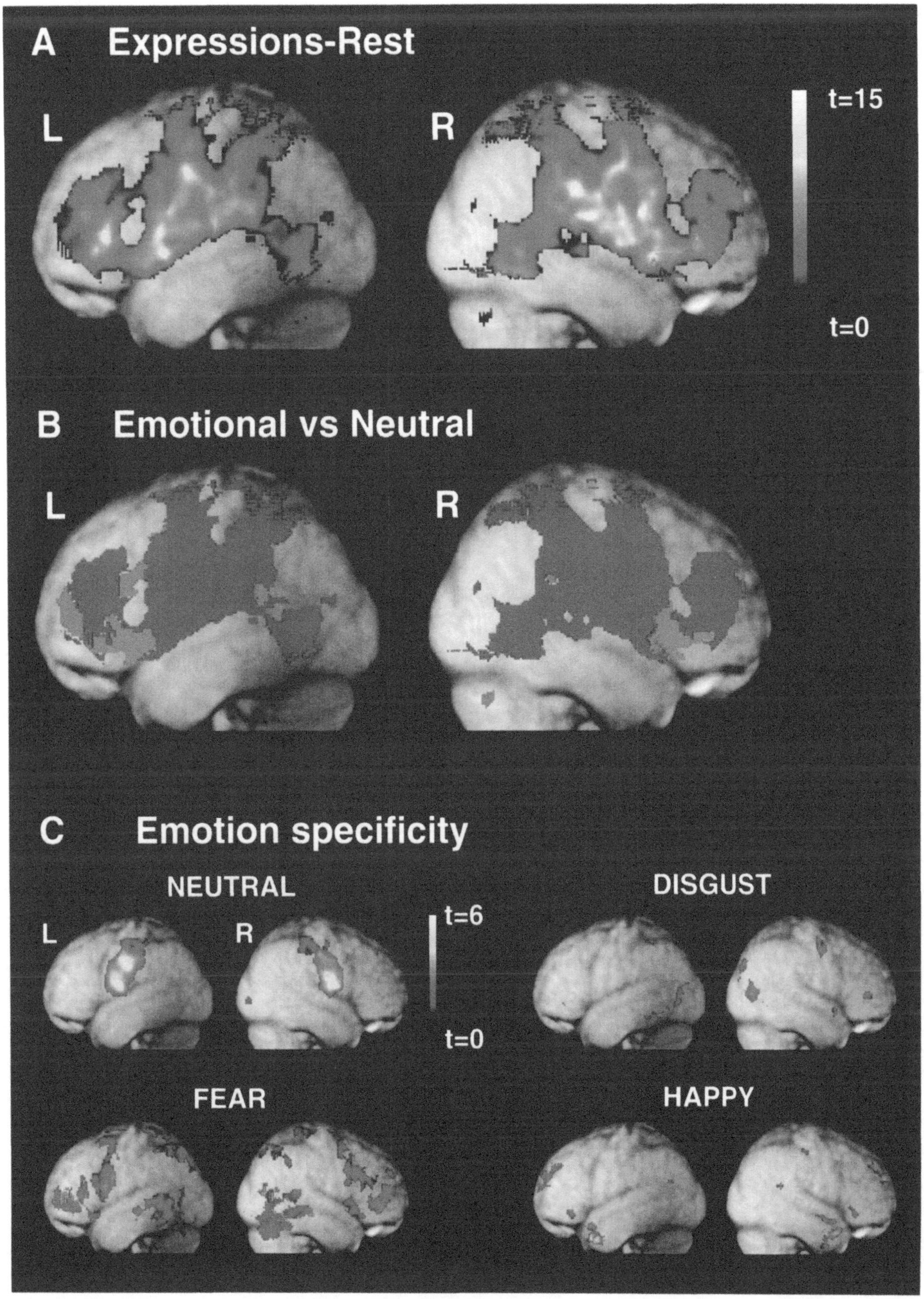

APPENDIX 3

Results of facial expression imitation (combining all facial conditions – rest, $p<.005$, $k=20$) as shown by a random effects analysis of 17 subjects. For each cluster, from left to right, we describe: the hemisphere containing the cluster; the anatomical and/or cytoarchitectonic description of the cluster, based on the anatomy toolbox provided by Eickhoff et al. (http://www.fz-juelich.de/ime/ime_brain_mapping); the number (k) of $2 \times 2 \times 2$ voxels contained in the cluster (if available); the uncorrected p-value (if available), corresponding t-value and the MNI coordinates of the peak voxel. Activations are presented per cluster and, if possible, are first shown for the regions of interest of this paper (premotor regions in inferior frontal cortex, BA 44/6 and BA 45, STS/MTG, posterior parietal cortex, motor regions and limbic regions like, insula, amygdala)

					MNI		
Hemisphere	*Anatomical description*	*k*	*p (uncorr.)*	*t*	*x*	*y*	*z*
Right	BA 44			13.31	56	8	18
Right	Area 6			12.03	12	−2	68
	Area 6			11.65	6	−4	64
	Area 6			10.85	12	4	72
Right	Area 6			8.59	28	−10	50
				6.55	32	0	50
Right	Area 6			8.15	22	−18	62
				6.15	22	−18	74
Right	Inferior frontal (p triangularis)			7.44	44	36	26
				7.14	42	36	18
Left	Inferior frontal (p triangularis)			12.58	−48	36	6
	Idem			11.09	−44	40	0
	Idem			8.79	−52	38	0
	BA 45			8.39	−42	46	18
Right	Middle temporal gyrus			8.35	58	−56	−2
	Inferior temporal gyrus			8.28	56	−58	−12
	Middle temporal gyrus			6.41	56	−62	6
Right	Middle temporal			7.49	46	−40	2
Right	Postcentral gyrus			10.67	24	−46	66
	Precuneus			10.36	8	−38	52
	Superior parietal			10.03	16	−60	60
	Superior parietal			9.25	18	50	54
Right	Supramarginal gyrus			14.90	58	−28	26
Left	Inferior parietal			7.73	−36	−48	40
	Middle occipital			7.39	−28	−56	36
	Inferior parietal			7.39	−36	−52	54
	Superior parietal			6.66	−28	−62	52
Left	Cuneus			7.60	−10	−82	34
	Idem			7.45	2	−80	38
	Idem			6.50	−6	−82	26
Left	Middle cingulate			9.70	−14	−32	42
	4p			9.49	−18	−30	60
	4p			8.77	−16	−46	62
	Superior parietal			7.98	−20	−54	62
Right	Insula			12.13	30	18	6
Right	Insula			10.57	40	−4	2
Right	Insula			9.91	40	2	6
Left	Superior temporal gyrus			12.48	−56	4	−2
	Pallidum			11.77	−24	0	−2
	Insula			11.50	−38	−6	−6
	Insula			11.30	−34	8	6
	Area 4 p			11.49	−54	−8	38
	L inferior frontal gyrus (opercularis)			11.40	−58	6	8

APPENDIX 3 (*Continued*)

					MNI		
Hemisphere	*Anatomical description*	*k*	*p (uncorr.)*	*t*	*x*	*y*	*z*
	OP1			11.45	−58	−22	14
	OP4			10.61	−56	−12	12
	Thalamus			11.45	−14	−14	2
Right	Middle cingulate cortex			11.53	6	20	30
Left	Middle cingulate cortex			13.82	−6	8	42
Right	Middle frontal			8.63	28	52	22
	Idem			8.00	32	40	28
Left	Middle frontal			6.59	−40	38	30
				6.54	−32	38	28
				6.57	52	−30	52
				6.35	14	−82	36
				6.26	18	−76	32
Right	Putamen			11.45	18	10	−2
Right/Left	Cerebellum	88 591	.000	15.62	28	−48	−34
	Areas 17, 18						
	Fusiform gyrus						

APPENDIX 4

Comparisons of emotional and neutral execution (combining all emotional facial conditions – neutral facial condition, $p<.005$, $k=20$) as shown by a random effects analysis of 17 subjects (for an explanation of the different columns see Appendix 3)

					MNI		
Hemisphere	*Anatomical description*	*k*	*p (uncorr.)*	*t*	*x*	*y*	*z*
Emot-N							
Right	Inferior frontal gyrus (p triangularis)/BA 45	1138	0	5.37	48	32	−2
	BA 44			4.98	56	20	6
	Insula			4.82	42	18	−6
Left	Inferior frontal gyrus (p orbitalis)	2574	0	5.96	−44	28	−6
	Middle frontal gyrus			5.10	−40	52	8
	Inferior frontal gyrus(p opercularis)			4.57	−40	14	18
	Middle orbital gyrus			3.93	−42	44	−8
	Insula			3.90	−26	22	−14
Right	Superior temporal gyrus	139	0	4.95	48	−24	−2
	Also MTG						
Left	MTG/STS	234	.001	3.98	−50	−62	12
Right	Supramarginal gyrus	31	.001	3.56	64	−40	26
Right	Precentral gyrus	25	.001	3.57	46	6	46
Right	Area 6	139	0	4.07	24	−26	64
	Area/3a/3b/4p/2						
Left	Area 6	72	.001	3.53	−18	−26	64
	Area 3b/4a/4p						
Right	Anterior cingulate cortex	121	0	4.42	2	36	0
Left	Anterior cingulate cortex	99	.001	3.89	−6	40	10
Right/Left	Anterior cingulate cortex	25	.001	3.58	6	12	26
Left	Middle cingulate cortex	1447	0	4.80	−4	30	32
Left	Middle cingulate cortex			5.17	−4	−20	32
Right	Pre-SMA			3.73	6	20	56
Left	Pre-SMA			4.19	2	20	48
Right/Left	Hypothalamus	120	0	4.39	−2	−10	−9
Right	Superior medial gyrus	53	0	3.51	10	60	34
Right	Middle frontal gyrus	47	.001	3.73	30	54	22
Left	Middle frontal gyrus	49	.002	3.46	−28	30	20
Left	Cerebellum	90	0	4.41	−22	−68	−40
Right/Left	Cerebellum	10642	0	7.28	2	−72	−30
	Area 17			6.2	−12	−52	2
Neutral-E							
Right	OP4	1323	0	7.11	60	−8	12
	1/2/3a/3b						
Left	Area 1	1709	0	7.29	−62	−16	34
	OP4/2/3a/3b						
Right	OP3	163	0	6.02	38	−6	16
Right	Cerebellum	50	0	4.32	20	−62	−22
Left	Cerebellum	39	0	4.13	−16	−64	−24
Right	Pallidum/amygdala	32	0	4.39	28	−8	−6
Right	Middle occipital gyrus/Area 18	28	0	5.08	34	−92	2

APPENDIX 5

Areas selectively involved in the production of one facial condition compared to the remaining facial conditions (execution specificity). Pairwise contrasts between the target emotion and the three remaining facial conditions were calculated ($p < .1$). In addition, the parameter estimate of the target emotion needed to be above zero at $p < .05$. Requiring that these four tests be significant results in a false positive rate under a global null hypothesis of $p_{global} < .1^3 \times .05 = .00005$, see also material and methods (for an explanation of the different columns see Appendix 3)

				MNI		
Hemisphere	*Anatomical description*	*k*	*t*	*x*	*y*	*z*
Neutral						
Right	Area 3b Also OP4/4a/4p/1/2	1945	4.89	58	−4	20
Left	Area 1 Also OP4/4a/4p/3b/2	2592	5.89	−60	−12	36
Right	Insula Also OP3	169	3.59	38	−8	8
Right	Cerebellum	98	3.65	20	−62	−22
Left	Cerebellum	98	3.47	−16	−64	−24
Right	Inferior/middle occipital gyrus	30	1.97	32	−94	−2
Right	Pallidum	28	2.34	28	−10	−4
Left	Putamen	24	1.92	−26	−8	−12
Disgust						
Right	Inferior frontal gyrus (p triangularis/opercularis)	58	2.15	42	16	8
Right	Area 6	124	2.48	40	−6	56
Left	Inferior frontal gyrus (orbital)	59	2.71	−30	38	−4
Right	Inferior parietal lobule	149	2.35	28	−50	52
Left	Superior parietal lobule	198	2.51	−32	−50	56
Right	Area 3a/4p	74	2.27	32	−20	40
Right	Insula	39	2.08	42	10	−12
Left	Insula	39	2.32	−36	−10	−4
Right	SMA/cingulate gyrus	52	2.11	10	14	46
Right	Putamen	23	2.00	16	16	−4
Left	Putamen	213	2.13	−18	6	−6
Right	Middle frontal	60	2.54	40	42	6
Right	Thalamus	212	2.01	18	−14	10
Right	Thalamus	31	2.26	6	−10	−2
Left	Inferior temporal gyrus/fusiform	62	2.61	−46	−42	−16
Left	Middle occipital gyrus	61	1.94	−38	−62	2
Left/Right	Lingual gyrus Also calcarine/17/18	4387	4.03	−18	−62	−6
Left/Right	Cerebellum	698	3.69	16	−66	−40
Fear						
Right	Inferior frontal gyrus (p triangularis) Also 44/45	625	2.62	44	12	24
Right	Inferior frontal gyrus (p opercularis) Also area 44	146	2.58	52	18	−6
Right	Area 6	126	2.69	14	−22	64
Left	Precentral gyrus Also 44/6/45	1537	3.63	−42	0	26

APPENDIX 5 (*Continued*)

				MNI		
Hemisphere	*Anatomical description*	*k*	*t*	*x*	*y*	*z*
Left	Inferior frontal gyrus (p triangularis)	248	3.06	−38	34	8
Left	BA 45	68	2.22	−52	32	16
Left	MTG	276	2.76	−48	−50	0
Left	Inferior parietal Also middle/superior occipital gyrus	558	2.05	−30	−72	44
Left/Right	Precuneus Also 4a/6/3b Inferior parietal cortex	1643	2.77	−8	−52	72
Left	Amygdala Also hippocampus	135	2.13	−18	2	−20
Right	Hippocampus	149	1.94	24	−22	−14
Right	Superior medial gyrus	121	2.05	6	32	48
Right	Middle frontal gyrus	120	2.10	42	8	58
Left/Right	Thalamus	36	1.75	2	−6	6
Right	Cerebellar vermis	1498	2.99	2	−50	2
Happy						
Right	STG	20	1.91	54	−46	18
Right	Middle temporal gyrus Also superior temporal gyrus	111	2.86	58	−2	−14
Right	3b Also 4a	37	1.95	42	−20	52
Right	Inferior frontal (orbital)	59	2.07	40	30	−10
Left	Inferior frontal (orbital)	50	2.95	−40	32	−10
Left/right	Mid orbital gyrus Also superior medial	1518	4.81	−4	50	−6
Right	Anterior cingulate	25	1.58	4	26	20
Right	Medial temporal pole	211	3.88	50	12	−28
Left	Medial temporal pole	134	4.48	−42	14	−34
Right	Caudate	57	2.55	16	18	12
Left	Caudate	40	1.86	−4	12	−2
Left	Putamen	22	1.46	−28	12	−6
Left	Thalamus	24	2.11	−22	−26	0
Left	Precuneus	75	2.51	0	−68	28
Left	Middle occipital gyrus	22	2.05	−38	−66	24

APPENDIX 6

MNS analyses during the different task conditions. Reported brain regions are active during facial expression viewing, all four facial conditions combined – rest, $t(16) > 2.92$, $p < .005$, $k = 20$, inclusively masked with facial expression execution, all four facial conditions combined – rest, $t(16) > 2.92$, $p < .005$. The same analysis was consecutively performed for (1) passive observation; (2) discrimination; and (3) viewing-before-imitation as viewing conditions. Moderate thresholds of both (viewing and production) group *t*-maps ($p < .005$) were combined, resulting in an overall false positive under a global null hypothesis of $p < .000025$ (for explanation of the different columns see Appendix 3)

				MNI		
Hemisphere	*Anatomical description*	*k*	*t*	*x*	*y*	*z*
(1) Observation						
Right	Middle frontal gyrus	2612	7.18	48	8	52
	Including: BA 44		6.72	50	16	26
	BA 44		6.71	52	14	14
	BA 45		6.70	48	22	−4
	BA 45		6.54	50	22	24
	BA 45		6.43	52	26	−2
	Insula					
Left	BA 45	1251	7.33	−52	38	6
	Including: BA 45		6.05	−54	30	8
	BA 45		5.18	−52	26	26
	BA 44		4.99	−46	12	28
	BA 44		4.93	−42	12	26
	Left inferior frontal gyrus (p orbitalis)		4.85	−48	20	−10
	Left insula					
Right	Inferior parietal lobule	671	6.10	34	−50	50
	Including: right superior occipital gyrus					
	Right angular gyrus					
	Right superior parietal lobule					
Left	Inferior parietal lobule	537	4.82	−34	−56	50
	Including: left superior parietal lobule					
	Left angular gyrus					
Right	Area 6	488	4.75	10	6	72
	Including: (Pre-) SMA		4.44	4	20	60
Left	Middle frontal gyrus	37	3.91	−46	6	52
	Including: area 6					
	Precentral gyrus					
Left/Right	Fusiform gyrus	15068	12.19	−20	−86	−18
	Also right MTG		11.14	62	−48	16
	Right fusiform		11.05	34	−76	−18
	Left cerebellum		10.58	−36	−60	−22
	Right amygdala		9.70	20	−2	−18
	Left amygdala		9.32	−14	−6	−16
	Thalamus					
	BA 17/18					
	Hippocampus					
Right	Area 1 (SI)	216	5.05	62	−16	42
	Including: Area 2 (SI)					
Left	Area 1 (SI)	23	5.75	−62	−12	28
Left	Area 17	77	4.38	−10	−76	10
Right	OP3 (SII)	50	4.32	36	−6	12
	Including: Right insula		3.86	40	−4	8
	OP4 (SII)					
Left	OP3 (SII)	28	6.32	−42	−6	18

APPENDIX 6 (*Continued*)

Hemisphere	*Anatomical description*	*k*	*t*	*MNI* *x*	*y*	*z*
(2) Discrimination						
Right	Inferior frontal gyrus (p opercularis)		7.20	48	16	28
Left	Inferior frontal gyrus (p triangularis)		7.96	−34	34	0
	BA 45		6.52	−54	24	26
	BA 44		6.42	−46	14	26
	BA 44/operculum		6.16	−50	10	4
Right	Area 6	85	4.63	18	−26	72
	Including: Area 4a		3.71	12	−30	76
Left	Area 6		6.80	−26	−24	66
Right	Inferior parietal lobule		6.63	34	−54	54
Left	Inferior parietal lobule		5.86	−32	−58	48
Right	Infcrior occipital gyrus	35 637	12.53	22	−28	−4
	Including: Area 6		11.50	58	6	38
	Inferior temporal gyrus		10.86	46	−52	−14
	Area 17					
	STS/MTG R		9.83	60	−38	18
	STS/MTG L		9.56	−54	−50	8
	Thalamus		5.54	−8	−16	0
	Putamen L		6.91	−22	14	−2
	Putamen R		5.97	20	12	−2
	Amygdala L		7.09	−18	−2	−18
	Amygdala R		8.95	26	0	−20
	Insula/operculum L		6.48	−46	16	−2
	Insula L		5.82	−40	6	−2
	Insula R		7.82	46	14	−4
	Area 1/3 L		6.06	−52	−12	50
	Area 2 L		5.79	−46	−32	46
	Area 1/2 R		7.00	52	−26	56
	Area 4p L		5.91	−38	−22	52
Left	Area 4a	62	4.96	−8	−44	64
	Including: Area 3a/3b		4.49	−8	−46	68
Left/Right	SMA L		7.93	−4	10	48
Right	Middle frontal gyrus	26	3.79	32	46	26
Left	Middle frontal gyrus	317	4.70	−26	42	28
Left	Precuneus	36	3.69	−16	−70	36
Right	Cerebellum		6.27	18	−70	−35
Left	Cerebellum		7.75	−12	−84	−37
(3) Viewing – Imitation						
Right	Inferior temporal gyrus	33 139	10.25	46	−52	−14
Right	Including: Area 17/18		6.46	12	−88	4
	STS/MTG R		8.96	58	−38	2
	STS/MTG L		6.67	−58	−44	14
	Thalamus L		8.84	−10	−18	2
	Thalamus R		7.73	18	−22	8
	Putamen L		7.55	−20	8	0
	Putamen R		7.04	20	10	6
	Amygdala L		9.74	−30	−4	−14
	Amygdala R		10.00	22	−2	−20
	Insula/operculum L					

APPENDIX 6 (*Continued*)

Hemisphere	*Anatomical description*	*k*	*t*	*MNI*		
				x	*y*	*z*
	Insula L		8.09	−28	22	4
	Insula L		8.12	−36	12	4
	Insula R		5.60	34	16	4
	Area 1/3 L					
	Area 2 L		6.18	−42	−32	40
	Area 1/2 R					
	Area 4a L/precentral		5.88	−50	−2	32
	Area 4a R/Area 6		6.56	52	−4	36
Left/Right	SMA L		6.56			
Right	Inferior parietal lobule		5.52	26	−52	44
Left	Inferior parietal lobule		6.63	−32	−50	42
	Superior parietal		8.34	−26	−60	46
Left	BA 44		10.12	−46	16	26
	Inferior frontal gyrus (p triangularis)		10.11	−32	32	−2
Left/Right	Area 6 L	2837	10.67	−21	0	48
	Including: Area 6 R		8.19	4	2	60
Left	Middle cingulate cortex		8.09	−6	14	40
Left	Middle frontal gyrus	254	5.48	−28	40	22
Right	Middle frontal gyrus	71	4.95	30	42	28

APPENDIX 7

Brain regions that significantly differentiate between facial viewing tasks, matched pair *t*-test, $t(16) > 2.92$, $p < .005$, $k = 20$, inclusively masked with facial expression production, all four facial conditions – rest, $t(16) > 2.92$, $p < .005$. The following pairwise comparisons were performed: (1) discrimination contrasted against observation; (2) viewing-before-imitation contrasted against observation; and (3) viewing-before-imitation contrasted against discrimination. Inclusive masking at $p < .005$ of a contrast thresholded at $p < .005$ resulted in an overall false positive rate of $p < .005^2 = p < .000025$ (for an explanation of the different columns see Appendix 3)

				MNI		
Hemisphere	*Anatomical description*	*k*	*t*	*x*	*y*	*z*
(1a) Observation > Discrimination						
No activations						
(1b) Discrimination > Observation						
Left/Right	Area 6/SMA Including: Inferior frontal gyrus (BA 44)/insula	14 472	9.84	58	4	40
Right	Insula	29	3.55	48	14	−4
Right	Amygdala	119	4.55	18	−2	−14
Right	Middle frontal gyrus	112	4.64	32	40	30
Left	Middle frontal gyrus	498	6.61	−36	44	16
Left	Thalamus	484	4.95	−10	−18	2
Right	Thalamus	68	4.07	12	−18	6
Right	Putamen	357	4.96	20	12	−2
Left	Putamen	340	6.68	−22	8	8
Right	Hippocampus	147	8.39	22	−28	−4
Right/Left	Cerebellum MTG/fusiform	10 005	7.69	26	−62	−18
(2a) Observation > Imitation						
Right	Superior frontal gyrus	45	5.57	22	54	32
(2b) Imitation > Observation						
Left/Right	SMA Including: Insula, inferior frontal gyrus (BA 44, BA 45), middle cingulate cortex, thalamus, putamen, Area 17/18	24 284	10.40	−4	8	46
Right	BA 44/45	512	5.11	50	30	14
Right	Precentral gyrus, Area 6, Area 3, 4p, BA 44, OP1, OP4	1495	5.90	50	−4	36
Right	Area 6, Area 4a/4p	82	5.13	18	−24	62
Right	Superior parietal	260	4.42	26	−52	42
Left	Area 3b, 2	31	4.05	−22	−36	64
Left	Amygdala	36	4.55	−30	−2	−32
Left	Middle frontal gyrus	437	5.80	−38	46	18
Right	Middle frontal gyrus	133	6.31	30	40	28
Left	Middle cingulate cortex	42	3.97	−6	−42	52
(3a) Discrimination > Imitation						
Left	Cerebellum	260	7.77	−32	−64	−36
Right	Cerebellum	107	3.99	24	−68	−34
Left	Area 6	56	3.66	−26	−24	70
Left	Area 4	29	3.86	−14	−34	72
	Cerebellar vermis	29	3.99	−2	−46	−10
Right	Cerebellum	24	3.67	8	−70	−38

APPENDIX 7 (*Continued*)

Hemisphere	*Anatomical description*	*k*	*t*	*MNI* *x*	*y*	*z*
Left	Area 6	23	3.49	−38	−14	64
Right	Supramarginal gyrus	20	3.75	50	−40	30
(3b) Imitation > Discrimination						
Right	Inferior frontal gyrus/BA 44	89	5.43	52	14	6
Left	Inferior frontal gyrus/BA 44	30	4.13	−60	6	16
Left	Inferior frontal gyrus/BA 44	163	3.34	−42	14	24
Left	Precuneus	47	4.68	−14	−66	32
Left	Precentral gyrus	25	3.98	−36	0	48
Right	Area 4a/p	288	4.22	58	−8	36
Left	Area 1/4p/6	167	4.05	−52	−8	32
Left	Pre-SMA	287	4.49	−8	14	46
	Middle cingulate cortex		3.36	−8	18	34
Left	Putamen	208	5.87	−22	10	4
	Insula		3.56	−30	22	4
Right	Pallidum	81	4.96	18	8	2
Right	Inferior temporal gyrus	23	3.96	48	−52	−14
Left	Cuneus	21	3.42	−8	−78	38
Left	Area 17/18	22	3.69	−8	−64	0
Left	Thalamus	41	3.91	−4	−10	2

APPENDIX 8

Brain regions that significantly differentiate between dynamic faces and dynamic patterns during: (1) passive observation; and (2) the discrimination task, inclusively masked with facial expression production, all four facial conditions – rest, $t(16) > 2.92$, $p < .005$. Brain regions that prefer the observation of dynamic faces over dynamic patterns, all four facial conditions combined-patterns, matched pair t-test, $t(16) > 2.92$, $p < .005$, $k = 20$, for the observation (1a) and discrimination task (2a), respectively. Brain regions that prefer the observation of dynamic patterns over facial expressions, matched pair t-test, $t(16) > 2.92$, $p < .005$, $k = 20$, for the observation (1b) and discrimination task (2b), respectively (for explanation of the different columns see Appendix 3)

				MNI		
Hemisphere	*Anatomical description*	*k*	*t*	*x*	*y*	*z*
(1a) Observation: Face > Pattern						
Right	BA 44/45	277	6.67	52	24	−4
Left	BA 44	20	3.18	−52	16	4
Right	Inferior frontal gyrus /BA 45	40	3.74	46	24	20
Left	Inferior frontal gyrus	88	5.11	−42	26	−12
Right	Area 6/pre-SMA	240	5.41	6	6	66
Right	Area 6	32	3.55	52	−2	50
Right	Superior temporal gyrus/MTG	1257	8.67	44	−36	4
Left	Superior temporal gyrus/MTG	47	3.73	−58	−48	14
Right	MTG	30	3.76	52	−58	14
Right	Amygdala/hippocampus	340	6.09	18	−4	−18
Left	Amygdala	238	5.60	−20	−4	−18
	Putamen		3.90	−26	6	−10
Right	Fusiform gyrus	86	4.93	38	−46	−26
Right/Left	Area 17/18	732	6.48	12	−90	8
Right	Cerebellum	51	7.26	18	−76	−38
Left	Cerebellum	215	5.05	−20	−80	−38
(1b) Observation: Pattern > Face						
Left	Precentral	308	5.48	−54	4	30
	Including: BA 44					
	Area 3a					
	OP4					
Right	Supramarginal gyrus	3694	7.43	38	−36	40
	Including: Superior parietal					
	Area 2					
	Precuneus					
	Inferior parietal					
Left	Superior parietal	2958	8.64	−12	−78	50
	Including: Inferior parietal					
Right	Superior frontal gyrus/precentral	618	6.29	26	−4	42
Left	Precentral/superior frontal gyrus	457	7.01	−28	−8	56
Left	Insula	52	4.62	−40	−2	6
Right	Inferior temporal gyrus	243	5.60	50	−50	−14
Left	Inferior temporal gyrus	357	7.88	−44	−60	−8
Right	Fusiform	86	6.13	26	−58	−14
Left	Fusiform	125	5.48	−24	−58	−14
Right	Middle cingulate gyrus	105	4.49	14	−34	42
Left	Superior temporal gyrus/OP1	75	4.63	−54	−32	10
Right	Middle occipital gyrus	51	6.92	42	−78	24

APPENDIX 8 *(Continued)*

Hemisphere	*Anatomical description*	*k*	*t*	*MNI* *x*	*y*	*z*
Left	Middle occipital gyrus	41	5.34	−38	−80	14
Left	Superior occipital gyrus	35	4.81	−20	−88	32
(2a) Discrimination: Face > Pattern						
Right	BA 45	35	3.71	46	26	−2
Right	Area 6/pre-SMA	54	4.17	4	8	68
Right	Area 6/SMA	26	4.07	6	−20	46
Right	MTG	1201	5.98	62	−40	6
Right	Amygdala/hippocampus	103	5.13	16	−8	−18
Right	Temporal lobe	33	4.90	32	6	−24
Right/Left	Area 17/18	643	7.59	10	−76	−6
Left	Cerebellum	26	3.52	−22	−76	−38
(2b) Discrimination: Pattern > Face						
Right	BA 44	462	8.76	50	6	28
Left	Precentral Including: BA 44	504	5.30 5.22	−50 −50	4 4	38 38
Right	BA 45/middle frontal	277	4.57	50	38	10
Left	BA 45/middle frontal	232	5.29	−44	50	4
Right	Superior occipital gyrus Including: Superior parietal Area 2 Inferior parietal	4359	11.33	28	−70	44
Left	Inferior parietal Including: Superior parietal Area 2	3996	9.41	−40	−46	42
Right	Insula	67	5.17	38	−4	8
Right	Superior frontal gyrus/middle frontal	749	7.66	24	6	60
Left	Superior frontal gyrus	403	5.14	−24	−6	56
Right	Middle cingulate gyrus/superior medial gyrus	159	6.00	6	26	42
Left	Anterior cingulate	67	5.12	−4	14	24
Left	Fusiform	206	7.52	−28	−56	−16
Right	Fusiform	132	6.47	26	−56	−14
Right	Inferior temporal gyrus	281	5.68	52	−50	−14
Left	Inferior temporal gyrus	421	7.01	−44	−60	−8
Left	Precuneus	28	3.88	−14	−44	46
Right	Middle occipital gyrus	27	5.86	42	−80	20
Left	Middle occipital gyrus	44	5.20	−38	−80	12
Left	BA 17/18	29	4.10	−6	−76	16
Left	Thalamus	51	4.37	−10	−24	8

APPENDIX 9

Emotional and neutral facial expression processing in the MNS. (1a) Direct-matching of emotional facial expressions: regions significantly more active during *emotional* facial expression viewing contrasted against *neutral* facial expression viewing, after combining the observation and discrimination task, average of three emotional facial conditions – neutral facial condition, $t(16) > 2.92$, $p < .005$, $k = 20$, inclusively masked with emotional facial expression production, three emotional facial conditions – neutral facial expressions, $t(16) > 2.92$, $p < .005$. (1b) Direct-matching of neutral facial expressions: regions significantly more active during *neutral* facial viewing contrasted against *emotional* facial expression viewing, after combining the observation and discrimination task, average of the neutral facial conditions – three emotional facial conditions, $t(16) > 2.92$, $p < .005$, $k = 20$, inclusively masked with neutral facial expression production, neutral facial conditions – emotional facial expressions, $t(16) > 2.92$, $p < .005$. (2) MNS analyses comparing the viewing of one specific emotional – with neutral facial expressions (combining observation and discrimination) for consecutively the disgusted, fearful and happy expressions, for disgust: disgust viewing – neutral viewing, $t(16) > 2.92$, $p < .005$, $k = 10$, inclusively masked with a congruent contrast for facial expression production, for disgust: disgust execution – neutral execution, $t(16) > 2.92$, $p < .005$. Inclusive masking at $p < .005$ of a contrast thresholded at $p < .005$ resulted in an overall false positive rate of $p < .005^2 = p < .000025$. (3) Specificity for individual facial conditions within the MNS. The target emotion (e.g., fear) against the other facial conditions needed to be significant in each case both during viewing and during production and that both during viewing and production the target emotion by itself exceeded zero, i.e., for fear for instance, we tested F–N, F–H, F–D for viewing and for execution at $p < .1$ and fear by itself for execution and viewing at $p = .05$. Combining these 8 tests leads to a false positive rate under the global null hypothesis of $p < .025 \times 10^{-8}$. Analyses were performed consecutively for neutral, disgust, fear and happy and reported using a spatial threshold of 10 voxels ($k = 10$)

				MNI		
Hemisphere	*Anatomical description*	*k*	*t*	*x*	*y*	*z*
(1a) Direct Matching: Emotional–Neutral (k = 20)						
Left	Inferior frontal gyrus (p triangularis)	121	4.34	−46	24	6
	Including: L inferior frontal (p opercularis)		4.28	−52	10	2
	BA 44		3.97	−46	24	10
	Left insula		3.29	−40	16	−2
Left	SMA	77	3.89	−4	14	60
Right	Insula	73	4.13	38	18	−8
	Including: BA 44		3.48	52	18	−4
	BA 45		3.18	54	26	−2
Right	STS	37	4.59	44	−26	−4
(1b) Direct Matching: Neutral – Emotional (k = 20)						
Right	SI	56	4.56			
	SII	78	3.89			
Left	SI	59	4.78			
(2) MN Regions: Emotion–Neutral (k = 10)						
Disgust						
Left	BA 44/BA 45	18	3.95	−44	24	14
Right	Frontal operculum	51	3.90	52	16	−4
	Insula		3.89	40	20	−6
Left	Operculum/insula	210	8.10	−44	20	0
	Operculum		3.55	−52	12	0
Left	Area 6	106	4.4	0	10	64
	Pre-SMA		4.26	−10	12	52
Left/Right	Middle cingulate gyrus	60	5.70	−6	28	32
Right	Middle occipital gyrus	17	3.53	38	−70	2
Left	Superior occipital gyrus	11	3.55	−22	−86	28
Right	Cerebellum	11	3.21	24	−72	−36
Fear						
Right	BA 44	12	3.40	56	16	2
Left	BA 44	27	3.68	−52	12	4
Left/Right	Pre-SMA	54	3.62	−2	18	46
Left	Cerebellum	34	5.38	−40	−66	−28
Happy						
Left	BA 44	13	3.69	−54	22	12
Right	Superior temporal gyrus	37	4.36	52	−24	−4

APPENDIX 9 (*Continued*)

				MNI		
Hemisphere	*Anatomical description*	*k*	*t*	*x*	*y*	*z*
Right	Superior medial gyrus	12	4.45	10	56	30
Left	Anterior cingulate gyrus	12	3.77	−2	40	10
(3) Emotion Specificity (k =10)						
Neutral						
Right	Area 6	15	1.84	60	2	28
Right	Area 2 (SI)	150	1.98	48	−34	56
Right	OP1 (SII)	177	2.73	62	−18	26
Left	Area 2 (SI)	256	2.81	−56	−26	44
	Including: Supramarginal gyrus		2.52	−60	−28	42
	OP1 (SII)		1.49	−56	−22	22
Disgust						
Left	Putamen/globus pallidus	18	2.00	−18	4	−6
Left	Lingual gyrus	31	1.96	−18	−74	−12
Right	Fusiform	28	2.02	26	−62	−10
Right	Middle occipital gyrus	210	2.41	40	−72	8
Left	Middle occipital gyrus	38	2.05	−34	−78	4
Right	Superior occipital gyrus	17	1.80	24	−84	32
Left	Superior occipital gyrus	20	1.91	−16	−88	28
Left	BA 17	30	1.92	−4	−68	−4
Right	Cerebellum	15	1.58	26	−58	−44
Fear						
Right	BA 44/BA 45	17	1.73	52	14	32
Left	BA 44	13	1.63	−48	12	10
Left	BA 44	13	1.59	−58	14	6
Right	Area 6	10	1.59	10	0	72
Left	Middle temporal gyrus	64	2.01	−52	−48	4
Right	Middle frontal gyrus	14	1.66	38	6	46
Happy						
Right	Superior medial gyrus	35	2.37	8	56	30
Right	Superior temporal gyrus	18	1.83	54	−46	18

APPENDIX 10 (overleaf)

Contrast values of the different viewing and execution conditions from remaining ROIs which were not shown in Figure 6. Contrast values from the different ROIs were extracted from individual subjects using the Marsbar toolbox (see Materials and methods section).

(A) Left column: Remaining ROIs from the congruent mapping analysis of the Emotional facial expression minus Neutral facial expression contrast, as depicted in Figure 5 and listed in Appendix 9. From top to bottom: left IFO, right STS and bilateral pre-SMA. Right column: Contrast values for the different stimuli during both viewing (combining observation and discrimination) (white bars) and execution (black bars) from our three ROIs. Error bars represent the *SEM*. The significant main effect of condition (see Results) was further explored using Newman-Keuls post-hoc comparisons: emotional facial conditions that significantly differed from the neutral condition during *viewing* ($p < .05$) are marked with (*), facial conditions that significantly differed from the pattern condition during *viewing* ($p < .05$) are marked with (†), emotional facial conditions that significantly differed from the neutral condition during *execution* ($p < .05$) are marked with (‡). Comparisons of the disgust condition with the remaining facial condition within the IFO (*a priori* hypothesis) is discussed further in the main text.

(B) Left column: Remaining ROIs from the congruent mapping analysis of the Neutral facial expression minus Emotional facial expression contrast, as listed in Appendix 9. From top to bottom: left SI and right SI. Right column: contrast values taken from these two ROIs during the different conditions. The same explanation for the graph and the symbols apply as described under (A). IFO, insula/frontal operculum; STS, superior temporal sulcus; pre-SMA, pre-supplementary motor area; SI, primary somatosensory cortex.

APPENDIX 10

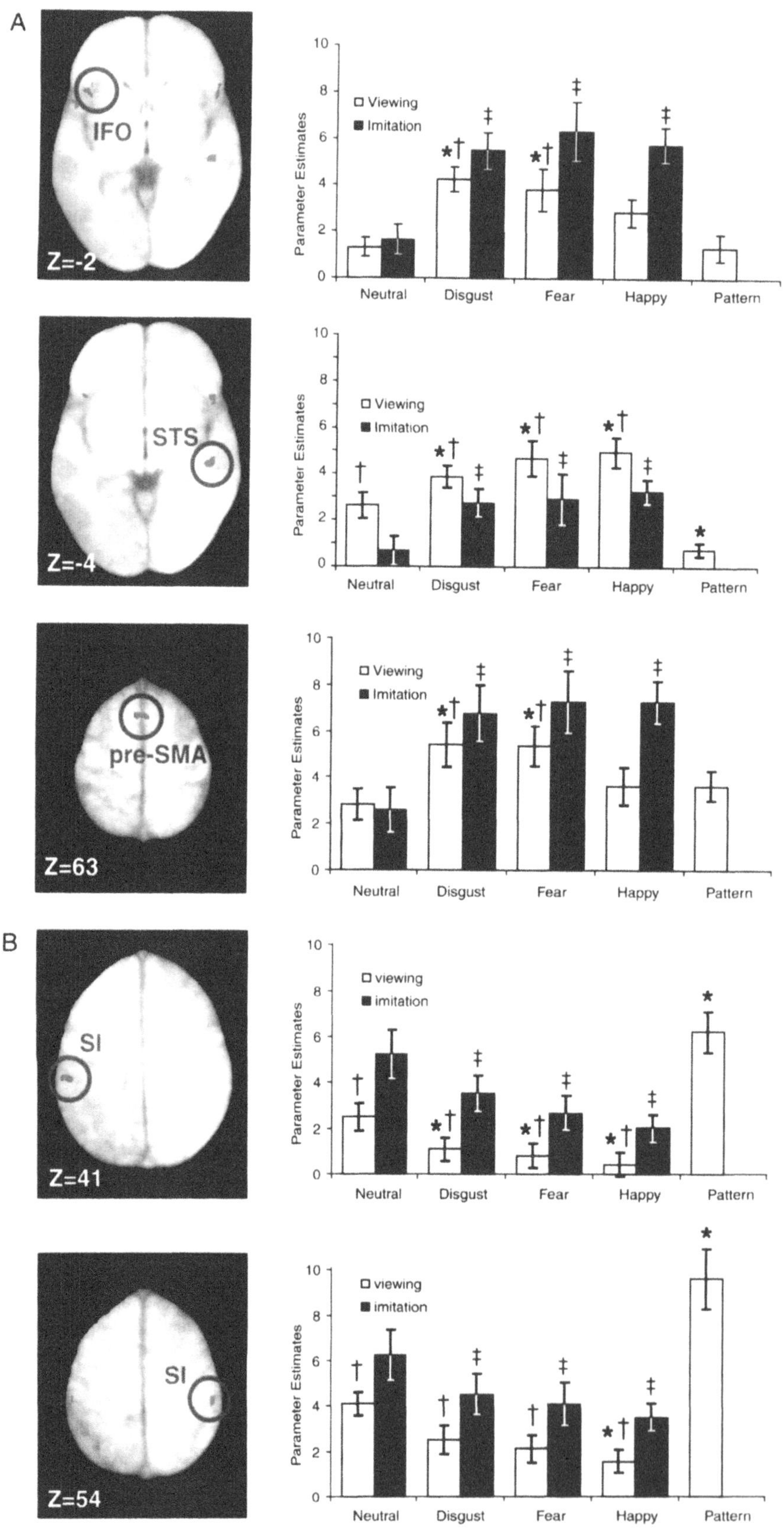
A
IFO
Z=-2
STS
Z=-4
pre-SMA
Z=63
B
SI
Z=41
SI
Z=54
Viewing
Imitation
viewing
imitation
Parameter Estimates
Neutral
Disgust
Fear
Happy
Pattern

APPENDIX 11

Signal increases for the contrast Emotional facial condition minus Neutral facial condition during the viewing condition (combining observation and discrimination) for consecutively the disgusted (A), fearful (B) and happy (C) expressions (for disgust: disgust viewing – neutral viewing, $t(16) > 2.92$, $p < .005$, $k = 10$). All regions displayed are inclusively masked with a congruent contrast for facial expression production (for disgust: disgust execution – neutral execution, $t(16) > 2.92$, $p < .005$). Inclusive masking at $p < .005$ of a contrast thresholded at $p < .005$ resulted in an overall false positive rate of $p < .005^2 = p < .000025$. Results are overlaid on an averaged T1-weigthed 3D render of the 17 subjects. See Appendix 9 for coordinates of local maxima. IFO, insula/frontal operculum; BA44, Brodmann Area 44; SMG, superior medial gyrus; Acc, Anterior Cingulate Gyrus.

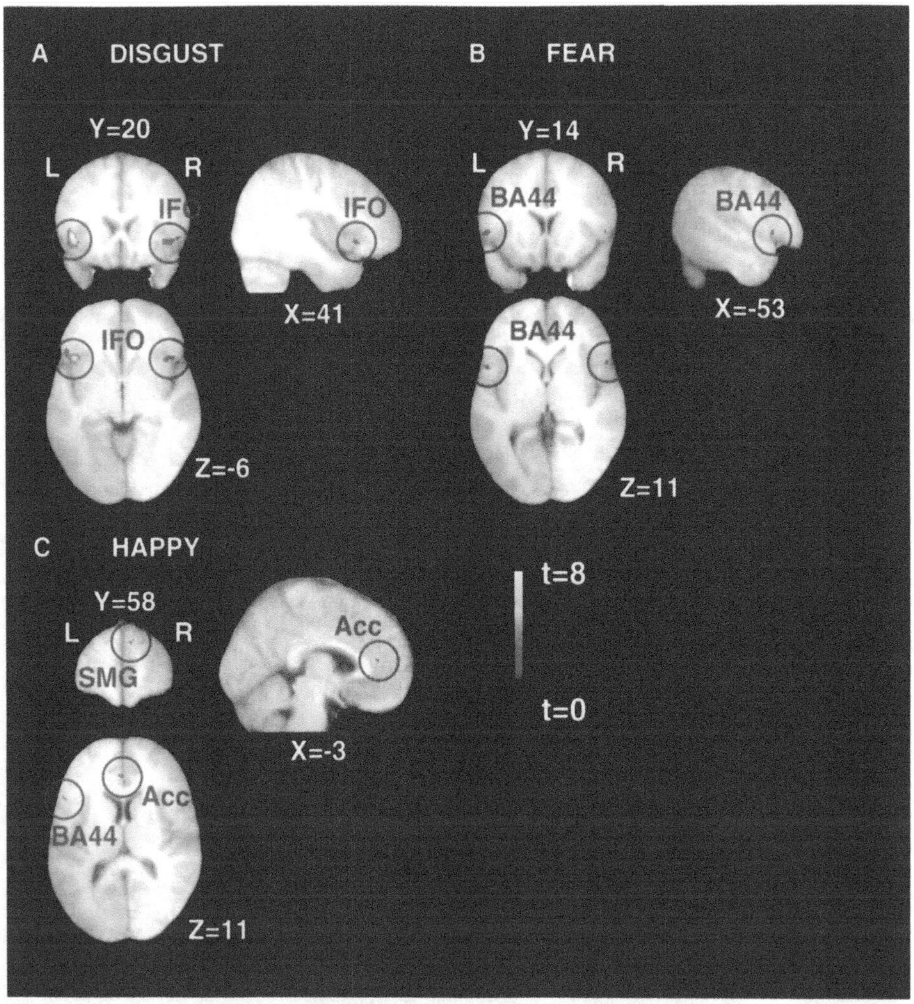

APPENDIX 12

Specificity for individual facial conditions within the MNS. The target emotion (e.g., fear) against the other facial conditions needed to be significant in each case both during viewing and during production and that both during viewing and production the target emotion by itself exceeded zero, i.e., for fear for instance, we tested F−N, F−H, F−D for viewing and for execution at $p<.1$ and fear for execution at .05. Combining these 8 tests leads to a false positive rate under the global null of $p<.025\times10^{-8}$. Analyses were performed consecutively for neutral (A), disgust (B), fear (C) and happy (D) and displayed using a spatial threshold of 10 voxels ($k=10$). Results are overlaid on an averaged T1-weigthed 3D render of the 17 subjects. See Appendix 9 for coordinates of local maxima. SI, primary somatosensory cortex; SII, secondary somatosensory cortex; Pt, putamen; BA44, Brodmann Area 44; STS, superior temporal sulcus; SMG, superior medial gyrus.

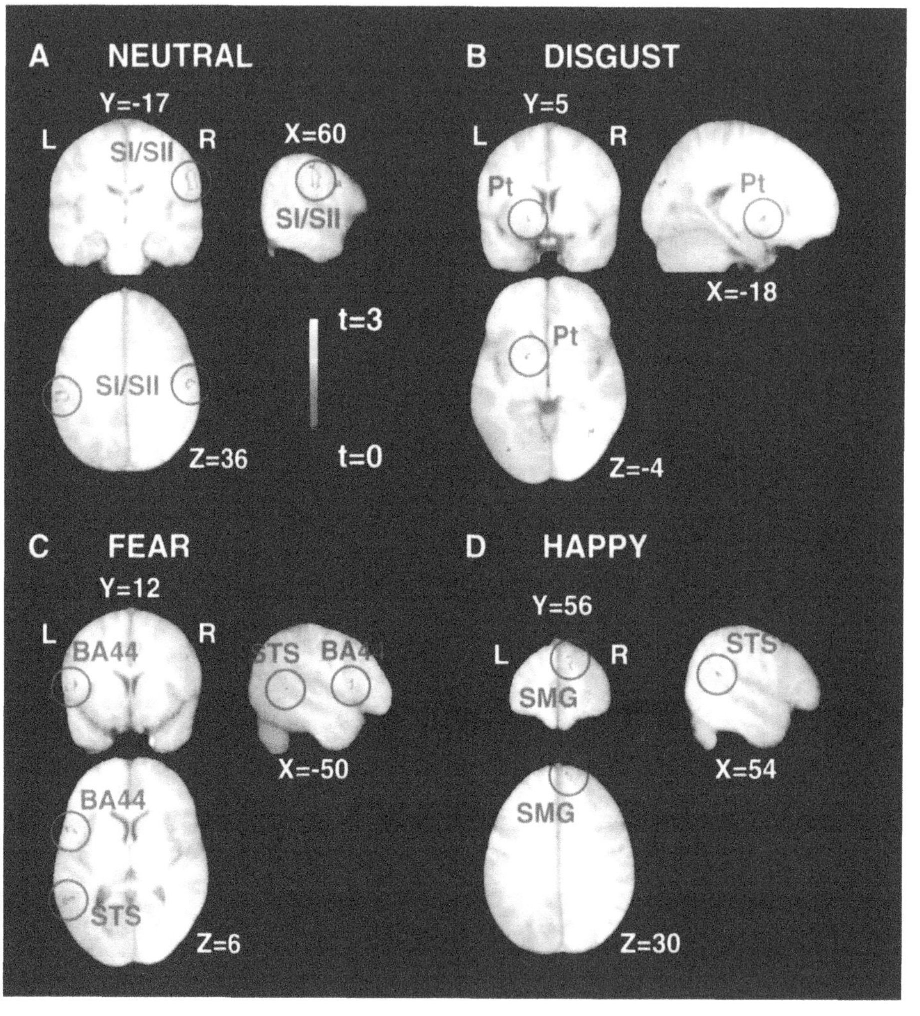

SOCIAL NEUROSCIENCE, 2007, 2 (3–4), 223–237

Psychology Press
Taylor & Francis Group

Imitation of in-group versus out-group members' facial expressions of anger: A test with a time perception task

Laurie Mondillon

University of Clermont-Ferrand, Clermont-Ferrand, France

Paula M. Niedenthal

CNRS and University of Clermont-Ferrand, Clermont-Ferrand, France

Sandrine Gil and Sylvie Droit-Volet

University of Clermont-Ferrand, Clermont-Ferrand, France

This research investigated the automatic imitation of facial expressions of anger by in-group and out-group members, using a temporal estimation task. Individuals typically overestimate duration represented by angry faces, probably due to increases in arousal (Droit-Volet, Brunot, & Niedenthal, 2004). Overestimation is not observed when imitation of the facial expressions is *inhibited*, suggesting that embodied simulation mediates the changes in arousal (Effron, Niedenthal, Gil, & Droit-Volet, 2006). This method thus provides an implicit measure of imitation and was used to test the hypothesis that individuals imitate in-group, but not out-group members' facial expressions of emotion. In separate studies Chinese and French Caucasian participants were presented with *short* (400 ms) and *long* (1600 ms) standard durations in a temporal bisection task. They then categorized standard and intermediate durations, represented by angry and neutral faces, in terms of similarity to the *short* and *long* standard durations. Half of the face stimuli were Chinese, and half Caucasian. Results revealed a bias in the temporal perception of emotion for the Caucasian participants when they were presented with Caucasian facial expressions and not Chinese ones. In contrast, this bias in time perception was observed when Chinese individuals imitated faces of both in- and out-group members. The results of the Chinese participants are interpreted in terms of familiarity with and motivations to understand the emotional expressions of members of a host culture.

How do we understand other people? The study of social cognition, which was for years impelled toward important advances by its endorsement of the computer metaphor and associated symbolic models of mental representation, is in the midst of a paradigm shift. Increasingly, understanding other people—judging their intentions and emotions, their goals, and making attributions for their actions—is proving to be importantly based on the ability to embody the physical and emotional states of the perceived. That is, embodied simulation in the motor system (e.g., Gallese, 2005a,b; Gallese & Metzinger, 2003; Glenberg, 1997), the perceptual systems (e.g., Barsalou, 1999), and the affect systems (e.g., Barsalou, Niedenthal, Barbey, & Ruppert, 2003; Decety & Jackson, 2003; Niedenthal, Barsalou, Ric, & Krauth-Grüber, 2005a) appear to ground the

Correspondence should be addressed to: Laurie Mondillon, Laboratoire de Psychologie Sociale et Cognitive (CNRS, UMR 6024), Université Blaise Pascal, F-63037 Clermont-Ferrand, France. E-mail: laurie.mondillon@univ-bpclermont.fr

Preparation of this article was supported by the Prosodie ACI Grant Pro-0052, Ministère de la Recherche (France) and by NSF grant BCS-0350687.

www.psypress.com/socialneuroscience DOI:10.1080/17470910701376894

higher-order cognitive processes of interest to students of social cognition. For example, and of specific interest in the present research, the imitation of facial expressions of others, which in turn reproduces at least in part the associated feeling state (Carr, Iacoboni, Dubeau, Mazziotta, & Lenzi, 2003; Gallese, Keysers, & Rizzolatti, 2004; Vaughan & Lanzetta, 1980, 1981), is importantly implicated in the recognition of facial expression and the understanding of the other's emotional state. To summarize, Meltzoff and Decety (2003) stated that imitation and its neural correlates provide the system by which empathy occurs in humans (see also Meltzoff, 2002).

Although showing that simulation in sensory, motor, and affect systems supports, rather than merely co-occurs with, social cognition is a scientific agenda that is well underway, a concurrent concern is the demonstration that individuals can distinguish the embodied simulations of *others'* experiences from their *own* experiences (e.g., Decety & Sommerville, 2003), and that this distinction is related to systematic constraints on embodied simulation. Indeed, people do not simulate the experiences of all of the people all the time. We are concerned here with the hypothesis that individuals embody the emotional expressions of in-group and not out-group members, or at least the former to a great degree. Secondarily we also examine the hypothesis that empathic individuals embody in-group members' emotional expressions more than do less-empathic individuals.

Imitation of in-group and out-group members

Most researchers agree that imitation of emotional gestures, but also of self-grooming, self-regulatory and other communicative and non-communicative nonverbal behaviors, serves the adaptive function of increasing affiliation, liking, and smoother rapport between people. (Anderson, Keltner, & John, 2003; Bernieri, 1988; Chartrand & Bargh, 1999; Friedlmeier, in press; LaFrance & Broadbent, 1976; Lakin & Chartrand, 2003; Lakin, Jefferis, Cheng, & Chartrand, 2003; Lanzetta & Englis, 1989; Neumann & Strack, 2000; Zajonc, Adelmann, Murphy, & Niedenthal, 1987; see Dijksterhuis, 2005, for a review of these studies). Van Baaren, Holland, Steenaert, and van Knippenberg (2003) showed the imitation-liking effect with a set of studies that took place in a restaurant. In these experiments, while some waitresses were instructed to repeat exactly the order of each customer (i.e., verbal behavior imitated condition), others had not to repeat any customer's words (i.e., control condition). The results showed that verbal mimicry significantly improved the tips assigned to the waitresses compared to the control condition in which there was no imitation.

However, people are not equally motivated to affiliate and have positive rapport with all individuals (LaFrance, 1979, 1985). According to primitive emotional contagion theory (Hatfield, Cacioppo, & Rapson, 1992, 1993), people spontaneously mimic the facial, bodily expressions, and speech duration of people with whom they have a close relationship of some kind (see also Gump & Kulik, 1997). Cheng and Chartrand (2003) showed that self-monitoring individuals, who are motivated and able to control their social appearance, were more likely to mimic a confederate's subtle gestures when they believed the confederate to be a peer or their superior, that is to say when their interaction partner was someone with whom it would have been beneficial to get along. Lakin and Chartrand (2005) showed that mimicry increases after an individual is socially excluded from his or her group, suggesting that those who are particularly motivated to be accepted tend to engage in more imitation. Relatedly, Smoski and Bachorowski (2003) showed that dyads composed of peers produced significantly more antiphonal laughter (i.e., "laughter that occurs during or immediately after a social partner laugh") than those composed of strangers.

Imitation of facial expression

The imitation of facial expression of emotion is of particular interest because recent neuroscientific evidence suggests that it is a mechanism underlying the recognition of facial expression, and thus a possible fundamental basis of empathy (Adolphs, Damasio, Tranel, Cooper, & Damasio, 2000; Decety & Chaminade, 2003; Dimberg, 1982, 1990; Gallese, 2003a,b; Gallese et al., 2004). Furthermore, there is evidence that people tend to imitate emotional facial expressions (whether positive or negative) in a relatively automatic and unintentional way, this processing taking place without awareness (Dimberg, Thunberg, & Elmehed, 2000; Hatfield et al., 1992), even if the emotional expression is not consciously perceived

(Stepper & Strack, 1993; Strack, Martin, & Stepper, 1988).

Despite the prevalent tendency to imitate facial expression, recent findings indirectly suggest constraints on this type of embodiment as well. Specifically, some recent research suggests that individuals are more accurate at recognizing emotions expressed by members of their own cultural group than by members of a different cultural group (Elfenbein & Ambady, 2003a,b; Elfenbein, Mandal, Ambady, Harizuka, & Kumar, 2004; Matsumoto, 2002). This relative performance difference in facial expression recognition has been called the "in-group advantage" effect (Elfenbein & Ambady, 2002b).

In the Elfenbein, and Ambady studies (2003c), Caucasian American participants were more accurate and also faster at recognizing emotions expressed by Caucasian American than by Chinese faces. Conversely, Chinese participants were better at identifying expressions displayed by individuals of their own culture than Caucasian American individuals. This in-group bias was also found when participants were simply told that the displayed faces corresponded to members of their own versus a different ethnic and socio-linguistic group (Hess, Sénécal, & Kirouac, 1996). Elfenbein and Ambady (2002a) propose a number of cultural differences to account for this effect such as cultural learning, differences in emotional concept, or cultural learning of emotional behavior. To us, however, these findings suggest the hypothesis that individuals do not imitate the facial expressions of others to the same degree. Specifically, based on such indirect evidence, we hypothesized that individuals tend to automatically imitate the facial expressions of in-group members to a greater degree than out-group members.

Finally, some studies have emphasized the moderator role of empathy on imitation. Zajonc et al.(1987) showed that couples who have been married for a long period of time resemble each other more than random couples of the same age or even than they did when they were first married. This finding was attributed to the fundamental role of facial imitation in empathizing with the spouse, and is consistent with embodied simulation theories of empathy (e.g., Bavelas, Black, Lemery, & Mullett, 1986; Decety & Jackson, 2004; Pacherie, 2004). In addition, however, that study showed that couples who grew more facially alike over their marriage also reported being happier in their marriage and more similar in their attitudes to their spouse. Relatedly, Chartrand and Bargh's (1999) study showed that empathic individuals exhibit more nonconscious mimicry of postures, mannerisms and facial expressions than other people. Such a finding suggests the hypothesis that empathic individuals imitate facial expressions more or more accurately than do less empathic individuals.

An implicit measure of automatic mimicry

To test the present prediction, we adapted a task recently used by Droit-Volet et al. (2004) to assess the effects of emotion on the perception of duration (e.g., Effron et al., 2006; Gil, Niedenthal, & Droit-Volet, 2007). Their findings showed that individuals tend to overestimate duration when it was represented by emotional faces, especially angry faces. The effect was observed in a temporal bisection task in which participants first learned to categorize two standard durations: a *short* (400 ms) and a *long* (1600 ms) one. The two standard durations were represented by a pink oval presented on a computer screen. After the initial learning phase, participants then had to categorize comparison durations of the same value as the two standard durations and of intermediate values in terms of similarity to the *short* or the *long* standard durations. In the test phase, duration was, however, represented by faces expressing one of three emotions, namely anger, joy, or sadness, or neutral emotion. For anger and joy expressions, compared to neutral ones, participants categorized intermediate durations more often as *long* durations, revealing that the duration of these high-arousal emotional stimuli was judged to be longer than that of the neutral emotion. Other crucial analyses of the data permitted the authors to conclude that the facial expressions augmented the arousal levels of the participants, and that this augmented arousal was responsible for the biased temporal perception. For instance, the internal clock model predicts a multiplicative effect on time judgment between the tested duration values and the emotions when the arousal speeds up the internal clock (Burle & Casini, 2001; Droit-Volet, Meck, & Penney, 2007; Maricq & Church, 1983; Meck, 1983; see Droit-Volet & Wearden, 2002, for complete discussion). Consistent with this predicted effect, the distortion in temporal judgment with the angry faces was more marked for the longer duration values.

Of interest for the present study, Effron and colleagues (2006) subsequently showed that facial expression imitation mediates the biased perception of duration represented by facial expressions of emotion. Specifically, these researchers replicated the study of Droit-Volet and colleagues (2004), using facial expressions of anger, joy, and neutral emotion (omitting sadness). Half of their participants performed the temporal bisection task with no other changes, allowing free facial movements, thus imitation. The remaining participants were instructed to hold a pencil lightly between both their teeth *and* their lips during the entire task. The pencil held in this manner significantly inhibits imitation (see Niedenthal, Brauer, Halberstadt, & Innes-Ker, 2001, for details). Results of the bisection task revealed that participants overestimated the duration of emotional faces relative to the neutral faces, but only when imitation was possible. That is, when facial expression could not be imitated, no significant effect of arousal on temporal perception was observed.

The Effron and colleagues' study suggests that inhibiting imitation eliminates the effect of emotion on the perception of duration. We thus decided to use the temporal bisection task to study the effect of imitation of facial expressions for in-group and out-group members. Using a temporal bisection task to measure imitation in this way has a number of advantages. One is the implicit nature of the task: imitation is not directly measured, so participants are not led inadvertently to generate correct experimental hypotheses and to adhere to experimental demand. This might occur if, for example, participants' imitative behavior were measured with electromyographic recoding (EMG) because the placement of electrodes could easily suggest an interest in facial expression. A second advantage is that the measure is not only implicit, but also subtle. Small changes in arousal such as those due to the imitation of facial expression can be expected to cause biased temporal perception. Thus the task is both implicit and sensitive.

Overview of the present experiments

Because our interest was in the automatic imitation of the facial expressions of in- and out-group members, and because recent research shows cultural differences in the recognition of facial expression, we chose to use the same in-group and out-group as in Elfenbein and Ambady's studies (2003b). In the temporal bisection task developed for this research, duration was therefore represented by faces that expressed anger or neutral emotion. Half of these faces were those of Caucasian individuals and half were those of Chinese individuals. All expressions had been FACS-coded for their authenticity (see Ekman & Friesen, 1978). In the first study, participants were all French Caucasians. In the second study, all participants were Chinese nationals studying in France. Our primary prediction for each study was that duration of presentations of facial expressions of anger, compared to neutral emotion, would be overestimated, and increasingly so over durations, and that this interaction would only be observed for in-group faces. Conversely, the duration of facial expressions displayed by out-group members would suggest no particular bias attributable to changes in arousal.

We also had a secondary interest. We measured individual differences in empathy in both of the following studies, and tested the secondary hypothesis that personal empathy levels moderate imitation such that biases in temporal perception of emotion are more marked for high-empathic individuals because they engage in more effective imitation.

STUDY 1

Method

Participants

Forty-seven female students participated in the study. All were first-year psychology students recruited from the University of Blaise Pascal in Clermont-Ferrand, France. The participants received course credit as compensation.

Materials

Bisection task. The participants were tested individually in a quiet laboratory room. Stimulus presentation and response collection were controlled by a program created in PsyScope for Macintosh (Cohen, MacWhinney, Flatt, & Provost, 1993). The stimuli used for the representation of

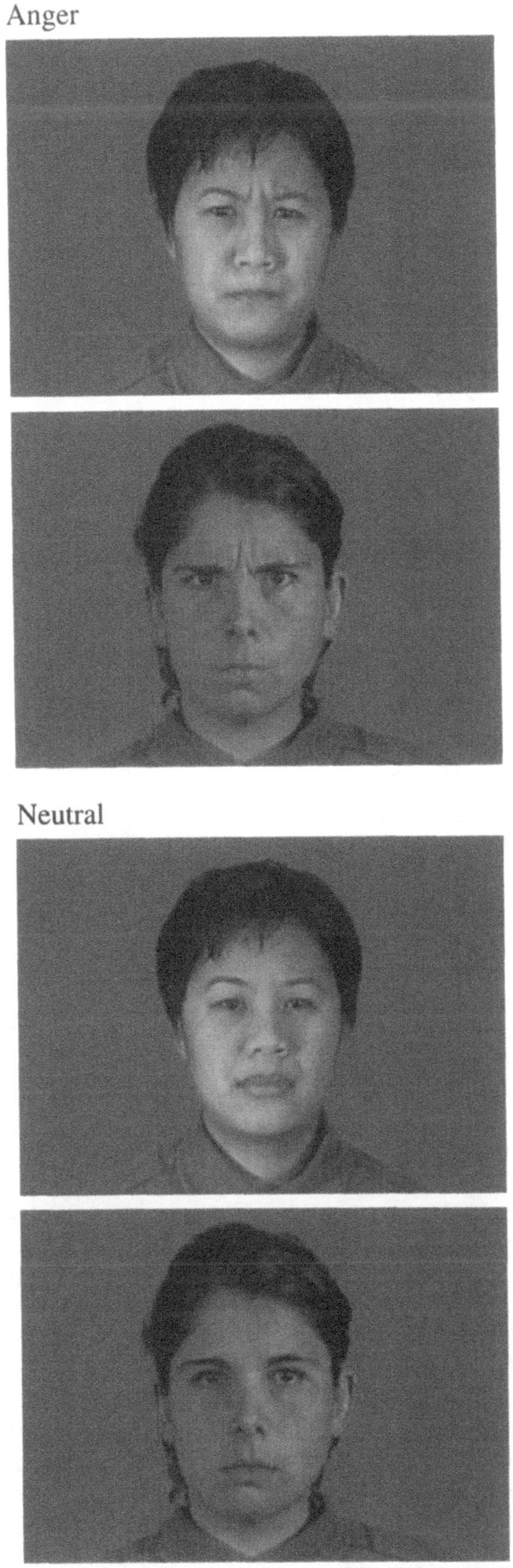

Figure 1. Examples of Caucasian and Chinese faces with angry and neutral expressions.

durations were a pink oval (12 × 16 cm) for the training phase, and 12 photographs of 6 different women (3 Chinese and 3 Caucasian) expressing anger or neutrality for the testing phase (from Beaupré & Hess, 2005; see Figure 1 for examples).

Empathy quotient questionnaire.. The Baron-Cohen Emotional Quotient Inventory (EQ-i; Baron-Cohen & Wheelwright, 2004) was used to measure individual differences in empathy. The EQ-i consists of 60 items (i.e., 40 empathy items and 20 fillers/control items) for use with adults of normal intelligence and gives an overall EQ score. The questionnaire was translated into French to produce a French-language version. A back-translation showed that the translation was acceptable.

Procedure

Participants first read and signed a statement of informed consent. Then, they performed a bisection task, which consisted of a training phase and a test phase. During the training phase, participants were shown a *short* (400 ms) and a *long* (1600 ms) standard stimulus duration, both represented by a neutral stimulus (i.e., a pink oval). The two standard durations were presented five times each in alteration. Participants pressed the *D* or the *K* key on the computer keyboard to indicate whether the duration was the *short* or the *long* one. The *D*/*K* response position was counterbalanced across participants. Participants then performed a block of 10 training trials in which the two standard durations had 50% probability of appearance.

The test phase maintained the conditions of training, except that the stimulus presented was a Chinese or Caucasian face rather than a pink oval. Each participant completed 3 blocks of 84 trials (i.e., 252 trials), in which the 12 photographs (3 different faces × 2 different expressions × 2 ethnicities) were presented three times for each of the seven durations in a random order. Each stimulus was presented at the two standard durations and five intermediated values (i.e., 400, 600, 800, 1000, 1200, 1400, and 1600 ms). The intertrial interval varied randomly between 1000 and 2000 ms.

After the test phase, participants completed the Baron Emotional Quotient Inventory, which

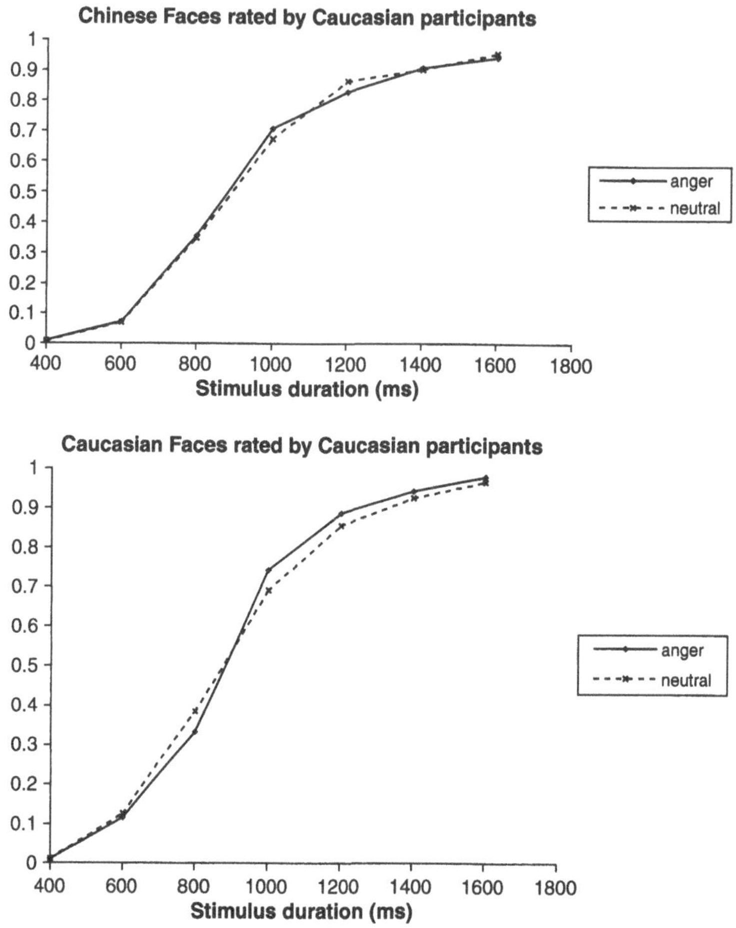

Figure 2. Average proportion of *long* responses plotted against stimulus durations, for the angry versus neutral faces as a function of stimuli's ethnicity, for Caucasian participants.

took about 15 minutes. Participants were then thanked for their participation.

Results

Four participants were excluded from analyses, including three due to their ethnicity (i.e., non-Caucasians) and one due to her tendency to strongly underestimate duration.

Time perception

Figure 2 shows the mean proportion of *long* responses plotted against the stimulus durations for the four experimental conditions: data from the two Stimulus Ethnicity conditions (Chinese faces vs. Caucasian faces) for the two types of Emotion (anger vs. neutral).

Because the observation of a three-way interaction in our 7 (Duration) × 2 (Emotion) × 2 (Ethnicity) model was judged as unnecessarily conservative,[1] to test our hypotheses we analyzed the data for the two stimulus Ethnicity conditions separately (Chinese faces and Caucasian faces), using stimulus Duration (400, 600, 800, 1000, 1200, 1400 and 1600 ms) and Emotion (angry vs. neutral) as within-subjects factors in an analysis of variance (ANOVA). The analysis on the proportion of *long* responses revealed a significant main effect of Duration for both Chinese and Caucasian stimuli, $F(6, 252) = 560.20, p < .001$ and $F(6, 252) = 562.91$, $p < .001$, respectively, indicating that participants were able to judge changes in duration length appropriately. Most importantly, an interaction between Duration and Emotion was observed for Caucasian stimuli, $F(6, 252) = 2.50$, $p < .05$, but not for Chinese stimuli,

[1] The three-way interaction involves the comparison of 28 conditions (7 2 2) and therefore lacks statistical power (e.g., participants sample, degree of freedom) for interpretation.

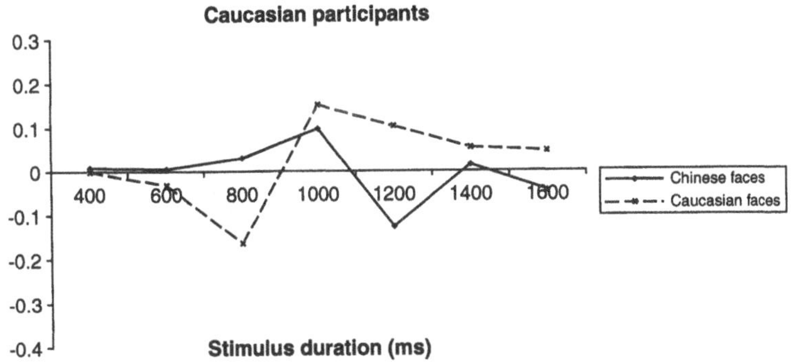

Figure 3. *D* scores plotted against stimulus durations, as a function of stimuli's ethnicity, for Caucasian participants.

$F(6, 252) = 1.19$, $p = .312$. The significant interaction for Caucasian faces was due to the expected tendency to overestimate the length of presentation of angry compared to neutral expressions, particularly at longer durations. No other theoretically meaningful results were obtained.

To further evaluate our predictions, the average proportion of trials that subjects categorized as *long* for each duration value was transformed using the *probit* function into *z*-scores. Then, a difference measure (*D*) was calculated at each duration by subtracting the *z*-scores for each angry face from the *z*-scores for the neutral face.[2] With 7 durations and 2 emotion pairs (happy–neural and sad–neutral), this yielded 14 *D* scores. Each score corresponded to the extent to which subjects' categorization of emotional faces differed from that of neutral faces; a positive score therefore represented an overestimation and a negative score represented an underestimation of duration relative to the neutral baseline. These scores were treated as the dependant variable in an ANOVA with Duration and Ethnicity as within-subjects factors.

The analysis revealed a marginal interaction between Ethnicity and Duration, suggesting a tendency for the magnitude of overestimation of stimulus durations for the anger faces compared to the neutral faces to be larger in the case of Caucasian faces ($M = 0.022$, $SD = 0.422$), relative to Chinese faces ($M = -0.002$, $SD = 0.416$), $F(6, 252) = 2.00$, $p = .066$. In order to further evaluate the marginal interaction, we examined the two types of Ethnicity separately. Consistent with the findings of the previous analyses, for Caucasian faces, *D* scores increased significantly as a function of time, $F(6, 252) = 2.31$, $p < .05$, going from negative to positive (overestimation) scores. In contrast, this was not the case for Chinese faces, $F(2, 252) = 1.20$, $p = .31$ (see Figure 3).

Empathy

Empathy Quotient scores varied from 27 to 62. They were subjected to a median split in order to obtain a categorical variable with 2 levels: low-empathic individuals ($M = 35.17$, $SD = 2.97$) and high-empathic individuals ($M = 44.95$, $SD = 5.47$). We then conducted the same overall ANOVA as before, including empathy as a between-subjects categorical variable. Results revealed an Empathy × Ethnicity interaction, $F(1, 41) = 5.4$, $p < .05$ (see Figure 4), which was further qualified by a significant Empathy × Ethnicity × Duration 3-way interaction, $F(6, 246) = 3.48$, $p < .01$.

The 3-way interaction was decomposed by conducting ANOVAs on the two sets of faces (i.e., Caucasian and Chinese) separately. Results revealed a main effect of empathy in the Caucasian faces condition, $F(1, 41) = 5.2$, $p < .05$, that was not present in the Chinese faces condition, $F(1, 41) = 1.37$, $p = .25$. Indeed, high-empathic participants ($M = 0.095$, $SD = 0.043$) overestimated duration represented by angry faces compared to neutral faces more than low-empathic participants ($M = -0.036$, $SD = 0.038$) when stimuli were Caucasian faces but this was not the case for Chinese faces ($M = -0.044$, $SD = 0.04$ and $M = 0.031$, $SD = 0.028$, respectively; see Figure 5).

Additionally, a series of *t*-test analyses showed that for high-empathic participants, *D* scores differed significantly from zero when duration was represented by Caucasian faces, $t(18) = 2.24$, $p < .05$, but not Chinese faces, $t(18) = 1.11$, $p = .28$. *D* differed from zero for low-empathic

[2] Note also that, in order for this transformation to be accomplished on proportions, values of 1 and 0 had to be replaced with a high and low decimal, respectively. We substituted 0.944 for 1, and 0.056 for 0.

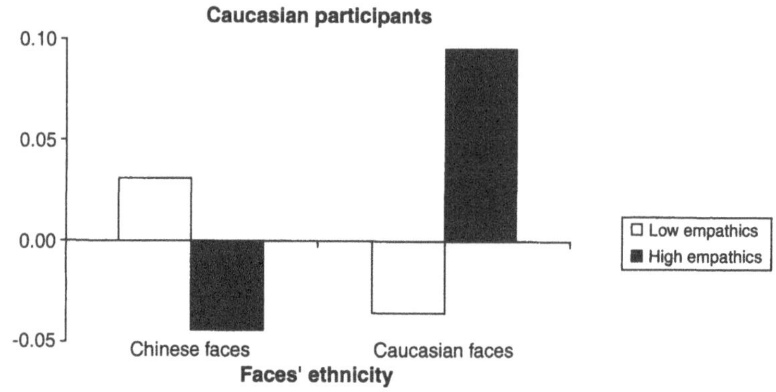

Figure 4. *D* scores' means for the Chinese versus Caucasian faces as a function of participants' empathy level, for Caucasian participants.

participants neither for Caucasian faces, $t(23) = -0.93$, $p = .36$, nor for Chinese ones, $t(23) = 0.65$, $p = .52$.

Discussion

The first study found support for the hypothesis that individuals imitate the facial expressions of individuals from their own group but not from an (ethnic) out-group. We used ethnicity to manipulate group membership in order to be able to inform and extend the current literature on possible cultural advantages in the recognition of facial expression of emotion, and because ethnicity constitutes a salient group membership. In our study, French Caucasian participants judged the duration of presentations of facial expressions (i.e., angry or neutral) on a computer screen, which were either displayed by other Caucasians or by Chinese individuals. Results of analyses of their performance on a temporal bisection task suggested that the participants tended to overestimate the duration of angry relative to neutral Caucasian faces, and that this overestimation had a propensity to increase over stimulus duration, thus producing a significant Emotion × Duration interaction. The same effect

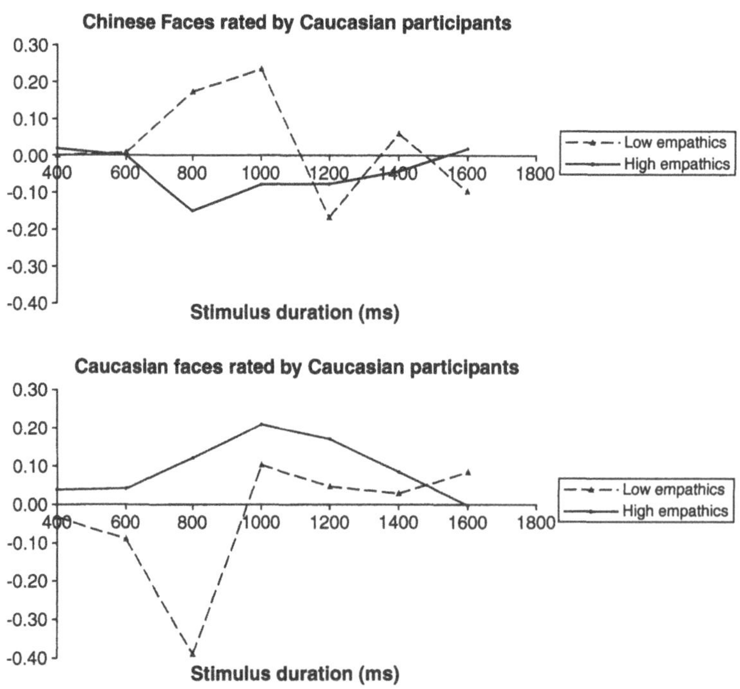

Figure 5. *D* scores plotted against stimulus durations, for the low- versus high-empathic participants as a function of stimuli's ethnicity, for Caucasian participants.

was not observed for Chinese faces. This suggests that the participants were imitating the anger expressions of the Caucasian faces, and that this had predictable effects on arousal (thus leading to overestimation of duration). Because there was no bias in judgments of the duration of angry compared to neutral expressions for the Chinese faces, we can suppose that the participants were not imitating these faces (e.g., Effron et al., 2006).

At a slightly later date, Chinese nationals were recruited for a similar study. Because the participants were run later in the academic year, and because historical events can affect performance on all tasks sensitive to psychological processes, we were obliged to consider this as constituting a separate study.

STUDY 2

Method

Participants, materials and procedure

Forty-one female students participated in the second study. All were Chinese nationals students residing and studying in France, and recruited from the University of Clermont-Ferrand, France. Because the participants were able to speak French, the French instructions from Study 1 were retained for use in this study.

The equipment, material, device model and procedure were the same as for Study 1 for both training and experimental tasks.

Results

Time perception

Figure 6 shows the mean proportion of *long* responses plotted against the stimulus duration for the four experimental conditions: data from the two Ethnicity conditions (Chinese faces vs. Caucasian faces) for the two types of emotion expressed (anger vs. neutral). Again, the scores expressed the extent to which the duration of emotional facial expressions was over- or underestimated compared to neutral ones.

As in Study 1, because of a non-conservative three-way interaction, we examined the proportion of *long* responses for the two Ethnicity conditions separately, using stimulus Duration

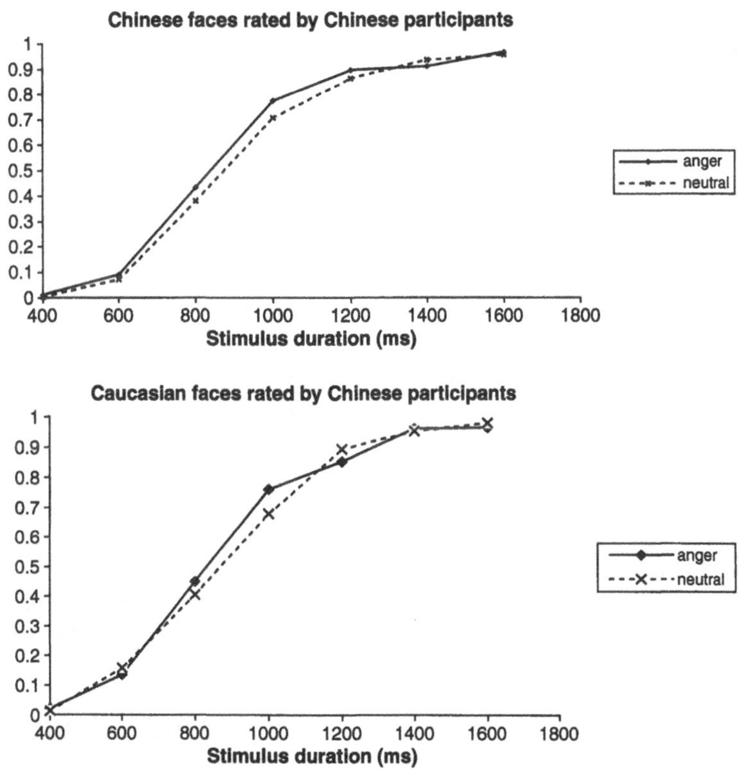

Figure 6. Average proportion of *long* responses plotted against stimulus durations, for the angry versus neutral faces as a function of stimuli's ethnicity, for Chinese participants.

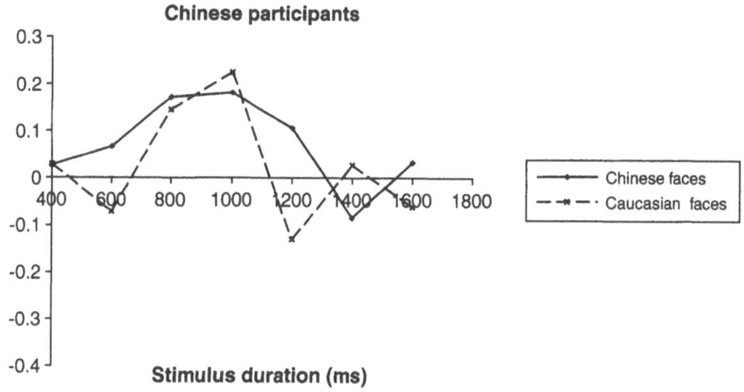

Figure 7. *D* scores plotted against stimulus durations, as a function of stimuli's ethnicity, for Chinese participants.

and facial expression of Emotion as within-subjects factors (Figure 7). Analyses revealed a significant main effect of Duration for both Chinese and Caucasian stimuli, $F(6, 240) = 476.52$, $p < .001$ and $F(6, 240) = 428.53$, $p < .001$, respectively, again suggesting that participants were able to appropriately judge duration.

A main effect of Emotion was observed for Chinese stimuli, $F(1, 40) = 5.84$, $p < .05$, but not for Caucasian stimuli, $F(1, 40) = 0.94$, $p = .34$. However, an interaction between Emotion and Duration was observed for Chinese stimuli, $F(6, 240) = 2.35$, $p < .05$, and for Caucasian stimuli, $F(6, 240) = 3.43$, $p \leq .01$. This suggests that the proportion of *long* responses for anger and neutral expressions differed with the duration value when Chinese participants saw Chinese and Caucasian faces.

As in the first study, we also computed difference (*D*) scores. Once more, the scores were treated as the dependant variable in a 7 × 2 ANOVA, with Duration (400, 600, 800, 1000, 1200, 1400 and 1600 ms) and Ethnicity (Chinese faces vs. Caucasian faces) as within-subjects factors. The analysis on difference score yielded only a main effect of Duration, revealing that the difference between angry and neutral faces varied as a function of the duration value, $F(6, 222) = 3.09$, $p < .01$. There was no interaction between Duration and Ethnicity, $F(6, 222) = 1.72$, $p = .119$, thus indicating that this overestimation occurred regardless of ethnicity (Figure 7).

Empathy

Empathy Quotient scores varied from 16 to 48 and were subjected to a median split in order to obtain a categorical variable with two levels: low-empathic individuals and high-empathic individuals. We then conducted the same ANOVA as before, including Empathy as a between-subjects categorical variable. Results revealed neither a main effect of Empathy, $F(1, 39) = 0.03$, $p = .87$, nor any interaction effect involving empathy as an interacting variable.

Discussion

Study 2 was a replication of the first study using Chinese participants. Interestingly, the Chinese participants overestimated duration when it was represented by angry facial expressions, compared to neutral, when the expressions were displayed by both Chinese and Caucasian faces. The interaction between Emotion and Duration for those faces suggests that the Chinese participants were imitating the faces and that attendant increases in arousal produced overestimation of the duration of angry faces. Indeed, there was a tendency to overestimate the duration represented by Chinese and Caucasian faces as suggested by the analyses on the proportion of *long* responses. However, as we will discuss, other processes, probably related to decisional or attentional processes, may have interfered with the arousal effect.

Unlike in the first study, empathy did not moderate the temporal distortions observed for either type of stimulus (i.e., Chinese vs. Caucasian). We take up the differences between the two studies in the general discussion.

GENERAL DISCUSSION

In the two studies reported here, we were concerned with the automatic imitation of emotional facial expressions as a function of group membership (i.e., in-group vs. out-group). We assessed processes of imitation using a temporal bisection task. In this type of task, when using relatively *short* durations (i.e., inferior to 2000 ms), it is well accepted that individuals typically overestimate durations of displayed angry faces (see Droit-Volet et al., 2004, for discussion). The effect of anger on temporal perception appears to be due to an increased arousal, which enhances the speed of the internal clock. In the framework of pacemaker-accumulator clock models, the arousal increases the number of pulses generated per unit of time by the pacemaker (Church, 1984; Gibbon, Church, & Meck, 1984). Such an increase in arousal seems to be mediated by embodied simulations (i.e., imitation) of the target's emotional expression (Effron et al., 2006). If perceivers do not imitate the expression displayed by a face representing duration, no distortion in temporal perception of emotional vs. neutral faces is observed (Effron et al., 2006). The task is thus ideal as an implicit measure of facial imitation.

We used ethnicity, or race, to represent in- and out-groups because of the now burgeoning literature on the recognition of facial expressions displayed by culturally (racially) similar and dissimilar others (e.g., Elfenbein & Ambady, 2002a,b). Such a design allowed us to test the limits of the imitation of facial expression of emotion.

Time perception

For both studies, one using French Caucasian and one Chinese individuals as participants, the basic prediction was, put statistically, that of an interaction between facial emotion (anger vs. neutral expression) and duration, and that this interaction would be greater (or exclusively observed) for in-group faces. For Caucasian participants, these results revealed the expected bias in the temporal perception of emotion when duration was represented by in-group members but not by out-group members. These results were consistent with the idea that the faces of out-group members were not imitated and therefore appeared to be quite consistent with Cheng and Chartrand's (2003) considerations about imitation of fellow in-group members.

For Chinese participants of Study 2, results were most supportive of the fact that both Caucasian and Chinese faces were imitated. Indeed, analyses revealed an interaction between emotion and stimulus duration on the proportion of *long* responses when faces were both Chinese or Caucasian. Specifically, the magnitude of the anger–neutral difference changed with the duration values, suggesting that the faces of out-group members were imitated. However, so far as the Caucasian faces are concerned, the results on the variation of *D* scores with the duration values, suggesting a pure arousal effect, are less clear. Indeed, the magnitude of the anger–neutral difference appeared to be particularly high at the point of subjective equality (i.e., between 800 and 1000 ms). In the mathematical model of temporal bisection performance, a distortion of the bisection function at the point of subjective equality (i.e., the stimulus duration giving rise to 50% of *long* responses) may be also explained by decisional processes (Delgado & Droit-Volet, in press; Droit-Volet & Wearden, 2001). Therefore, we cannot exclude that decision processes have interfered with earlier imitation-related processes for the final temporal judgment of Chinese participants. Attentional processes may also have interfered with the arousal effect. According to attentional models of time perception, when attentional resources are distracted away from the processing of time, the perceived time shortens (Fortin, 1999; Lejeune, 1998). However, this shortening of judged time is independent of the judged duration value, which is inconsistent with our results showing both a temporal overestimation of emotional stimuli compared to neutral ones and that this overestimation changed with the different duration values.

In order to interpret the finding that the Chinese seemed to be imitating the Caucasian faces, it is important to note, as mentioned above, that the Chinese participants were all students residing in France since the start of their studies. This raises the possibility that the Chinese were motivated to imitate the out-group, but that this same motivation was not present for the French Caucasian participants. We mentioned in the introduction that the imitation of emotional expressions improves affiliation and relationships between people (Dijksterhuis, 2005). Thus, it is in fact likely that individuals living in another country are not only motivated to imitate the facial expressions of peers, but also the majority ethnic group, with the general aim of assimilating to the culture.

This idea is suggested by a study by Bourgeois, Herrera, and Hess (submitted). The contribution of their study was to show that Asian Canadians were more accurate at identifying intense expressions displayed by Caucasian Canadians than expressions displayed by Asians. Furthermore, Beaupré and Hess (2006) recently showed, using the same base of normed facial expressions used here, that Chinese immigrants to Canada were as confident in their judgments of facial expressions of Caucasians as in their judgments of facial expressions of Chinese. Also relevant to the present study, Elfenbein and Ambady (2003c) provided evidence for the role of cultural familiarity in accuracy and speed at recognizing facial expressions of emotion. Their results demonstrated that participants were faster and more accurate when evaluating emotions expressed by hosts versus initial in-group members. That is, Chinese participants residing in the United States (for an average of 2.4 years) could better recognize facial expressions of American individuals than fellow in-group members, with a large increase across time.

The contribution of the present study, then, is to highlight imitation effects that are consistent with recently published recognition effects. In our view, our results are more consistent with the possibility that imitation is the mechanism that is responsible for the types of cultural group effects observed in the literature on facial expression recognition. However, given the fact that decisional effects may interfere with the arousal effects, it will be necessary to further investigate this distinction.

However, another explanation of the present findings is suggested by the recent studies of van Baaren, Maddux, Chartrand, de Bouter, and van Knippenberg (2003). The authors found that Asian (i.e., Japanese) participants, who have a chronic interdependent self-construal, were more likely to engage in mimicry than chronic independent self-construal individuals such as American participants. That is, van Baaren et al. (2003) did not find an in-group bias in imitation, but a self-construal orientation effect on mimicry. As concerns the present study, it may be that Chinese participants are more interdependent than French Caucasian ones, and thus show more sensitivity to context cues, including others' facial expressions. This interpretation may account for the lack of in-group bias effect for Chinese participants in Study 2. Participants' self-construals should therefore be addressed in further research dealing with inter-group or interpersonal differences in mimicry.

Empathy. Our second prediction was that personal empathy levels moderate imitation such that empathic individuals imitate facial expressions more competently than do less empathic individuals (see also Cheng & Chartrand, 2003). In both studies, we thus measured the participants' empathy quotient, and assessed the hypothesis that bias in temporal perception of emotion displayed by others is pronounced to a greater extent for high-empathic individuals, because they engage more in effective imitation.

Regarding Caucasian participants, the pattern of the results clearly supports the "empathy–imitation" assumption. The overestimation of time presentation of emotional expressions relative to neutral expressions revealed that the original in-group advantage effect was effective for empathic participants, whereas less-empathic participants did not show differences in time perception between emotional and neutral facial expressions for both Caucasian and Asian faces. Empathic individuals may exhibit more, and more efficiently, mimicry of others' postures and facial expressions than do less empathic individuals. This skill should confer a greater ability to understand and be sympathetic with others (Chartrand & Bargh, 1999). In addition, it recently appeared that imitation and empathy processes may share a common neural location involving mirror neurons (Gazzola, Aziz-Zadeh, & Keysers, 2006; Iacoboni, 2005). As Decety (2002) has noted, "by imitation we may feel what another person felt, which is the very definition of human empathy". That is empathy plays a central role via imitation in understanding accurately others' emotions, and thus in behaving properly.

The same finding was not obtained for the Chinese participants. However, a number of experimenter observations and methodological considerations leads us to be very cautious in drawing any conclusions from that finding. Despite the fact that the Chinese participants did speak French, experimenters noted that they showed difficulty in comprehending the ideas and words used in the questionnaire. Participants took a long time to complete the questionnaire, and required the use of a French–Chinese dictionary. Thus, it does not appear to us that these results constitute a hindrance in previous effects' reading in terms of empathy skills. Consequently, conducting a replication using participants living in their country of origin would allow us to assess the classical in-group effect appropriately. In addition, this may provide further evidence for the crucial role of embodiment in social cognition.

In sum, the present study stands in line with the mounting evidence described earlier in suggesting a central role of embodiment in social perception (Niedenthal, Barsalou, Winkielman, Krauth-Grüber, & Ric, 2005b). In agreement with Effron et al. (2006), it suggests that the imitation of emotional faces modifies our subjective experience of time that speeds up our internal clock. However, this does not exclude the fact that other cognitive processes (decision, memory) may also affect the final temporal judgment in a given task, the time judgment resulting from a complex processing of information.

REFERENCES

Adolphs, R., Damasio, H., Tranel, D., Cooper, G., & Damasio, A. R. (2000). A role for somatosensory cortices in the visual recognition of emotion as revealed by three-dimensional lesion mapping. *The Journal of Neuroscience*, *20*, 2683–2690.

Anderson, C., Keltner, D., & John, O. P. (2003). Emotional convergence between people over time. *Journal of Personality and Social Psychology*, *84*, 1054–1068.

Baron-Cohen, S., & Wheelwright, S. (2004). The Empathy Quotient (EQ). An investigation of adults with Asperger syndrome or high functioning autism, and normal sex differences. *Journal of Autism and Developmental Disorders*, *34*, 163–175.

Barsalou, L. W. (1999). Perceptual symbol system. *Behavioral and Brain Sciences*, *22*, 577–609.

Barsalou, L. W., Niedenthal, P. M., Barbey, A. K., & Ruppert, J. A. (2003). Social embodiment. *The Psychology of Learning and Motivation*, *43*, 43–92.

Bavelas, J. B., Black, A., Lemery, C. R., & Mullett, J. (1986). Motor mimicry as primitive empathy. In N. Eisenberg & J. Strayer (Eds.), *Empathy and its development* (pp. 317–338). New York: Cambridge University Press.

Beaupré, M. G., & Hess, U. (2005). Cross-cultural emotion recognition among Canadian ethnic groups. *Journal of Cross-Cultural Psychology*, *26*, 355–370.

Beaupré, M. G., & Hess, U. (2006). An ingroup advantage for confidence in emotion recognition judgments: The moderating effect of familiarity with the expressions of outgroup members. *Personality and Social Psychology Bulletin*, *32*, 16–26.

Bernieri, F. J. (1988). Coordinated movement and rapport in teacher–student interactions. *Journal of Nonverbal Behavior*, *12*, 120–138.

Bourgeois, P., Herrera, P., & Hess, U. (submitted). *Impact of cultural factors on the judgments of emotional facial expressions.*

Burle, B., & Casini, L. (2001). Dissociation between activation and attention effects in time estimation: Implications for internal clock models. *Journal of Experimental Psychology: Human Perception and Performance*, *27*, 195–205.

Carr, L., Iacoboni, M., Dubeau, M. C., Mazziotta, J. C., & Lenzi, G. L. (2003). Neural mechanisms of empathy in humans: A relay from neural systems for imitation to limbic areas. *Proceedings of the National Academy of Sciences of the USA*, *100*, 5497–5502.

Chartrand, T. L., & Bargh, J. A. (1999). The chameleon effect: The perception–behavior link and social interaction. *Journal of Personality and Social Psychology*, *76*, 893–910.

Cheng, C. M., & Chartrand, T. L. (2003). Self-monitoring without awareness: Using mimicry as a nonconscious affiliation strategy. *Journal of Personality and Social Psychology*, *85*, 1170–1179.

Church, R. M. (1984). Properties of the internal clock. In J. Gibbon & L. Allan (Eds.), *Timing and time perception* (pp. 566–582). New York: Annals of New York Academy of Sciences.

Cohen, J., MacWhinney, B., Flatt, M., & Provost, J. (1993). PsyScope: An interactive graphic system for designing and controlling experiments in the psychological laboratory using Macintosh computers. *Behavioral Research Methods, Instruments & Computers*, *25*, 257–271.

Decety, J. (2002). *Neuroscientists searching for roots of empathy find brain regions involved in learning by imitation.* (Available at: http://www.empathogens.com/empathy – accessed February 2007)

Decety, J., & Chaminade, T. (2003). Neural correlates of feeling sympathy. *Neuropsychologia*, *41*, 127–138.

Decety, J., & Jackson, P. L. (2003). Le corps acteur de l'esprit. *La Recherche*, *366*, 79–86.

Decety, J., & Jackson, P. L. (2004). The functional architecture of human empathy. *Behavioral and Cognitive Neuroscience Reviews*, *3*, 71–100.

Decety, J., & Sommerville, J. A. (2003). Shared representations between self and other: A social cognitive neuroscience view. *Trends in Cognitive Science*, *7*, 527–533.

Delgado, M., & Droit-Volet, S. (in press). Testing the representation of time in reference memory in the bisection and the generalization task: The utility of a developmental approach. *Quarterly Journal of Experimental Psychology.*

Dijksterhuis, A. (2005). Why we are social animals: The high road to imitation as social glue. In S. Hurley & N. Chater (Eds.), *Perspectives on imitation: From cognitive neuroscience to social science* (Vol. 2, pp. 207–220). Cambridge, MA: MIT Press.

Dimberg, U. (1982). Facial reactions to facial expressions. *Psychophysiology*, *19*, 643–647.

Dimberg, U. (1990). Facial electromyography and emotional reactions. *Psychophysiology*, *27*, 481–494.

Dimberg, U., Thunberg, M., & Elmehed, K. (2000). Unconscious facial reactions to emotional facial expressions. *Psychological Science*, *11*, 86–89.

Droit-Volet, S., Brunot, S., & Niedenthal, P. M. (2004). Perception of the duration of emotional events. *Cognition and Emotion, 18*, 849–858.
Droit-Volet, S., Meck, W. H., & Penney, T. B. (2007). Sensory modality and time perception in children and adults. *Behavioural Processes, 74*, 224–250.
Droit-Volet, S., & Wearden, J. (2002). Speeding up an internal clock in children? Effects of visual flicker on subjective duration. *The Quarterly Journal of Experimental Psychology, 55B*, 193–211.
Droit-Volet, S., & Wearden, J. H. (2001). Temporal bisection in children. *Journal of Experimental Child Psychology, 80*, 142–159.
Effron, D. A., Niedenthal, P. M., Gil, S., & Droit-Volet, S. (2006). Embodied temporal perception of emotion. *Emotion, 6*, 1–9.
Ekman, P., & Friesen, W. V. (1978). *Facial Action Coding System: A technique for the measurement of facial action.* Palo Alto, CA: Consulting Psychologists Press.
Elfenbein, H. A., & Ambady, N. (2002a). Is there an ingroup advantage in emotion recognition? *Psychological Bulletin, 128*, 243–249.
Elfenbein, H. A., & Ambady, N. (2002b). On the universality and cultural specificity of emotion recognition: A meta-analysis. *Psychological Bulletin, 128*, 205–235.
Elfenbein, H. A., & Ambady, N. (2003a). Cultural similarity's consequences: A distance perspective on cross-cultural differences in emotion recognition. *Journal of Cross-Cultural Psychology, 34*, 92–110.
Elfenbein, H. A., & Ambady, N. (2003b). Universals and cultural differences in recognizing emotions. *Current Direction in Psychological Science, 12*, 159–164.
Elfenbein, H. A., & Ambady, N. (2003c). When familiarity breeds accuracy: Cultural exposure and facial emotion recognition. *Journal of Personality and Social Psychology, 85*, 276–290.
Elfenbein, H. A., Mandal, M. K., Ambady, N., Harizuka, S., & Kumar, S. (2004). Hemifacial differences in the in-group advantage in emotion recognition. *Cognition and Emotion, 18*, 613–629.
Fortin, C. (1999). Short-term memory in temporal interval production. In I. Neath, G. D. A. Brown, M. Poirier, & C. Fortin (Eds.), *Short-term/working memory: A special issue of the International Journal of Psychology* (pp. 308–316). Hove, UK: Psychology Press.
Friedlmeier, W. (in press). Empathy. In J. R. Miller, R. M. Lerner, & L. B. Schiamberg (Eds.), *Human ecology: An encyclopedia of children, families, communities, and environments.* Santa Barbara, CA: ABC-Clio.
Gallese, V. (2003a). *Intentional attunement. The mirror neuron system and its role in interpersonal relations.* (Available at: http://www.interdisciplines.org/mirror/papers/1 – accessed May 2006)
Gallese, V. (2003b). The roots of empathy: The shared manifold hypothesis and the neural basis of intersubjectivity. *Psychopathology, 36*, 171–180.
Gallese, V. (2005a). Being like me. Self–other identity, mirror neurons and empathy In S. Hurley & N. Chater (Eds.), *Perspectives on imitation: From neuroscience to social science* (Vol. 1, pp. 101–118). Cambridge, MA: MIT Press.
Gallese, V. (2005b). Embodied simulation: Form neurons to phenomenal experience. *Phenomenology and the Cognitive Science, 4*, 23–48.
Gallese, V., Keysers, C., & Rizzolatti, G. (2004). A unifying view of the basis of social cognition. *Trends in Cognitive Sciences, 8*, 396–402.
Gallese, V., & Metzinger, T. (2003). Motor ontology: The representational reality of goals, actions, and selves. *Philosophical Psychology, 16*, 365–388.
Gazzola, V., Aziz-Zadeh, L., & Keysers, C. (2006). Empathy and the somatotopic auditory mirror system in humans. *Current Biology, 16*, 1824–1829.
Gibbon, J., Church, R. M., & Meck, W. H. (1984). Scalar timing in memory. In J. Gibbon & L. Allan (Eds.), *Timing and time perception* (pp. 58–87). New York: Annals of New York Academy of Sciences.
Gil, S., Niedenthal, P. M., & Droit-Volet, S. (2007). Anger and temporal perception in children. *Emotion, 7*, 219–225.
Glenberg, A. M. (1997). What memory is for. *Behavioral & Brain Sciences, 20*, 1–55.
Gump, B. B., & Kulik, J. A. (1997). Stress, affiliation, and emotional contagion. *Journal of Personality and Social Psychology, 72*, 305–319.
Hatfield, E., Cacioppo, J. T., & Rapson, R. L. (1992). Primitive emotional contagion. In M. S. Clark (Ed.), *Review of personality and social psychology. Vol. 14: Emotions and social behavior* (pp. 151–177). Newbury Park, CA: Sage.
Hatfield, E., Cacioppo, J. T., & Rapson, R. L. (1993). Emotional contagion. *Current Direction in Psychological Science, 2*, 96–99.
Hess, U., Sénécal, S., & Kirouac, G. (1996). Recognizing emotional facial expressions: Does perceived sociolinguistic group make a difference? *International Journal of Psychology, 31*, 3–4.
Iacoboni, M. (2005). Understanding others: Imitation, language, empathy. In S. Hurley & N. Chater (Eds.), *Perspectives on imitation: From neuroscience to social science* (Vol. 1, pp. 77–99). Cambridge, MA: MIT Press.
LaFrance, M. (1979). Nonverbal synchrony and rapport: Analysis by the cross-lag panel technique. *Social Psychology Quarterly, 42*, 66–70.
LaFrance, M. (1985). Postural mirroring and intergroup relations. *Personality and Social Psychology Bulletin, 11*, 207–217.
LaFrance, M., & Broadbent, M. (1976). Group rapport: Posture sharing as a nonverbal indicator. *Group and Organization Studies, 1*, 328–333.
Lakin, J., & Chartrand, T. (2005). Exclusion and nonconscious behavioral mimicry. In K. D. Williams, J. P. Forgas, & W. von Hippel (Eds.), *The social outcast: Ostracism, social exclusion, rejection, and bullying.* New York: Psychology Press.
Lakin, J. L., & Chartrand, T. L. (2003). Using nonconscious behavioral mimicry to create affiliation and rapport. *Psychological Science, 14*, 334–339.
Lakin, J. L., Jefferis, V. E., Cheng, C. M., & Chartrand, T. L. (2003). The chameleon effect as social glue: Evidence for the evolutionary significance of

nonconscious mimicry *Journal of nonverbal behavior, 27*, 145–162.

Lanzetta, J. T., & Englis, B. G. (1989). Expectations of cooperation and competition and their effects on observers vicarious emotional responses. *Journal of Personality and Social Psychology, 56*, 543–554.

Lejeune, H. (1998). Switching or gating? The attentional challenge in cognitive models of psychological time. *Behavioural Processes, 44*, 127–145.

Maricq, A. V., & Church, R. M. (1983). The differential effects of haloperidol and methamphetamine on time estimation in the rat. *Psychopharmacology, 79*, 10–15.

Matsumoto, D. (2002). Methodological requirements to test a possible in-group advantage in judging emotions across cultures: Comment on Elfenbein and Ambady (2002) and evidence. *Psychological Bulletin, 128*, 236–242.

Meck, W. H. (1983). Selective adjustment of the speed of internal clock and memory processes. Journal of Experimental Psychology. *Animal Behavior Processes, 9*, 171–201.

Meltzoff, A. N. (2002). Imitation as a mechanism of social cognition: Origins of empathy, theory of mind, and the representation of action. In U. Goswami (Ed.), *Handbook of childhood cognitive development* (pp. 6–25). Oxford, UK: Blackwell Publishers.

Meltzoff, A. N., & Decety, J. (2003). What imitation tells us about social cognition: A rapprochement between developmental psychology and cognitive neuroscience. *Philosophical Transactions of the Royal Society of London, Series B, Biological Sciences, 358*, 491–500.

Neumann, R., & Strack, F. (2000). Mood contagion: The automatic transfer of mood between persons. *Journal of Personality and Social Psychology, 79*, 211–223.

Niedenthal, P. M., Barsalou, L. W., Ric, F., & Krauth-Grüber, S. (2005a). Embodiment in the acquisition and use of emotion knowledge. In L. Feldman Barrett, P. M. Niedenthal, & P. Winkielman (Eds.), *Emotion: Conscious and unconscious*. New York: Guilford Press.

Niedenthal, P. M., Barsalou, L. W., Winkielman, P., Krauth-Grüber, S., & Ric, F. (2005b). Embodiment in attitudes, social perception, and emotion. *Personality and Social Psychology Review, 9*, 184–211.

Niedenthal, P. M., Brauer, M., Halberstadt, J. B., & Innes-Ker, A. H. (2001). When did her smile drop? Contrast effects in the influence of emotional state on the detection of change in emotional expression. *Cognition and Emotion, 15*, 853–864.

Pacherie, E. (2004). L'empathie et ses degrés. In A. Berthoz & G. Jorland (Eds.), *L'Empathie* (pp. 149–181). Paris: Editions Odile Jacob.

Smoski, M. J., & Bachorowski, J. A. (2003). Antiphonal laughter between friends and strangers. *Cognition and Emotion, 17*, 327–340.

Stepper, S., & Strack, F. (1993). Proprioceptive determinants of emotional and nonemotional feelings. *Journal of Personality and Social Psychology, 64*, 211–220.

Strack, F., Martin, L. L., & Stepper, S. (1988). Inhibiting and facilitating conditions of the human smile: A nonobstrusive test of the facial feedback hypothesis. *Journal of Personality and Social Psychology, 54*, 768–777.

Van Baaren, R. B., Holland, R. W., Steenaert, B., & van Knippenberg, A. (2003). Mimicry for money: Behavioral consequences of imitation. *Journal of Experimental Social Psychology, 39*, 393–398.

Van Baaren, R. B., Maddux, W. W., Chartrand, T. L., de Bouter, C., & van Knippenberg, A. (2003). It takes two to mimic: Behavioral consequences of self-construals. *Journal of Personality and Social Psychology, 84*, 1093–1102.

Vaughan, K., & Lanzetta, J. (1980). Vicarious investigation and conditioning of facial expressive and autonomic responses to a models expressive display of pain. *Journal of Experimental Social Psychology, 38*, 909–923.

Vaughan, K., & Lanzetta, J. (1981). The effect of modification of expressive display on vicarious emotional arousal. *Journal of Experimental Social Psychology, 17*, 16–30.

Zajonc, R. B., Adelmann, P. K., Murphy, S. T., & Niedenthal, P. M. (1987). Convergence in the physical appearance of spouses: An implication of the vascular theory of emotional efference? *Motivation and Emotion, 11*, 335–346.

SOCIAL NEUROSCIENCE, 2007, 2 (3–4), 238–253

Psychology Press
Taylor & Francis Group

The face of rejection: Rejection sensitivity moderates dorsal anterior cingulate activity to disapproving facial expressions

Lisa J. Burklund
University of California, Los Angeles, California, USA

Naomi I. Eisenberger
UCLA Cousins Center for Psychoneuroimmunology, Los Angeles, California, USA

Matthew D. Lieberman
University of California, Los Angeles, California, USA

Previous research has examined neural responses to threatening facial expressions such as those displaying anger, fear, and disgust. Here, we examined neural responses to a different type of threatening facial expression that primarily signifies a threat to social connection, namely a "disapproving" facial expression. We hypothesized that neural responses to disapproving facial expressions would be moderated by individual differences in rejection sensitivity. Using functional magnetic resonance imaging (fMRI), we scanned participants while they viewed brief video clips of facial expressions depicting disapproval, anger, and disgust. As expected, all three expressions yielded bilateral amygdala activation relative to a resting baseline. Additionally, individuals who scored higher on a measure of rejection sensitivity exhibited greater dorsal anterior cingulate cortex activity in response to disapproving facial expressions, but not in response to anger or disgust facial expressions. Results suggest that, at the neural level, individuals high in rejection sensitivity may be more sensitive to facial expressions signaling potential rejection, but not to threatening facial expressions in general. Results also suggest that disapproving facial expressions convey a distinct type of threat and should be considered in future studies of socially threatening facial expressions.

Facial expressions are a powerful and efficient source of social information (Darwin, 1872/1998). They are a reflection of how we feel and they help us communicate these feelings to others. As highly social beings, we are constantly exposed to a variety of different facial expressions, each providing valuable information about our physical and social surroundings and how to respond appropriately to them (Darwin, 1872/1998; Ekman, 2003). Negative or threatening

Correspondence should be addressed to: Lisa J. Burklund, Department of Psychology, Franz Hall, University of California, Los Angeles, CA 90095-1563 USA. E-mail: burklund@ucla.edu

This research was supported by a National Institutes of Mental Health (NIMH) postdoctoral research fellowship (T32MH-019925) to NIE, a UCLA Health Psychology Program university graduate fellowship to LJB, a predoctoral research fellowship from NIMH to LJB as part of the UCLA Health Psychology Program (MH15750), and an NIMH grant to MDL (MH071521). We also appreciate the support provided by the Brain Mapping Medical Research Organization, Brain Mapping Support Foundation, Pierson-Lovelace Foundation, the Ahmanson Foundation, Tamkin Foundation, Jennifer Jones-Simon Foundation, Capital Group Companies Charitable Foundation, Robson Family, William M. and Linda R. Dietel Philanthropic Fund at the Northern Piedmont Community Foundation, Northstar Fund, and National Center for Research Resources grants RR12169, RR13642 and RR08655.

www.psypress.com/socialneuroscience DOI:10.1080/17470910701391711

facial expressions, which signal to the target individual that "something is wrong" or that he or she is in some sort of danger, are especially potent. For example, an angry scowl can signal that we have crossed someone and should prepare to either fight or flee, and a grimace of disgust can warn us to steer clear of certain foods, helping us avoid food poisoning. Using tools such as functional magnetic resonance imaging (fMRI), we can examine the underlying neural responses to these negative facial expressions, and thus, deepen our understanding of how these expressions affect us.

Dozens of studies have investigated the brain's response to various threatening expressions, such as fear, anger, and disgust (fear: Morris et al., 1996; Phillips et al., 1997; Whalen et al., 1998, 2001; anger: Fitzgerald, Angstadt, Jelsone, Nathan, & Phan, 2006; Whalen et al., 2001; disgust: Fitzgerald et al., 2006; Phillips et al., 1997), with most studies finding activity in the amygdala,[1] a limbic structure theorized to play a role in the processing of threatening as well as socioemotional information from faces (Adolphs, 2002, 2003; Adolphs, Baron-Cohen, & Tranel, 2002; Whalen et al., 1998, 2001). Numerous studies have also observed insula and prefrontal activity in response to these negative emotional cues (Britton, Taylor, Sudheimer, & Liberzon, 2006; Kesler-West et al., 2001; Kilts, Egan, Gideon, Ely, & Hoffman, 2003; Phillips et al., 1997, 1998; Sprengelmeyer, Rausch, Eysel, & Przuntek, 1998; see Murphy, Nimmo-Smith, & Lawrence, 2003; Phan, Wager, Taylor, & Liberzon, 2002; Wager, Phan, Liberzon, & Taylor, 2003, for reviews). Despite this wealth of research, there is another type of threatening facial expression whose neural signature has yet to be examined. To date, the neural response to an expression that specifically conveys social rejection or disapproval is unknown. Thus, the goal of this study was to use fMRI to examine the neural responses to disapproving facial expressions, how these responses may resemble or differ from responses to other previously studied threatening facial expressions, and how they may vary across individuals based on differences in rejection sensitivity.

What is a disapproving facial expression? A disapproving facial expression is an indication of a negative evaluation. It is a signal that the target has done or said something socially undesirable and, consequently, has potentially damaged a social connection. A disapproving facial expression can convey a qualitatively different message than other threatening facial expressions such as fear, anger, or disgust. Fear and anger expressions often communicate that the target may be in danger of physical harm, and disgust expressions typically signal that the target may be in danger of physical contamination. In fact, the primary evolved purpose of anger, fear, and disgust expressions is assumed to be to signal a threat to physical safety (Darwin, 1872/1998). In contrast, disapproving facial expressions *only* signify a threat to social connection and have no other connotation.

Of course, facial expressions, such as anger or disgust, which can convey the threat of physical harm, may also convey the threat of social harm. One factor that may determine whether an expression is interpreted as a physical threat, a social threat, or both, is the familiarity of the individual displaying the expression. When the individual is familiar, the target individual has additional information from past interactions to help process the facial expression. For example, if the individual has a history of violence, the target individual will likely perceive an anger expression as primarily a physical threat. If the individual is known to be a peaceful person, the anger expression may not be interpreted as signaling physical danger, but rather, as indicating a purely social threat. Additionally, if the individual is a close other (e.g., spouse, parent, friend), expressions of anger and disgust will likely be interpreted, to some extent, as a social threat capable of damaging the relevant relationship. On the other hand, if the individual is unfamiliar, it may be difficult to definitively eliminate the possibility of physical danger associated with an anger or disgust expression and thus may be adaptive to prepare for the possibility of physical harm. Thus, we suggest that, at least with unfamiliar others, disapproving facial expressions are more likely to be interpreted as *purely* social threats than other threatening expressions. In the present study, we focused exclusively on responses to expressions conveyed by unfamiliar others.

We operationalized a disapproving facial expression as one that involves raising one side of the upper lip, lowering the inner corners of the brow in a fashion similar to that displayed when expressing "confusion," and slightly tilting or

[1] Amygdala activity has been reported less consistently for disgust expressions than for either fear or anger expressions (Fitzgerald et al., 2006; Phillips et al., 1997; see Murphy, Nimmo-Smith, & Lawrence, 2003, for a meta-analysis).

pulling the head backwards. Figure 1 shows a still frame from a video clip of a volunteer demonstrating a disapproving expression. The disapproving facial expressions used in our study are somewhat similar to the "contempt" faces described by Darwin (1872/1998) and Matsumoto and Ekman (2004) in that both convey dissatisfaction with some aspect of the target individual, and both involve a unilateral lip raise and sometimes a slight tilting or pulling back of the head. However, unlike the disapproval expressions we used, contempt expressions do not usually involve brow movement and are further described as involving a "wrinkling of the nose ... commonly accompanied by a slight snort ... half-closing [the] eyelids, or turning away ... " (Darwin, 1872/1998, pp. 252–254). Additionally, contempt expressions may involve more disgust or hatred than disapproving facial expressions.

Understanding the impact of disapproving facial expressions on targets has important consequences for mental and physical health outcomes. For example, social rejection has been shown to be associated with such negative outcomes as decrements in self-regulation (Baumeister, DeWall, Ciarocco, & Twenge, 2005), increases in negative affect, and increased physiological stress responses such as elevated blood pressure (Stroud, Tanofsky-Kraff, Wilfley, & Salovey, 2000) and cortisol (Dickerson & Kemeny, 2004). Facial expressions signaling potential social rejection may have similar effects. Additionally, while facial expressions such as fear and anger have been used to assess negative affective processes in several clinical disorders, including social phobia (Amir et al., 2005; Birbaumer et al., 1998; Phan, Fitzgerald, Nathan, & Tancer, 2006), posttraumatic stress disorder (Rauch et al., 2000), depression (Lawrence et al., 2004; Sheline et al., 2001), and borderline personality disorder (Donegan et al., 2003), disapproving facial expressions may be more relevant and potent for understanding clinical disorders that predominantly involve a sensitivity to negative social experiences (e.g., social phobia, depression).

Whether at the clinical or subclinical level, individuals likely differ in the extent to which they may be affected by cues of potential rejection, such as disapproving facial expressions. "Rejection sensitivity" refers to the extent to which an individual anxiously expects, readily perceives, and overacts to social rejection (Downey & Feldman, 1996). In other words, an individual high in rejection sensitivity is more likely to interpret ambiguous stimuli as rejecting, subsequently overreact, and consequently feel greater distress, compared to individuals low in rejection sensitivity (Downey & Feldman, 1996). Since rejection sensitivity can moderate social, cognitive, and behavioral responses to potentially socially threatening information, it follows that it may also moderate neural responses to cues of potential social rejection, such as disapproving facial expressions. Thus, we expected that individuals high in rejection sensitivity would be particularly sensitive to disapproving facial expressions at the neural level.

Although the neural response to disapproving facial expressions has not yet been examined, we predicted that certain neural regions would be involved based on related research. First, we expected to see amygdala activity in response to viewing disapproving facial expressions because

Figure 1. Still frames from video clips of (from left to right) disapproval, anger, and disgust facial expressions. The video clips presented all facial expressions in full color.

the amygdala has been shown to be activated in response to a variety of other threatening facial expressions such as anger, fear, and disgust (Fitzgerald et al., 2006; Morris et al., 1996; Phillips et al., 1997; Whalen et al., 1998, 2001). However, because the threat displayed in a disapproving facial expression is fundamentally more social in nature than the threat conveyed by anger, fear, or disgust expressions, we also expected to find neural activity unique to the disapproving facial expressions. The dorsal anterior cingulate cortex (dACC) was a primary candidate. Prior work has shown that dACC is activated in response to an episode of social rejection and that the magnitude of dACC activity correlates with the magnitude of self-reported social distress (e.g., "I felt rejected") following rejection (Eisenberger, Lieberman, & Williams, 2003; Eisenberger, Way, Taylor, Welch, & Lieberman, 2007). Based on this finding, it is possible that facial stimuli that indicate social rejection (i.e., disapproving expressions) may activate the dACC as well, either (a) because these expressions elicit social distress in the target and thus activate distress-related dACC activity or (b) because dACC activity may be sensitive to cues of social rejection, in addition to experiences of social rejection. Thus, we hypothesized that dACC would be responsive to disapproving facial expressions, but not to other threatening expressions. Most importantly, we hypothesized greater dACC activity in response to the disapproving facial expressions in individuals who are high in rejection sensitivity.

Here we present the findings of an fMRI study that supports the hypothesis that rejection-sensitive individuals show a greater "threat response" to disapproving facial expressions than non rejection-sensitive individuals. In this study, we scanned healthy participants while they viewed dynamic disapproval, anger, and disgust faces.[2] We examined neural activity in response to the presentation of each of these expressions (i.e., main effects) as well as the activity that was correlated with participants' self-reported scores of rejection sensitivity.

METHODS

Participants

Twenty-one participants (15 females; mean age = 27.10 years, $SD = 9.04$), who were recruited from a UCLA undergraduate psychology class and the UCLA community, were paid $30 to complete the study. Data from two participants (both female) were excluded due to excessive head movement during scanning, and therefore, main effects analyses included 19 participants. In addition, only 16 of these 19 participants (11 females) completed the self-report measure of rejection sensitivity and thus all regression analyses using this variable involved these 16 participants.

Stimuli

The stimuli consisted of brief, 3-second video clips of seven volunteers (four female) demonstrating anger, disgust, disapproving, and neutral[3] facial expressions. The video clips were recorded by the experimenters using a Sony digital video recorder and edited on a Macintosh G4 iMac with iMovie software. The video clips were validated by having a separate group of participants ($n = 17$) rate them, using two different methods. First, participants were instructed to label the emotion or feeling conveyed by each facial expression using a free-response format. Subsequently, participants were instructed to identify the emotion or feeling conveyed by each facial expression using a forced-choice response sheet with the following options: anger, disgust, disapproval, sadness, confusion, and neutral. Using the free-response format, participants identified the expressions at a rate significantly higher than chance (84% correct overall; 85% correct for disapproval faces specifically). Examples of free responses (to the question "What emotion or feeling is being conveyed?") that were coded as "disapproval" included phrases like: "disapproving," "disappointed," and "shouldn't have done

[2] Video clips were used in lieu of still photographs because the videos more accurately captured the facial expressions as they would be seen in real-life settings. Furthermore, dynamic expressions have been shown to improve recognition of the emotional content of facial expressions (Frijda, 1953; Wehrle, Kaiser, Schmidt, & Scherer, 2000) and have been associated with greater amygdala responses (Sato, Kochiyama, Yoshikawa, Naito, & Matsumura, 2004).

[3] The neutral facial expression condition was originally included to serve as a comparison condition for the other facial expressions. However, because this condition produced greater limbic and prefrontal activity than the other facial expressions, a finding consistent with recent research (Fitzgerald et al., 2006), it was not used as a control condition in the current analyses. All results pertaining to the neutral condition will be presented in a separate report.

that." Examples of free responses that were coded as "anger" included "angry" and "mad." Examples of free responses that were coded as "disgust" included "disgusted," "repulsed," and "grossed out." Using the forced-choice method, participants accurately labeled the expressions at a level significantly higher than chance (86% correct overall; 79% correct for disapproval faces specifically; chance = 16.7%). These recognition rates are consistent with those observed for emotional expressions (i.e., anger, disgust, fear) in standardized stimulus sets (Biehl et al., 1997; Tottenham, Borscheid, Ellertsen, Marcus, & Nelson, 2002).

Procedure

To assess neural activity in response to viewing the dynamic facial expressions, participants viewed the video clips of these expressions within the fMRI scanner. Participants were fitted with LCD goggles that enabled them to see the stimuli presented on a Macintosh G4 Powerbook computer. Each participant viewed a total of eight blocks of video clips. This included two blocks each of anger, disgust, disapproval, and neutral facial expressions. Each block consisted of ten 3-second video clips of different individuals demonstrating the particular expression type for that block, for a total of 30 seconds per block. Participants were instructed to simply watch the video clips. Experimental blocks were separated by 10-second rest blocks, during which time, the participants were instructed to focus on a cross-hair, thereby establishing a baseline comparison condition ("fixation").

Outside the scanner, participants completed a self-report measure of rejection sensitivity—the Rejection Sensitivity Questionnaire (RSQ-Short Form; Downey & Feldman, 1996). This measure consisted of 8 items in which participants were given social scenarios (e.g., "You ask a friend to do you a big favor") and then asked about the extent to which they would feel anxious about rejection ("How concerned or anxious would you be over whether or not your friend would do this favor?") and the extent to which they would expect rejection ("I would expect that he/she would willingly do this favor for me"), given the scenario. Ratings of anxiety were made on a 6-point scale from 1 (*very unconcerned*) to 6 (*very concerned*). Ratings of expectation of rejection were made on a 6-point scale from 1 (*very unlikely*) to 6 (*very likely*). The RSQ-Short Form scores were obtained by multiplying the anxiety and expectation rating for each of the eight scenarios, and then taking the average of these eight products. Most participants completed the measure in the same session as the fMRI task, prior to being scanned; however, some completed the measure several weeks following the scanning session.

Image acquisition and analysis

Data were acquired on a Siemens Allegra 3T scanner at the UCLA Ahmanson-Lovelace Brainmapping Center. Head movements were restrained with foam padding and surgical tape that was attached to the goggles and scanner bed. High-resolution structural T2-weighted echo-planar images (spin-echo; TR = 5000 ms; TE = 33 ms; matrix size 128 × 128; 36 saggital slices; FOV = 20 cm; 3 mm thick, skip 1 mm) were acquired coplanar with the functional scan. One functional scan lasting 5 minutes and 30 seconds was acquired (echo planar $T2^*$-weighted gradient-echo, TR = 3000 ms, TE = 25 ms, flip angle = 90°, matrix size 64 × 64, 36 axial slices, FOV = 20 cm; 3 mm thick, skip 1 mm).

The imaging data were analyzed using SPM'9 (Wellcome Department of Cognitive Neurology, Institute of Neurology, London, UK). Images for each participant were realigned to correct for head motion, normalized into a standard stereotactic space as defined by the Montreal Neurological Institute, and smoothed with an 8 mm Gaussian kernel, full width at half maximum, to increase the signal-to-noise ratio. The design was modeled using a boxcar function convolved with a canonical hemodynamic response function. Linear contrasts were used to assess neural activity during each of the experimental conditions (anger, disgust, disapproval) compared to the baseline rest condition (crosshair fixation), as well as during each experimental condition compared directly to the others (e.g., disapproval vs. anger). Random effects group analyses were computed using participants' individual contrast images.

To assess correlations between rejection sensitivity and neural activity in response to the disapproving expressions, self-report scores of the rejection sensitivity measure were entered as a regressor into a whole-brain random effects group analysis for the contrast comparing the disapproval condition to fixation. Similar whole-brain regression analyses were also completed for the contrasts comparing the anger condition to fixation and the disgust condition to fixation in order to assess correlations between rejection sensitivity and neural activity in response to the anger and disgust expressions.

To examine functional connectivity between those neural regions hypothesized a priori to be moderated by rejection sensitivity (i.e., dACC) and other regions associated with rejection sensitivity in response to disapproving faces, parameter estimates were extracted at the region of interest and entered as a regressor into a whole-brain random effects group analysis to reveal regions correlated with the region of interest.

To correct for multiple comparisons in the whole-brain analyses, we used an uncorrected *p*-value of .005 combined with a cluster size threshold of 10 voxels (Forman et al., 1995) for all limbic (e.g., amygdala, insula, ACC) and prefrontal cortical regions, as these regions have been shown to be involved when viewing emotional facial expressions (Fitzgerald et al., 2006; Morris et al., 1996; Phillips et al., 1997; Wager et al., 2003; Whalen et al., 1998, 2001). For all other regions, we used a *p*-value of .05, corrected, with a 10-voxel extent threshold. All coordinates are reported in Montreal Neurological Institute (MNI) format.

RESULTS

Rejection sensitivity

The mean of participants' scores on the self-report measure of rejection sensitivity ($M = 7.53$) was slightly lower than the measure's reported mean ($M = 9.69$; Downey & Feldman, 1996). The observed standard deviation ($SD = 3.05$) was similar to the measure's reported standard deviation ($SD = 3.07$; Downey & Feldman, 1996). The reliability of this scale was strong ($\alpha = .73$, standardized item $\alpha = .76$).

Neural activations common to disapproval, anger, and disgust

As predicted and shown in Table 1, each condition (disapproval, anger, disgust), compared to fixation, resulted in significant activity in bilateral amygdala ($p < .005$, 10-voxels), similar to previous studies of neural activity to negative facial expressions (Fitzgerald et al., 2006; Morris et al., 1996; Philips et al., 1997; Whalen et al., 1998, 2001). In addition, each condition, compared to fixation, resulted in activity in several regions of the prefrontal cortex, including the medial prefrontal cortex (MPFC; Brodmann's area (BA) 10), dorsomedial prefrontal cortex (DMPFC; BA 8 & 9), ventromedial prefrontal cortex (VMPFC; BA 11), right and left ventrolateral prefrontal cortex (RVLPFC, LVLPFC; BA 47), and right and left inferior frontal gyrus (RIFG, LIFG; BA 45). Contrary to our prediction, however, there was no significant ACC activity (dorsal or rostral) in response to any of the conditions relative to fixation. Table 1 lists activations for regions found to be activated in response to all three types of facial expressions relative to fixation.

Activations that differentiate disapproval from anger and disgust

Table 2 shows regions differentially active when comparing the disapproval condition directly with the anger and disgust conditions. Interestingly, there was significantly greater activity in the left amygdala during disapproval expressions than during anger expressions ($p < .005$, 10-voxels; see Figure 2). In addition, consistent with prior work on disgust (Phillips et al., 1997, 1998; Sprengelmeyer et al., 1998), there was significantly greater insula activity when subjects viewed the disgust faces, compared to when they viewed the disapproval faces ($p < .005$, 10-voxels; see Figure 3). Again, there were no significant differences in dACC activity when comparing the disapproval condition with the other conditions.

Correlations with rejection sensitivity

Positive correlations. To examine correlations between rejection sensitivity and neural responses to the facial expressions, we regressed rejection sensitivity scores into whole-brain random effects group analyses, comparing each facial expression to fixation. There were no limbic or prefrontal regions that correlated positively with rejection sensitivity during the anger condition relative to fixation or during the disgust

TABLE 1

Common neural regions that showed increased activity during the viewing of each facial expression compared to fixation

	Disapproval				*Anger*				*Disgust*			
Region	*x*	*y*	*z*	*t-statistic*	*x*	*y*	*z*	*t-statistic*	*x*	*y*	*z*	*t-statistic*
Left Amygdala	−22	−6	−16	6.42	−22	−4	−16	6.87	−22	−6	−18	6.55
Right Amygdala	16	−6	−14	4.12	30	0	−24	5.09	28	−4	−20	7.08
DMPFC (BA 8)	−8	50	44	7.52	−4	56	44	3.07	−8	50	48	4.36
DMPFC (BA 9)	−6	60	30	5.52	−6	60	34	4.28	−6	64	30	5.15
									−6	52	40	5.92
MPFC (BA 10)	−6	64	20	4.77	−6	68	18	3.38	−8	66	18	3.72
					−10	64	26	3.89	−8	66	24	5.94
VMPFC	2	44	−20	6.00	−2	44	−20	5.26	−2	46	−22	4.42
	2	50	−20	5.92	2	46	−20	5.26				
LVLPFC	−46	28	−4	5.89	−52	40	−2	3.19	−46	28	−4	3.38
	−40	38	−14	3.63					−32	26	−20	4.43
									−32	36	−20	3.77
RVLPFC	44	28	−8	6.64	46	26	−4	5.20	46	28	−8	3.79
	46	30	−6	6.70	42	34	−16	3.55				
	46	50	−12	3.96								
	50	26	−12	6.05								
LIFG (BA 45)	−56	22	8	6.56	−48	24	16	8.70	−54	24	14	4.74
RIFG (BA 45)	56	22	20	7.52	56	22	8	3.09	56	28	10	3.54
Occipital (BA 17/18)*	−8	−98	6	16.52	−14	−98	12	16.60	12	−100	10	22.22

Note: All activations: $p < .005$, uncorrected, 10-voxel extent threshold except $^*p < .05$, corrected, 10-voxel extent threshold. BA = Brodmann's area.

TABLE 2
Regions that show differential neural activity while viewing disapproving facial expressions compared to other facial expressions

Region	*x*	*y*	*z*	*t-statistic*
Disapproval > anger				
Left Amygdala	−22	−6	−12	4.39
LVLPFC (BA 47)	−52	32	−4	3.77
RVLPFC (BA 47)	48	26	−14	4.14
MTG (BA 21)*	56	−34	0	7.76
Anger > disapproval				
MPFC	12	62	2	3.77
ROFC (BA 11/47)	20	24	−16	3.63
Disapproval > disgust				
No significant activity				
Disgust > disapproval				
Insula	38	−18	−2	4.12
MPFC	12	66	8	3.76
RVLPFC (BA 11/47)	28	26	−20	4.24
	28	30	−12	4.16
LDLPFC (BA 9)	−36	42	36	3.69

Note: All activations: $p < .005$, uncorrected, 10-voxel extent threshold except *$p < .05$, corrected, 10-voxel extent threshold. BA = Brodmann's area; MTG = middle temporal gyrus; ROFC = right orbital frontal gyrus.

condition relative to fixation. In contrast, during the disapproval condition, relative to fixation, rejection sensitivity was positively correlated with activity in the dACC ($r = .70$, $p < .005$; see Table 3 and Figure 4) as well as with activity in RVLPFC and right dorsolateral prefrontal cortex (RDLPFC; see Table 3). Thus, as predicted, to the extent that an individual scored higher in rejection sensitivity, that individual exhibited increased activity in the dACC and other prefrontal regions in response to the disapproving facial expressions relative to fixation, but not in response to the anger or disgust facial expressions relative to fixation.

To further explore the specificity of the relationship between rejection sensitivity and dACC activity in response to the disapproving expressions, we completed two sets of post hoc analyses. First, we lowered our statistical threshold to further examine the relationship between rejection sensitivity and dACC activity in response to the anger and disgust expressions relative to fixation. Even when the statistical threshold was lowered to $p < .3$, there were no regions of dACC that were positively correlated with rejection sensitivity in response to the anger faces relative to fixation or in response to the disgust faces relative to fixation. Next, we completed additional analyses examining the relationship between dACC activity and rejection sensitivity in response to the disapproving expressions compared directly with anger and disgust faces. As shown in Table 4, using dACC ROIs based on the activity correlating with rejection sensitivity in response to the disapproving faces compared to fixation, rejection sensitivity was also positively correlated with dACC activity in response to the disapproving facial expressions when compared directly with either anger or disgust expressions ($p < .05$), suggesting that rejection sensitivity correlated with dACC activity to a significantly greater extent while viewing disapproving faces than while viewing disgust or anger faces.

Negative correlations. When examining the neural regions that correlated negatively with

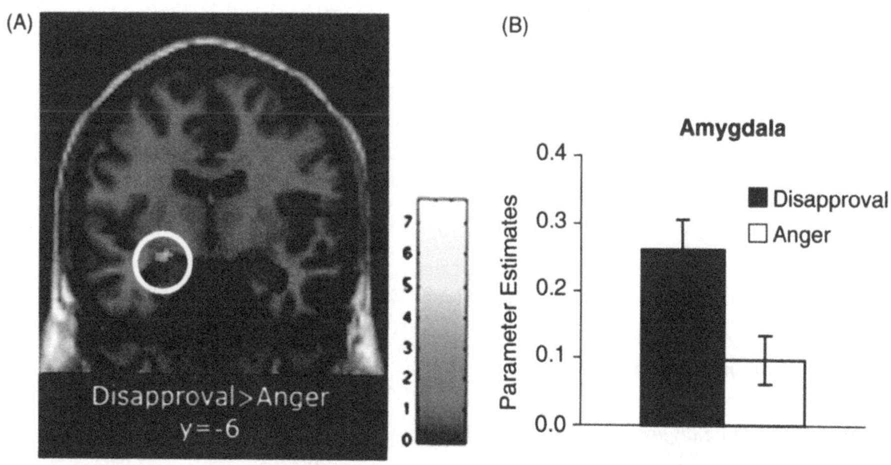

Figure 2. (A) Coronal slice (at $y = -6$); (B) bar graph showing greater amygdala activity in response to disapproval than anger ($p < .005$, 10-voxels).

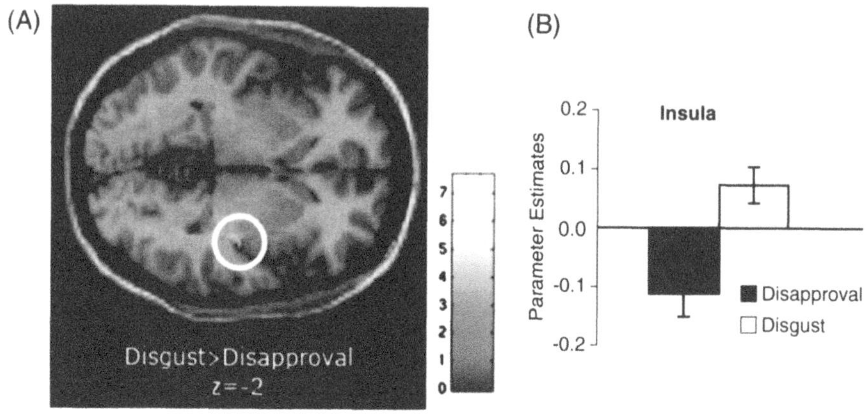

Figure 3. (A) Axial slice (at $z = -2$); (B) bar graph showing greater insula activity in response to disgust than disapproval ($p < .005$, 10-voxels).

rejection sensitivity during disapproval versus fixation, rejection sensitivity correlated negatively with activity in a single region, the subgenual ACC (subACC/VMPFC), ($r = -.75$, $p < .005$, 10-voxels; see Table 3 and Figure 5). Thus, individuals who scored higher in rejection sensitivity also exhibited less activity in subACC/VMPFC in response to the disapproval expressions. The fact that rejection-sensitive individuals showed reduced subACC/VMPFC activity to disapproving facial expressions fits with previous work demonstrating a role for the subACC/VMPFC in the extinction of conditioned fear responses in humans (Phelps, Delgado, Nearing, & LeDoux, 2004) as well as in signaling a less-threatening interpretation of a negative stimulus (Kim, Somerville, Johnstone, Alexander, & Whalen, 2003). Thus, it is possible that rejec-

TABLE 3
Neural regions that correlate with rejection sensitivity while viewing facial expressions compared to fixation

Region	*x*	*y*	*z*	*r*	*t-statistic*
Positive correlation with RS during disapproval > fixation					
dACC	0	24	30	.68	3.43
	8	30	22	.66	3.31
	14	24	40	.70	3.68
	14	32	24	.65	3.18
	6	38	36	.66	3.30
	16	26	42	.69	3.54
RVLPFC	26	22	−18	.75	4.24
RDLPFC	22	52	42	.71	3.74
Negative correlation with RS during disapproval > fixation					
SubACC/VMPFC	−4	26	−12	−.75	4.18
Positive correlation with RS during anger > fixation					
No significant activity					
Negative correlation with RS during anger > fixation					
dACC	0	4	32	−.68	3.44
	−8	6	48	−.70	3.69
SubACC/VMPFC	8	26	−14	−.75	4.18
MPFC (BA 10)	2	46	20	−.65	3.24
Left Amygdala	−20	−2	−12	−.68	3.46
Positive correlation with RS during disgust > fixation					
No significant activity					
Negative correlation with RS during disgust > fixation					
Right Amygdala	22	−8	−10	−.64	3.08

Note: All activations $p < .005$, uncorrected, 10-voxel extent threshold. BA = Brodmann's area; RS = rejection sensitivity.

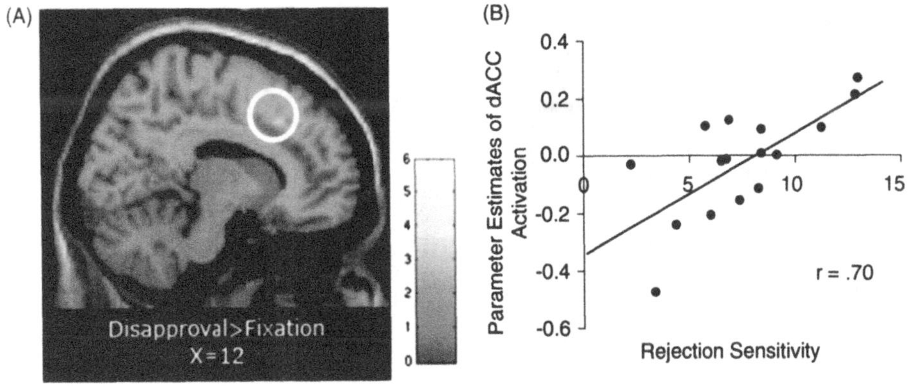

Figure 4. (A) dACC activity (14, 24, 40) in response to disapproval expressions compared to fixation that correlates positively with self-reported rejection sensitivity scores ($r = .70$; $p < .005$, 10-voxels); (B) scatterplot showing the relationship between rejection sensitivity scores and dACC activity (14, 24, 40) in response to disapproving expressions compared to fixation.

tion-sensitive individuals may show reduced subACC/VMPFC activity to disapproving faces and thus are less able to regulate the negative responses to these stimuli.

Somewhat unexpectedly, rejection sensitivity was also negatively correlated with activity in a variety of regions during the anger and disgust conditions relative to fixation (see Table 3). During anger versus fixation, higher levels of rejection sensitivity were associated with lower levels of activity in the dACC, subACC/VMPFC, left amygdala, and MPFC ($p < .005$, 10-voxels). During disgust versus fixation, higher levels of rejection sensitivity were associated with lower levels of activity in the right amygdala ($p < .005$, 10-voxels). This result can alternatively be interpreted as individuals high in rejection sensitivity showing *more* activity in these limbic and prefrontal regions during the *rest* (i.e., fixation) period, relative to viewing the anger or disgust expressions.

Functional connectivity during the disapproval condition relative to fixation

Connectivity analyses were completed to examine the relationship between dACC, which was hypothesized a priori to be moderated by rejection sensitivity, and other regions of the brain during the viewing of disapproving expressions compared to fixation. Parameter estimates were extracted from each of the regions of the dACC that correlated with rejection sensitivity and then entered as regressors into whole-brain random effects group analyses, comparing neural activity

TABLE 4
dACC ROI activity that is positively correlated with rejection sensitivity while viewing disapproving facial expressions compared directly to anger and disgust facial expressions

Region	*x*	*y*	*z*	*r*	*t-statistic*
Positive correlation with RS during disapproval > anger					
dACC	0	24	30	.59	2.73
	8	30	22	.61	2.91
	14	24	40	.50	2.18
	14	32	24	.55	2.49
	6	38	36	.54	2.43
	16	26	42	.45	1.91
Positive correlation with RS during disapproval > disgust					
dACC	0	24	30	.66	3.26
	8	30	22	.62	2.95
	14	24	40	.59	2.74
	14	32	24	.48	2.04
	6	38	36	.48	2.07
	16	26	42	.59	2.71

Note: All activations $p < .05$, uncorrected, 10-voxel extent threshold. RS = rejection sensitivity.

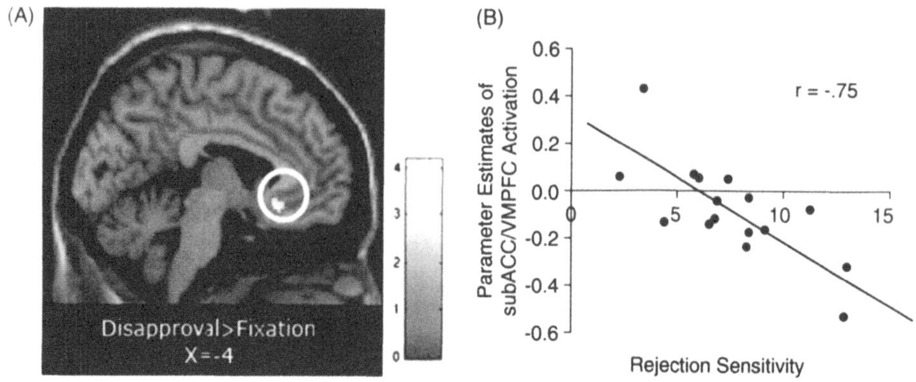

Figure 5. (A) SubACC/VMPFC activity (−4, 26, −12) in response to disapproval expressions compared to fixation that correlates negatively with self-reported rejection sensitivity scores ($r = -.75$; $p < .005$,10-voxels); (B) scatterplot showing the relationship between rejection sensitivity scores and subACC/VMPFC activity (−4, 26, −12) in response to disapproving expressions compared to fixation.

during disapproval to neural activity during fixation. There were no regions of activity that correlated with dACC at $p < .005$, 10-voxels; however, when reducing the extent threshold, results revealed that activity in the dACC ($x = 14$, $y = 24$, $z = 40$) was negatively correlated with activity in subACC/VMPFC ($x = -4$, $y = 26$, $z = -12$; $r = -.63$; $p < .005$, 6-voxels). Thus, individuals who showed greater activity in the dACC while viewing disapproving faces, compared to fixation, also showed a corresponding reduction in subACC/VMPFC activity. Interestingly, this is the precise region of subACC/VMPFC that was negatively correlated with rejection sensitivity.

DISCUSSION

The purpose of the present study was to: (a) examine the neural signature associated with viewing disapproving facial expressions, which have not been investigated with neuroimaging methods, and (b) to examine how neural responses to disapproving faces are moderated by rejection sensitivity. To accomplish these goals, participants completed a measure of rejection sensitivity and were scanned while viewing disapproval, anger, and disgust facial expressions.

As expected, all three facial expressions (disapproval, anger, disgust) resulted in significant bilateral amygdala activity compared to fixation. This extends prior work showing amygdala activity in response to viewing negative emotional expressions, such as anger, fear, or disgust (Fitzgerald et al., 2006; Morris et al., 1996; Phillips et al., 1997; Whalen et al., 1998, 2001), and indicates that a disapproving facial expression may be interpreted as an inherently threatening emotional expression as well. Interestingly, when comparing neural activity to disapproving facial expressions versus anger expressions, there was significantly greater left amygdala activation, suggesting that disapproving faces may represent a more potent threat at the neural level than some other previously studied negative expressions. In addition, the finding of greater insula activity in response to disgust expressions relative to disapproval expressions provides additional evidence that distinguishes disapproval faces from other threatening facial expressions, highlighting the notion that disapproval may not just be a variant of disgust (cf. Ekman & Friesen, 1986).

Contrary to our predictions, however, individuals did not show greater dACC activity while viewing disapproving faces compared to fixation. In hindsight, this lack of a main effect for dACC may not be entirely surprising. The dACC is a region that has been shown to be involved in responding to personal threats to the self such as the distress associated with social exclusion (Eisenberger et al., 2003, 2007) or the distressing or unpleasant experience associated with physical pain (Rainville, Duncan, Price, Carrier, & Bushnell, 1997). In the present study, participants watched video clips of strangers demonstrating disapproving expressions, which may not have been experienced as personally threatening.

Nonetheless, as predicted, dACC activity was moderated by rejection sensitivity in response to viewing disapproving facial expressions. Specifically, individuals high in rejection sensitivity exhibited greater dACC activity in response to viewing the disapproving facial expressions than individuals low in rejection sensitivity. This was true even when disapproving facial expressions were compared directly with anger and disgust facial expressions. Thus, for individuals who are dispositionally sensitive to rejection, stimuli that signal social rejection (i.e., video clips of strangers demonstrating disapproving expressions) elicited activity in some of the same neural regions that are involved in the experience of social rejection. Because individuals high in rejection sensitivity are more likely to interpret ambiguous stimuli as rejecting (Downey & Feldman, 1996), these individuals may have interpreted these disapproving facial expressions as more personally threatening. It is also possible that individuals high in rejection sensitivity were more likely to become distressed by these disapproving facial expressions, either because they are more sensitive to these types of rejecting expressions or because these rejecting expressions are more likely to trigger memories of past episodes of rejecting experiences. Because we did not assess distress levels in response to viewing the disapproving faces, we cannot determine whether the greater dACC activity observed in rejection-sensitive individuals is related to greater experiences of distress or to a more sensitive detection of cues that predict social rejection. Future studies will be needed to disentangle these two alternatives.

Of particular interest is that the positive relationship between rejection sensitivity and dACC activity was not seen in response to the anger or disgust faces, suggesting that individuals sensitive to rejection exhibited a greater dACC response specifically to facial expressions indicating possible rejection, but not to threatening facial expressions in general. Although previous researchers have suggested that disgust expressions may actually represent social rejection (Amir et al., 2005; Rozin, Haidt, & McCauley, 1999) the results here qualify this claim. The selective dACC response to disapproving facial expressions by high rejection sensitive individuals is also consistent with previous work showing that high-rejection-sensitive individuals showed heightened autonomic reactivity in response to rejection-themed cues, but not to other negative (but non-rejection-themed) stimuli (Downey, Mougios, Ayduk, London, & Shoda, 2004).

Another interesting point is that rejection sensitivity correlated specifically with dACC activity to disapproving faces, but not with other limbic system activity (e.g., amygdala, insula), suggesting that dACC activity, rather than more general limbic system activity, is specifically responsive to these disapproving faces in high, relative to low, rejection-sensitive individuals. Again, this finding suggests that disapproving facial expressions may represent a distinct type of threat that has not been examined previously in fMRI studies and that the dACC may play a unique role in responding to this expression.

In light of these findings, disapproving face stimuli may be more appropriate to use than other face stimuli when assessing negative affective processes in clinical disorders that involve a heightened sensitivity to negative social experiences. For example, fear of rejection has been shown to be the primary cognition for individuals with social anxiety (Turner, Johnson, Beidel, Heiser, & Lydiard, 2003), and hypersensitivity to social threat cues is a key feature of depressed states (Allen & Badcock, 2003; Mathews, Ridgeway, & Williamson, 1996). Thus, a socially threatening facial expression that specifically conveys potential social rejection may be more effective in engaging the maladaptive affective processes present in these populations.

In addition to its role in personal threats to the self, dACC has also been theorized to play a role in conflict monitoring, in which the dACC monitors for conflicting response tendencies (Botvinick, Cohen, & Carter, 2004; Carter et al., 2000; MacDonald, Cohen, Stenger, & Carter, 2000). Indeed, some have suggested that the dACC activity observed in response to the Cyberball social exclusion paradigm (Eisenberger et al., 2003) is due to the fact that the exclusion episode is unexpected and thus conflicts with prior expectations to be included (Somerville, Heatherton, & Kelley, 2006). Based on these accounts of dACC function, it may be argued that the dACC activity seen in response to the disapproving facial expressions is the consequence of disapproving faces being more "unexpected" than anger or disgust faces. However, if this were the case, we would expect to see a main effect of greater dACC activity in response to the disapproving facial expressions compared to the anger or disgust faces. Instead, dACC activity in response to the disapproving facial expressions correlated with rejection sensitivity. Because rejection-sensitive individuals are presumably more likely to expect to see disapproving faces (Downey & Feldman, 1996), it is unlikely that the enhanced dACC activity observed in rejection-sensitive individuals is due to expectancy violations (cf. Somerville et al., 2006). Furthermore, rejection-sensitive individuals have previously been shown to have an emotional response to rejection-related stimuli (Downey et al., 2004). In that study, high-rejection-sensitive individuals exhibited amplified startle responses in response to rejection-themed images. As the startle response is a measure of autonomic nervous system activity, this finding supports the idea that rejection sensitive individuals' responses to rejection cues have an affective component and are not strictly reducible to cognitive expectation effects. Thus, rejection sensitivity is associated with emotional responses to rejection-related cues. It follows that neural responses that differ by level of rejection sensitivity while viewing rejection-related cues may be involved in this emotional process.

Rejection sensitivity also correlated negatively with subACC/VMPFC activity in response to the disapproving facial expressions. The fact that rejection-sensitive individuals showed reduced subACC/VMPFC activity to disapproving facial expressions is consistent with previous work demonstrating a role for the subACC/VMPFC in the extinction of conditioned fear responses in both animals and humans (Morgan, Romanski, & LeDoux, 1993; Phelps et al., 2004; Quirk, Russo, Barron, & Lebron, 2000). For example, conditioned fear responses, which normally decrease during extinction trials (when the conditioned stimulus is presented without the unconditioned stimulus), persist among rats with lesions to the VMPFC, suggesting that the VMPFC may be essential for inhibiting the conditioned response. Moreover, in humans, extinction of a conditioned fear response has been associated with reduced amygdala activation and increased subACC/VMPFC activation (Phelps et al., 2004). Thus, it is possible that decreased subACC/VMPFC activity to disapproving faces in rejection-sensitive individuals may reflect a diminished ability to regulate the negative responses to these disapproving facial expressions.[4]

Consistent with this idea, in a functional connectivity analysis, we found dACC activity to be negatively correlated with subACC/VMPFC activity. This result is also similar to previous findings showing an inverse relationship between subACC/VMPFC and amygdala activity when assessing the valence of certain stimuli (Kim et al., 2003). In that study, to the extent that surprised facial expressions were interpreted more negatively, participants showed reduced subACC/VMPFC activity and greater amygdala activity; conversely, to the extent that surprised facial expressions were interpreted more positively, participants showed greater subACC/VMPFC activity and reduced amygdala activity. Additionally, these two regions were negatively correlated with each other, suggesting that subACC/VMPFC inputs to the amygdala may be involved in down-regulating the activity of the amygdala, leading to more positive interpretations of certain stimuli. In a similar manner, the present findings may suggest that individuals who interpret the disapproving facial expressions more negatively (i.e., those high in rejection sensitivity) show reduced subACC/VMPFC and greater dACC activity, whereas individuals who interpret the disapproving facial expressions less negatively (i.e., those low in rejection sensitivity) show greater subACC/VMPFC and reduced dACC activity. Consistent with these results, Somerville and colleagues (Somerville et al., 2006) recently observed that the receipt of gestures of social acceptance were associated with increased activity in the subACC/VMPFC.

Thus, it is possible that the reduced subACC/VMPFC activity observed in rejection-sensitive individuals in the present study may be indicative of biased appraisal processes, such that rejection-sensitive individuals are less able to interpret these disapproving facial expressions in a non-threatening way. The diminished ability to re-interpret these disapproving facial expressions as non-threatening may then be related to the increased dACC activity that was evidenced among those high in rejection sensitivity.

One possible confound in our study may have been the level of realism present in each of the facial expressions. Like the classic Ekman-style photographs of emotions, the anger and disgust expressions may have seemed exaggerated and thus less ecologically valid than the disapproving facial expressions. Future studies should involve more realistic dynamic facial expressions of anger and disgust in addition to disapproval.

Additionally, future studies should include a direct comparison of disapproval faces with contempt faces, in order to determine whether or not there are meaningful differences in the responses to these expressions at the neural level. Darwin's original description of the contempt expression focused on its role in displaying hatred

[4] These results should be interpreted with caution as rejection sensitivity also correlated negatively with subACC/VMPFC while viewing anger faces relative to fixation.

for another person, specifically when that person is considered insignificant (Darwin, 1872/1998, p. 234). Accordingly, a contempt expression is dismissive, signaling a lack of interest in establishing or maintaining any sort of social relationship. In contrast, a disapproving face is more akin to the facial expression that a parent would direct at a child who is misbehaving, a signal, not of hatred, but of disappointment in the child's behavior. The disapproving facial expression acts as a signal or motivational cue to correct another's behavior and, perhaps, to improve the state of the social relationship. Thus, these facial expressions may trigger different types of neural activity that reflect being rejected with the inability to reconcile (contempt) versus the ability to make amends (disapproval).

In future studies, it would also be interesting to examine how neural responses to disapproving and other facial expressions would differ from the results presented here if participants were given more specific instructions for viewing the video clips, such as to "imagine the facial expressions are directed at you," or "imagine what this person is thinking." It is likely that participants in our study used a variety of cognitive processes while watching the video clips. Constraining what participants are thinking while viewing the video clips may provide stronger, more uniform neural responses to the expressions presented.

Finally, future studies should also include a "baseline" control condition other than a crosshair fixation to more tightly control for the viewing of facial stimuli. Future studies would also benefit by including a baseline condition other than "neutral" facial expressions, as these expressions have been shown to elicit amygdala activity (Fitzgerald et al., 2006) and can be interpreted as threatening (Somerville, Kim, Johnstone, Alexander, & Whalen, 2004).

In conclusion, the current study is the first to examine neural responses to disapproving facial expressions. In addition to finding significant bilateral amygdala activity to this threatening facial expression, we also found that individuals who scored higher in rejection sensitivity showed greater dACC and reduced subACC/VMPFC activity to these disapproving faces. Not only did individuals sensitive to rejection show a greater threat response to the facial expressions indicating possible rejection, but they also did so selectively, suggesting that different types of threatening facial expressions convey specific types of threats, with disapproving facial expressions primarily signifying a threat to social connection.

Over a hundred years ago, Darwin observed how different emotional facial expressions convey unique information, with each expression adapted for a specific purpose. Today, we are extending these observations to the neural level. Clearly, facial expressions other than those representing "basic emotions" can have a profound effect on our functioning and well-being. Further research in this area will enable us to better understand these effects. As Ovid observed centuries ago, "Often a silent face has voice and words" (*Ars Amatoria*, Bk. I, 574).

REFERENCES

Adolphs, R. (2002). Recognizing emotion from facial expressions: Psychological and neurological mechanisms. *Behavioral and Cognitive Neuroscience Review, 1*, 21–61.

Adolphs, R. (2003). Is the human amygdala specialized for processing social information? *Annals of the New York Academy of Sciences, 985*, 326–340.

Adolphs, R., Baron-Cohen, S., & Tranel, D. (2002). Impaired recognition of social emotions following amygdala damage. *Journal of Cognitive Neuroscience, 14*(8), 1264–1274.

Allen, N. B., & Badcock, P. B. T. (2003). The social risk hypothesis of depressed mood: Evolutionary, psychosocial, and neurobiological perspectives. *Psychological Bulletin, 129*, 887–913.

Amir, N., Klumpp, H., Elias, J., Bedwell, J. S., Yanasak, N., & Miller, L. S. (2005). Increased activation of the anterior cingulate cortex during processing of disgust faces in individuals with social phobia. *Biological Psychiatry, 57*, 975–981.

Baumeister, R. F., DeWall, C. N., Ciarocco, N. J., & Twenge, J. M. (2005). Social exclusion impairs self-regulation. *Journal of Personality and Social Psychology, 88*(4), 589–604.

Biehl, M., Matsumoto, D., Ekman, P., Hearn, V., Heider, K., Kudoh, T., et al. (1997). Matsumoto and Ekman's Japanese and Caucasian Facial Expressions of Emotion (JACFEE): Reliability data and cross-national differences. *Journal of Nonverbal Behavior, 21*(1), 3–21.

Birbaumer, N., Grodd, W., Diedrich, O., Klose, U., Erb, M., Lotze, M., et al. (1998). fMRI reveals amygdala activation to human faces in social phobics. *Neuroreport, 9*, 1223–1226.

Botvinick, M. M., Cohen, J. D., & Carter, C. S. (2004). Conflict monitoring and anterior cingulate cortex: An update. *Trends in Cognitive Science, 8*(12), 539–546.

Britton, J. C., Taylor, S. F., Sudheimer, K. D., & Liberzon, I. (2006). Facial expressions and complex IAPS pictures: Common and differential networks. *NeuroImage, 31*(2), 906–919.

Carter, C. S., Macdonald, A. M., Botvinick, M., Ross, L. L., Stenger, V. A., Noll, D., et al. (2000). Parsing executive processes: Strategic vs. evaluative functions of the anterior cingulate cortex. *Proceedings of the National Academy of Sciences*, *97*(4), 1944–1948.

Darwin, C. (1998). *The expression of the emotions in man and animals.* New York: Oxford University Press. (Originally published 1872)

Dickerson, S. D., & Kemeny, M. E. (2004). Acute stressors and cortisol responses: A theoretical integration and synthesis of laboratory research. *Psychological Bulletin*, *130*(3), 355–391.

Donegan, N. H., Sanislow, C. A., Blumberg, H. P., Fulbright, R. K., Lacadie, C., Skudlarski, P., et al. (2003). Amygdala hyperreactivity in borderline personality disorder: Implications for emotional dysregulation. *Biological Psychiatry*, *54*(11), 1284–1293.

Downey, G., & Feldman, S. I. (1996). Implications for rejection sensitivity for intimate relationships. *Journal of Personality and Social Psychology*, *70*(6), 1327–1343.

Downey, G., Mougios, V., Ayduk, O., London, B. E., & Shoda, Y. (2004). Rejection sensitivity and the defensive motivational system: Insights from the startle response to rejection cues. *Psychological Science*, *15*(10), 668–673.

Eisenberger, N. I., Lieberman, M. D., & Williams, K. (2003). Does rejection hurt? An fMRI study of social exclusion. *Science*, *302*, 290–292.

Eisenberger, N. I., Way, B. M., Taylor, S. E., Welch, W. T., & Lieberman, M. D. (in press). Understanding genetic risk for aggression: Clues from the brain's response to social exclusion. *Biological Psychiatry.*

Ekman, P. (2003). *Emotions revealed: Recognizing faces and feelings to improve communication and emotional life.* New York: Times Books.

Ekman, P., & Friesen, W. V. (1986). A new pan-cultural facial expression of emotion. *Motivation and Emotion*, *10*(2), 159–168.

Fitzgerald, D. A., Angstadt, M., Jelsone, L. M., Nathan, P. J., & Phan, K. L. (2006). Beyond threat: Amygdala reactivity across multiple expression of facial affect. *NeuroImage*, *30*, 1441–1448.

Forman, S. D., Cohen, J. D., Fitzgerald, M., Eddy, W. F., Mintun, M. A., & Noll, D. C. (1995). Improved assessment of significant activation in functional magnetic resonance imaging (fMRI): Use of a cluster-size threshold. *Magnetic Resonance in Medicine*, *33*(5), 636–47.

Frijda, N. H. (1953). The understanding of facial expressions of emotion. *Acta Psychologica*, *9*, 294–362.

Kesler-West, M. L., Andersen, A. H., Smith, C. D., Avison, M. J., Davis, C. E., Kryscio, R. J., et al. (2001). Neural substrates of facial emotion processing using fMRI. *Brain Research. Cognitive Brain Research*, *11*(2), 213–226.

Kilts, C. D., Egan, G., Gideon, D. A., Ely, T. D., & Hoffman, J. M. (2003). Dissociable neural pathways are involved in the recognition of emotion in static and dynamic facial expressions. *NeuroImage*, *18*(1), 156–168.

Kim, H., Somerville, L. H., Johnstone, T., Alexander, A. L., & Whalen, P. J. (2003). Inverse amygdala and medial prefrontal cortex responses to surprised faces. *Neuroreport*, *14*, 2317–2322.

Lawrence, N. S., Williams, A. M., Surguladze, S., Giampietro, V., Brammer, M. J., Andrew, C., et al. (2004). Subcortical and ventral prefrontal cortical neural responses to facial expressions distinguish patients with bipolar disorder and major depression. *Biological Psychiatry*, *55*, 578–587.

MacDonald, A. W., III, Cohen, J. D., Stenger, V. A., & Carter, C. S. (2000). Dissociating the role of the dorsolateral prefrontal and anterior cingulate cortex in cognitive control. *Science*, *288*(5472), 1835–1838.

Mathews, A., Ridgeway, V., & Williamson, D. A. (1996). Evidence for attention to threatening stimuli in depression. *Behaviour Research and Therapy*, *34*, 695–705.

Matsumoto, D., & Ekman, P. (2004). The relationship among expressions, labels, and descriptions of contempt. *Journal of Personality and Social Psychology*, *87*(4), 529–540.

Morgan, M. A., Romanski, L. M., & LeDoux, J. E. (1993). Extinction of emotional learning: Contribution of medial prefrontal cortex. *Neuroscience Letters*, *163*, 109–113.

Morris, J. S., Frith, C. D., Perrett, D. I., Rowland, D., Young, A. W., Calder, A. J., et al. (1996). A differential neural response in the human amygdala to fearful and happy facial expressions. *Nature*, *383*(6603), 812–815.

Murphy, F. C., Nimmo-Smith, I., & Lawrence, A. D. (2003). Functional neuroanatomy of emotions: A meta-analysis. Cognitive. *Affective & Behavioral Neuroscience*, *3*(3), 207–233.

Phan, K. L., Fitzgerald, D. A., Nathan, P. J., & Tancer, M. E. (2006). Association between amygdala hyperactivity to harsh faces and severity of social anxiety in generalized social phobia. *Biological Psychiatry*, *59*(5), 424–429.

Phan, K. L., Wager, T., Taylor, S. F., & Liberzon, I. (2002). Functional neuroanatomy of emotion: A meta-analysis of emotion activation studies in PET and fMRI. *NeuroImage*, *16*(2), 331–348.

Phelps, E. A., Delgado, M. R., Nearing, K. I., & LeDoux, J. E. (2004). Extinction learning in humans: Role of the amygdala and vmPFC. *Neuron*, *43*, 897–905.

Phillips, M. L., Young, A. W., Scott, S. K., Calder, A. J., Andrew, C., Giampietro, V., et al. (1998). Neural responses to facial and vocal expressions of fear and disgust. *Proceedings. Biological Sciences/The Royal Society*, *265*(1408), 1809–1817.

Phillips, M. L., Young, A. W., Senior, C., Brammer, M., Andrews, C., Calder, A. J., et al. (1997). A specific neural substrate for perceiving facial expressions of disgust. *Nature*, *389*(6650), 495–498.

Quirk, G. J., Russo, G. K., Barron, J. L., & Lebron, K. (2000). The role of ventromedial prefrontal cortex in the recovery of extinguished fear. *Journal of Neuroscience*, *20*, 6225–6231.

Rainville, P., Duncan, G. H., Price, D. D., Carrier, B., & Bushnell, M. C. (1997). Pain affect encoded in human anterior cingulate but not somatosensory cortex. *Science, 277*, 968–970.

Rauch, S. L., Whalen, P. J., Shin, L. M., McInerney, S. C., Macklin, M. L., Lasko, N. B., et al. (2000). Exaggerated amygdala response to masked facial stimuli in posttraumatic stress disorder: A functional MRI study. *Biological Psychiatry, 47*, 769–776.

Rozin, P., Haidt, J., & McCauley, C. (1999). Individual differences in disgust sensitivity: Comparison and evaluations of paper-and-pencil versus behavioral measures. *Journal of Research in Personality, 33*, 330–351.

Sato, W., Kochiyama, T., Yoshikawa, S., Naito, E., & Matsumura, M. (2004). Enhanced neural activity in response to dynamic facial expressions of emotion: An fMRI study. *Cognitive Brain Research, 20*(1), 81–91.

Sheline, Y. I., Barch, D. M., Donnelly, J. M., Ollinger, J. M., Snyder, A. Z., & Mintun, M. A. (2001). Increased amygdala response to masked emotional faces in depressed subjects resolves with antidepressant treatment: An fMRI study. *Biological Psychiatry, 50*, 651–658.

Somerville, L. H., Heatherton, T. F., & Kelley, W. M. (2006). Anterior cingulate cortex responds differentially to expectancy violation and social rejection. *Nature Neuroscience, 9*(8), 1007–1008.

Somerville, L. H., Kim, H., Johnstone, T., Alexander, A. L., & Whalen, P. J. (2004). Human amygdala responses during presentation of happy and neutral faces: Correlations with state anxiety. *Biological Psychiatry, 55*, 897–903.

Sprengelmeyer, R., Rausch, M., Eysel, U. T., & Przuntek, H. (1998). Neural structures associated with recognition of facial expressions of basic emotions. *Proceedings. Biological Sciences/The Royal Society, 265*(1409), 1927–1931.

Stroud, L. R., Tanofsky-Kraff, M., Wilfley, D. E., & Salovey, P. (2000). The Yale Interpersonal Stressor (YIPS): Affective, physiological, and behavioral responses to a novel interpersonal rejection paradigm. *Annals of Behavioral Medicine, 22*(3), 204–213.

Tottenham, N., Borscheid, A., Ellertsen, K., Marcus, D. J., & Nelson, C. A. (2002). *Categorization of facial expressions in children and adults: Establishing a larger stimulus set*. Poster presented at the Cognitive Neuroscience Society annual meeting, San Francisco, CA.

Turner, S. M., Johnson, M. R., Beidel, D. C., Heiser, N. A., & Lydiard, S. B. (2003). The Social Thoughts and Beliefs Scale: A new inventory for assessing cognitions in social phobia. *Psychological Assessment, 15*, 384–391.

Wager, T. D., Phan, K. L., Liberzon, I., & Taylor, S. F. (2003). Valence, gender, and lateralization of functional brain anatomy in emotion: A meta-analysis of findings from neuroimaging. *NeuroImage, 19*(3), 513–531.

Wehrle, T., Kaiser, S., Schmidt, S., & Scherer, K. R. (2000). Studying the dynamics of emotional expression using synthesized facial muscle movements. *Journal of Personality and Social Psychology, 78*(1), 105–119.

Whalen, P. J., Rauch, S. L., Etcoff, N. L., McInerney, S. C., Lee, M. B., & Jenike, M. A. (1998). Masked presentations of emotional facial expressions modulate amygdala activity without explicit knowledge. *Journal of Neuroscience, 18*(1), 411–418.

Whalen, P. J., Shin, L. M., McInerney, S. C., Fischer, H., Wright, C. I., & Rauch, S. L. (2001). A functional MRI study of human amygdala responses to facial expressions of fear versus anger. *Emotion, 1*(1), 70–83.

SOCIAL NEUROSCIENCE, 2007, 2 (3–4), 254–275

Psychology Press
Taylor & Francis Group

The neural substrates of cognitive empathy

Stephanie D. Preston
University of Michigan, Ann Arbor, Michigan, USA

Antoine Bechara
University of Southern California, Los Angeles, California, USA, and University of Iowa, Iowa City, Iowa, USA

Hanna Damasio
University of Southern California, Los Angeles, California, USA

Thomas J. Grabowski
University of Iowa, Iowa City, Iowa, USA

R. Brent Stansfield
University of Michigan Medical School, Ann Arbor, Michigan, USA

Sonya Mehta
University of Iowa, Iowa City, Iowa, USA

Antonio R. Damasio
University of Southern California, Los Angeles, California, USA

Neuroscientific research has consistently found that the perception of an affective state in another activates the observer's own neural substrates for the corresponding state, which is likely the neural mechanism for "true empathy." However, to date there has not been a brain-imaging investigation of so-called "cognitive empathy", whereby one "actively projects oneself into the shoes of another person," imagining someone's personal, emotional experience as if it were one's own. In order to investigate this process, we conducted a combined psychophysiology and PET and study in which participants imagined: (1) a personal experience of fear or anger from their own past; (2) an equivalent experience from another person as if it were happening to them; and (3) a nonemotional experience from their own past. When participants could relate to the scenario of the other, they produced patterns of psychophysiological and

Correspondence should be addressed to: Stephanie D. Preston, Department of Psychology, University of Michigan, 530 Church Street, Ann Arbor, MI 48109, USA. E-mail: prestos@umich.edu

This project was funded primarily by a grant from the Mathers Foundation; but funding was also received from NIH NINDS P01NS19632. Partial support to SP was also provided by a research grant from the Institute for Research on Unlimited Love and a fellowship from Eli Lilly and Company.

The authors would like to thank Emily Recknor for help with data collection and processing, Nasir Naqvi for help with psychophysiology data collection, Kathy Jones for help with subject recruitment and testing, Jocelyn Cole for processing the MR data, Joel Bruss for help with scan visualization, and the staff at the University of Iowa Department of Radiology for administration of the PET scans.

www.psypress.com/socialneuroscience DOI:10.1080/17470910701376902

neuroimaging activation equivalent to those of personal emotional imagery, but when they could not relate to the other's story, differences emerged on all measures, e.g., decreased psychophysiological responses and recruitment of a region between the inferior temporal and fusiform gyri. The substrates of cognitive empathy overlap with those of personal feeling states to the extent that one can relate to the state and situation of the other.

Empathy is an elusive process that has stirred centuries of debate regarding its mechanisms (e.g., Preston & de Waal, 2002). From a neuroscientific perspective, investigators have consistently found neural evidence that perception of another's emotional state, be it a basic emotion (Carr, Iacoboni, Dubeau, Mazziotta, & Lenzi, 2003; Decety & Chaminade, 2003; Wicker, Perrett, Baron-Cohen, & Decety, 2003) or pain (Decety & Grèzes, 2006; Jackson, Brunet, Meltzoff, & Decety, 2006; Lamm, Batson, & Decety, 2007; Singer et al., 2004), engage brain areas related to the representation of one's own feeling states such as the anterior insula and the cingulate cortices. These data generally support the application of mirror-neuron (Gallese, 2001; Iacoboni & Lenzi, 2002), simulation (Adolphs, 1999; Gallese & Goldman, 1998) or perception–action theories (Blakemore & Decety, 2001; Lipps, 1903; Merleau-Ponty, 1962/1970; Preston & de Waal, 2002; Prinz, 1987) of empathy, although brain imaging cannot be used to identify common representations for perception and action within a single cell, as was demonstrated in mirror neurons in the monkey premotor cortex (di Pellegrino, Fadiga, Fogassi, Gallese, & Rizzolatti, 1992).

Studies comparing first- and third-person perspective taking related to emotion (Ruby & Decety, 2004) or pain (Jackson et al., 2006; Jackson, Meltzoff, & Decety, 2005; Lamm et al., 2007) also find great overlap in the feeling structures activated by the two conditions; however, there are also differences as third-person perspective taking engages to a greater extent the posterior cingulate/precuneus (Jackson et al., 2005, 2006; Ruby & Decety, 2004) and inferior parietal cortex (Decety & Chaminade, 2003; Jackson et al., 2006; Ruby & Decety, 2004).

Taken together, neural substrates have been discovered in relation to both the emotion-matching aspects of empathy as well as the cognitive, perspective-taking aspects. This supports long-standing, developmental theories of empathy that postulate a sequential development of empathic abilities in evolution and ontogeny, beginning with emotional resonance and developing into more complex forms such as cognitive empathy and perspective taking (e.g., Eisenberg, 1986; Feshbach, 1975; Hoffman, 1978, 1982). These studies further suggest that empathic processes rely heavily on regions that subserve one's own feeling states, but also recruit areas specific to the task of adopting another's perspective.

Because the empathy literature is particularly plagued by confusion and disagreement regarding the definition of terms such as empathy, cognitive empathy and perspective taking (Wispé, 1987), we provide the following working definitions to clarify our usage throughout (after Preston & de Waal, 2002). We define *empathy* broadly, comparably to Hoffman (2000), as *any process where the attended perception of the object's state generates a state in the subject that is more applicable to the object's state or situation than to the subject's own prior state or situation.* This is largely a bottom-up process that occurs directly from attending to the state of an object, magnified by the extent to which the subject is similar to or familiar with the object or the object's situation. In contrast, this study investigated *cognitive empathy*, which is a top-down process whereby *the subject effortfully tries to represent the state of the object*; also referred to as "putting oneself in the place of another" or "imaginatively projecting oneself into the situation of another" (Allport, 1937; Buchheimer, 1963; Demos, 1984; Goldie, 1999; Smith, 1989). While cognitive empathy is often used interchangeably with *perspective taking*, perspective taking is a higher-order concept that includes both emotional and nonemotional forms of transitioning into the situation of another; in this experiment we are not investigating the cognitive process of transitioning between perspectives, but rather the state of being in the shoes of the other, after the switch has taken place, which we refer to as cognitive empathy.

In everyday situations, cognitive empathy often occurs as one tries to actively imagine what it would be like to be another person in a particular, difficult situation. Existing data suggest that subjects do not engage their own feeling substrates to the same extent when imagining an event from another's perspective (Jackson et al., 2005, 2006; Ruby & Decety, 2004). According to most models of empathy, the emotional experience of subjects are expected to be quantitatively reduced during cognitive empathy because subjects must maintain awareness of themselves as distinct individuals, lest they become personally distressed by the state of the other and confuse the object's plight with their own (e.g., Decety & Sommerville, 2003; Eisenberg et al., 1989). Additionally, because the aforementioned studies investigated perspective taking per se, comparing first- and third-person perspectives, they focused on activation related to the process of switching from first- to third-person perspectives. It is to be expected that subjects would be distracted from fully feeling the state of the other during this perspective-switching process. However, once a subject has successfully adopted another's perspective, there is no a priori reason why subjects should feel less emotion when imagining their own or another's experience. For example, according to simulation theory (Gallese & Goldman, 1998; Goldman, 1992), which provides a very useful model of cognitive empathy, one could hypothetically access all of the necessary sensory information such as sights, sounds, and even feeling states needed to construct the other's scenario in an online, top-down manner, resulting in a successful image. However, a perception–action model of empathy (Preston & de Waal, 2002) and Damasio's view on emotion (Damasio, 1994, 2003) specifically postulate that subjects commonly access existing representations from past experience when imagining emotional states; thus, the intensity of the subject's feeling, as well as the accuracy of the empathic process, depend crucially upon the extent to which the subject has existing representations for the object, state, and situation. Given this, even after successfully switching perspectives, the extent to which a subject feels the state of the other is directly correlated with the extent to which they have existing representations for the state. This study was designed to test these predictions.

We compared the feeling states of participants while imagining a personal scenario from their own past versus a nonpersonal scenario from the past of another participant, *as if it were happening to them*. Psychophysiological data and functional neuroimaging data using positron emissions tomography (PET) were obtained during emotional imagery; a nonemotional personal imagery condition was also included as a reference. We did not focus on the process of transitioning from one's own perspective to the perspective of the other; rather, we were interested in what happens in the brain and body while the participant maximally experiences the state of the other, reliving the other's experience as if it is happening to them. Thus, data were collected only after participants indicated that they were feeling emotion from the imagined experience in the other's perspective, in the other's shoes. In this way, regions associated with perspective taking in former studies such as the inferior parietal cortex and the frontal pole (Decety & Chaminade, 2003; Decety, Chaminade, Grèzes, & Meltzoff, 2002; Ruby & Decety, 2001) were not expected to be significantly different between conditions. That is, since both the personal and nonpersonal imagery conditions were in the first person, and were measured after the subject successfully imagined the object's situation, no additional cognitive processes were expected to be necessary for the nonpersonal imagery. In contrast, we predicted that there would remain great overlap between personal and nonpersonal imagery in brain areas associated with the representation of feeling states such as the orbitofrontal, insula, anterior cingulate, somatosensory (SII), and mesial parietal cortices, brainstem autonomic areas and cerebellum (Damasio, 2003).

According to our view of empathy (cf. Damasio, 1994, 2003; Preston & de Waal, 2002), even if subjects activate the same neural regions to sustain personal and nonpersonal imagery, the success of the nonpersonal imagery and the amount of brain activation still depend on the extent to which subjects have an existing representation for the object's state. To investigate this further, Experiment 1 compared personal scenarios to nonpersonal scenarios to which participants could relate *the most* (maximizing the possibility for overlap between conditions), while Experiment 2 compared personal scenarios to nonpersonal scenarios to which subjects could relate *the least* (minimizing the possibility for overlap between conditions). We predicted that the nonpersonal scenario would create almost identical levels of psychophysiological and brain activation when the participants could relate to the scenario well in Experiment 1 but that the nonpersonal scenario would produce reduced levels of psychophysiological and brain activation when participants could not relate well to the scenario in Experiment 2, and this is where it is more likely that additional areas may be recruited to guide the imagery of an unfamiliar situation.

EXPERIMENT 1

Method

The [^{15}O]H_2O PET experiment was approved by the University of Iowa's Institutional Review Board and Medical Radiation Protection Committee. The emotional imagery was induced and the PET scans were administered as in a prior investigation (Damasio et al., 2000). The major differences were that we only investigated anger and fear imagery and did not investigate sad or happy imagery; instead, we added a nonpersonal emotional episode to compare to the personal emotional episodes.

Two emotions were included to determine if the correlates of cognitive empathy generalized across emotions, as they would if there was a region associated with the process of being in the shoes of another, or if the patterns of activation were specific to the state being imagined, as they would if one simply recruits neural systems associated with imagining the respective emotion (cf. Damasio et al., 2000).

Participants. Participants were recruited through advertisements in a local newsletter, a newspaper, and campus posting. Prescreening on the phone was used to exclude participants with a history of neurological or psychiatric illness or currently on medications that affect the central nervous system. The screening session in the laboratory (detailed below) was used to exclude participants who were not dominantly right-handed according to the Oldfield–Geschwind questionnaire (scores $< +85$) or were not able to recall a specific, profound, emotional memory of either fear or anger to be used as their personal imagery scenario. This latter criterion excluded abstract emotional memories like anger stories about hating a particular person, or fear stories about phobias and excluded routine emotional events such anger caused by a typical dispute with a roommate or fear caused by a recent bad dream. Four volunteers were excluded on the basis of this criterion. Five subjects dropped out of the study after the screening session in the lab and before the PET scan. A total of sixteen participants were studied, half male and half female, mean age $= 31.81$ ($SD = 6.53$) ranging from 25 to 44; half of each gender was assigned before testing to one of the two emotion conditions (fear or anger) and to an order of imagery types (personal first, or nonpersonal first; Table 1). All participants gave informed consent in compliance with federal and institutional guidelines, and they were all approved by the IRB at the University of Iowa.

Screening session. After verifying that the subject could indeed recall an appropriate personal imagery scenario and recording the event, participants selected a nonpersonal imagery scenario from a list of seven scenarios that were collected from participants in a prior investigation of emotional imagery (Damasio et al., 2000). All fourteen nonpersonal stories (seven for each emotion) are included in full in Table 2 as a reference, along with the number of subjects who selected each story by condition and experiment. To operationalize the construct of "strongly relating" to another's experience, we gave participants the complete text of all seven scenarios and asked them to select the one to which they could relate the *most.* Lastly, the experimenter collected a scenario for the neutral imagery condition by asking participants to recall a nonemotional event from a specific day (the current day unless it was emotional), from getting up in the morning and preparing breakfast to getting ready to leave for school or work.

After all scenarios were compiled, participants were instructed by the experimenter to begin each imagery scenario in turn while lying back with eyes closed in a darkened testing room. The complete verbal instructions given to subjects to induce emotional imagery are included below; no other instructions were given regarding the method of induction or imagery. Based on their assigned order, half of the participants imagined neutral, personal, neutral, nonpersonal while the other half imagined neutral, nonpersonal, neutral,

TABLE 1
Participants for Experiments 1 and 2 broken down by emotion, degree to which subjects related to their nonpersonal imagery scenario, and gender

Experiment	*Relatability*	*Emotion*	*Female*	*Male*
1	High	Fear	4	4
		Anger	4	4
2	Low	Fear	5	4
		Anger	4	4

TABLE 2

Scenarios used for nonpersonal imagery. All scenarios were generated as personal imagery scenarios by subjects in a prior investigation (Damasio et al., 2000). Subjects were given the complete text of all seven scenarios in their emotion condition (fear or anger) and were asked to select the scenario to which they could relate the most (Experiment 1) or the least (Experiment 2). This exact text was also read to the subjects just before imagery in the prescreening and PET imaging sessions along with the instructions presented in the methods section. The word count and number of subjects who selected each scenario is also provided

Emotion	*Story*	*Word count*	*N, high relate*	*N, low relate*	*Full text of scenario given to subjects*
Fear	1	63	0	4	You are traveling in Russia when three cops stop you. Before you know it, the cops take you away and start beating you up. After a while they let you go, but they are soon chasing after you again. You are very scared and are trying to run away from them. Picture yourself in that situation and feel as scared as you can.
	2	114	2	0	You are driving on a highway in the middle lane with a lot of traffic. Your son is sitting in the front seat next to you, and your two daughters are in the back. You see a semi truck in the right lane suddenly pull in front of you. You slam on the breaks, but discover that the car has no breaks! You try to escape to the left lane, but a car appears on your left. You are cornered, approaching the semi truck at high speed. Your kids are screaming and you are scared that you are all going to die. Picture yourself in that situation and feel as scared as you can.
	3	108	0	1	You are climbing a mountain over a very high cliff with a river underneath. You are almost at the top when your foot slips, and you slide down towards the cliff. Just before you go off of the cliff, you grab onto a small tree with one hand. All of your body is hanging over the cliff and you are holding on with only one hand. Your friend is coming close to save you, but your arm is giving up and you are about to let go and fall. You life is hanging by a thread. Picture yourself in that situation and feel as scared as you can.
	4	80	3	0	You are riding in the car with your friend, crossing over a train track. You see that a train is coming closer and closer. You are not driving and have no control over the wheel and cannot run away. Your heart is pumping and you are scared. You see the train approaching. The train actually hits the car and it is smashing in your side of the car. Picture yourself in that situation and feel as scared as you can.
	5	75	1	2	You are white water rafting in Colorado. You are knocked out of the raft. Your kids were with you and now you don't know where they are. You spouse was also with you and you realize that your spouse's leg is caught in a rock. The water is drifting you downstream and you don't know where you are all going to end up. Picture yourself in that situation and feel as scared as you can.
	6	67	1	2	You were younger and playing with fireworks with your brother. The next firecracker you shoot off hits your sister in the chest. Your sister is knocked out, unconscious. You are screaming and freaking out, fearing that your sister is dead. You are running around, trying to find some way to get her to a hospital. Picture yourself in that situation and feel as scared as you can.
	7	50	1	0	You are walking at night to a friend's house. You see two strangers approaching you, holding a knife. They come towards you and hold you at knifepoint. You are freaking out and think you are about to die. Picture yourself in that situation and feel as scared as you can.

TABLE 2 (*Continued*)

Emotion	*Story*	*Word count*	*N, high relate*	*N, low relate*	*Full text of scenario given to subjects*
Anger	1	87	5	0	You need to study for an important exam that is tomorrow. You were studying late in the library, but the library closed, so you went home to continue your studying. You arrive home and discover your roommate having a party in your room. This party was never discussed between the two of you. The exam is tomorrow and now you cannot study in your room. You are extremely angry and want to kill your roommate. Picture yourself in that situation and feel as angry as you can.
	2	88	0	2	You are sitting in the hospital next to your dad who just came out of surgery. You mom is sitting across the room. You sister arrives and immediately starts to put you down and tell stories that make you look bad in front of your parents. Your sister always does these kinds of things, and you hate her for it. You are especially mad that she is doing it now and you feel like killing her. Picture yourself in that situation and feel as angry as you can.
	3	76	1	0	You are driving and another guy hits your car. He admitted that it was his fault, so you agreed not to call the police. Later on, he started denying that it was his fault and was avoiding you so he wouldn't have to pay the damage. You keep trying to get him to pay, but he keeps avoiding you. You are angry and frustrated. Picture yourself in that situation and feel as angry as you can.
	4	69	0	2	You are walking with your mother and she gets mugged. You see the guy getting out of the car, approaching your mother from behind, and snatching her purse. You become very mad and angry with him. Now you are running after the guy; you are so mad and you want to beat the hell out of him. Picture yourself in that situation and feel as angry as you can.
	5	57	2	3	You found out from your daughters that the babysitter was hitting them. The babysitter finds out that your daughters told you and then tries to force your kids to say that they originally lied about the hitting. This makes you very angry with the sitter. Picture yourself in that situation and feel as angry as you can.
	6	73	0	1	Your spouse always makes excuses about having to work late and you are becoming angry and suspicious. One day you decide to look into it and discover that your spouse is not working late, but instead is going to a coworker's place. You confront your spouse but s/he denies it. You get into a big argument and you are very angry. Picture yourself in that situation and feel as angry as you can.
	7	93	1	0	A woman guaranteed you a truck rental. You go to the rental place on the day of your move and there are no trucks available. She tells you to come back later, which you do, and there are still no trucks. Meanwhile, you need to move out of your apartment and there are people waiting to move in who are pressuring you. The situation is getting desperate and you become very angry and furious at the woman from the rental company. Picture yourself in that situation and feel as angry as you can.

personal, in that order. After participants indicated that they were successfully engaging in the imagery (with a hand signal), timing began, and participants imagined the scenario for 90 seconds. To verify that the participant experienced the assigned emotion in the lab during imagery, self-report ratings of the primary emotion experienced and its intensity were collected after each scenario. All subjects reported feeling the intended emotion with a minimum subjective rating of 3 (from 0–4) for the intensity of that emotion (fear or anger). On this basis, all subjects tested in prescreening were included in the PET study.

In this study, we were specifically interested in investigating the feeling states and neural correlates of cognitive empathy once subjects were already in the shoes of the other person. Thus, we did not manipulate, measure, or try to control for possible differences in the way subjects chose to achieve this emotional imagery by instructing them to use a specific method of recall. Instead, we chose to maximize the validity of the construct of emotional imagery by allowing each participant to use the recall strategy they regarded as the most effective. The possible variability in content of the personal scenarios was exchanged for the effectiveness of autobiographical memories in inducing an intense emotion. The variability in cognitive strategies used by participants was compensated by the fact that data were collected only after participants reported feeling the target emotion.

Imagery induction instructions

> I want you to close your eyes and pay attention to me. When you hear me say the word "begin", I want you to picture in your mind that time when [description of story], and you felt very [scared or angry]. I want you to focus on that image as much as you can, and feel as [scared or angry] as you can. As you are trying to re-experience that emotional event, your mind may wander sometimes, and you may lose concentration. Whenever you lose that concentration, the important thing is to return to thinking again about that experience, bring back these images again, and focus again on that emotional experience until you hear me asking you to stop. Before we start this, I want you to remember the following: Close your eyes and listen to the description of the story. Wait for me to say "begin." After I tell you to begin, start bringing that emotional experience into focus, and as soon as you start feeling that emotion give me a hand signal [lift your hand], to indicate to me that you are starting to feel the emotion. You may hear walking, whispering, and noise around you. Don't pay any attention to anything that goes around you. You have only one thing to do: to focus on that emotional experience, feel it, and hold it as much as you can. Feel free to express emotion, just don't move your head or body too much. Continue to feel the emotion until I tell you to stop, it will be a few minutes.

Self-report measures. Self-report data ratings of overall emotional intensity from 0 (*none*) to 4 (*most intense*) were collected after each trial in the screening and PET imaging sessions to verify that participants had felt the target emotion during their personal, emotional imagery, and to compare how emotional they felt during their imagery of a personal event compared to a nonpersonal or neutral event.

Psychophysiology. Heart rate and skin conductance were recorded from electrodes attached to the participants at the beginning of both the screening and PET imaging sessions. Heart rate was collected using lead II EKG, with one electrode attached inferior to the costal margin, and another attached anterior to the sternocleidomastoid muscle; SCR electrodes were attached to the thenar and hypothenar areas on the palms of the left hand. Psychophysiological measures were recorded in an Astro-Med polygraph (Warwick, Rhode Island, USA) at 1000Hz and simultaneously converted from analog to digital format and transferred to a computer using an MP100WS system (Biopac systems, Santa Barbara, California, USA). The recorded data were analyzed with AcqKnowledge III software (Biopac Systems).

MR imaging. After the lab screening session, participants who qualified for inclusion in the PET study returned later to complete a T_1-weighted MR scan to plan slice orientation for use in the PET scanner, and to provide an anatomical reference for interpreting activation. For each subject, 3 MR scans were acquired on a 1.5T GE Signa scanner (SPGR 30, TR 24, TE 7, Nex 1, FOV 24cm, matrix 256 × 192). The scans consisted of 124 contiguous coronal slices (interslice distance 1.5–1.7mm, interpixel distance of 0.94) covering the whole brain. Images were co-registered using Automated Image Registration (AIR) version 3.03 (Woods, Cherry, & Mazziotta, 1992; Woods, Grafton, Watson, Sicotte, & Mazziotta, 1998; Woods, Mazziotta, & Cherry, 1993) and averaged to produce a single data set of enhanced higher quality and higher in-plane resolution (0.7mm interpixel distance; Holmes et al., 1998) that was used to plan slice orientation for use in the PET scanner using Brainvox (Grabowski et al., 1995).

PET imaging. PET imaging was conducted in a GE 4096B tomograph (Uppsala, Sweden), yielding 15 slices covering a 10cm axial field of view, with an in-plane resolution of 10mm. Each participant received eight injections of 50mCi $[^{15}O]H_2O$, with a minimum inter-interjection interval of 15 minutes. All participants first imagined two neutral scenarios, followed by four emotional scenarios, followed by two neutral scenarios. The order of the emotional scenarios was counterbalanced, with half of the participants in each emotion condition (fear or anger) imagining two nonpersonal scenarios followed by two personal scenarios, and the other half imagining two personal scenarios followed by two nonpersonal scenarios. The same personal and nonpersonal scenarios were used for each subject's PET scan as were used in the screening session, but because the neutral scenario depicted getting ready for school or work on a particular nonemotional day, the participants would not vividly remember the day from the prescreening many weeks prior; thus, neutral scenarios were updated for the PET session to a more recent day to control for the vividness of the emotional images.

Participants were cued to recall and re-experience the appropriate scenario as described above and signaled with a right-hand movement when they felt the intended emotion. The radiotracer was then injected within five seconds into the already attached IV. Because the injection was given through an IV, subjects did not feel a needle injection at this point, but may have experienced a warm sensation as the tracer entered the arm. Dynamic image acquisition enabled reconstruction of images based on the 40 seconds of data collection following clearance of the arterial blood pool, as judged from time-activity curves in ROIs over the major vessels. Our injection technique results in arrival of the bolus of labeled water in the brain about 15 seconds after injection (Hichwa, Ponto, & Watkins, 1995). Heart rate and SCR were monitored as during the screening session, and ratings of emotional intensity (0–4) were obtained after each injection.

PET data were spatially normalized to a Talairach-compatible averaged brain through a series of co-registration steps (see Damasio, Tranel, Grabowski, Adolphs, & Damasio, 2004; Grabowski et al., 1995, for details). All registrations were performed using AIR 3.03 (Woods et al., 1993). First, scans from each injection were co-registered to each other and then a mean PET image was registered to the MRI scan using the PET-Brainvox fiducials (Damasio et al., 1993; Grabowski et al., 1995). Using a fifth-order nonlinear warp, the MRI data were registered to a Talairach-compatible atlas constructed by iteratively averaging 50 normal brains in Talairach space (Woods, Dapretto, Sicotte, Toga, & Mazziotta, 1999). Finally, the displacement fields from each registration step were concatenated and applied to the PET data. After spatial normalization, the PET data were smoothed with a 16mm FWHM Gaussian kernel using complex multiplication in the frequency domain. The final calculated voxel resolution was 17.9 × 17.9 × 17.9mm. Each injection was normalized to a global mean of 1000 counts per voxel.

Statistical analysis

Ancillary data. Self-report and psychophysiology data were analyzed only for the PET imagery session and are not provided from the screening session. Screening session psychophysiology data were collected only in the event that self-report or observable referents of emotion were not informative or reliable. All subjects reported feeling the intended emotion with at least moderate intensity, and the experimenter observed visible changes in respiration (from the rise and fall of the chest) and muscle tone (e.g., clenched fists) in all subjects during personal imagery; thus the prescreening psychophysiology data were not additionally analyzed here. All PET imagery measures were analyzed using separate general linear models (GLMs) in two stages. First, a GLM was used to test whether the first two neutral imagery trials differed from the last two, modeling subject identity as random factor (reduced by iterative residual maximum likelihood estimation; REML) and using a dummy code for neutral imagery trials (neutral 1 and 2 = 1, neutral 3 and 4 = 2). Assuming there were no differences between the neutral imagery trials before the emotional imagery trials and the neutral imagery trials after the emotional imagery trials, all neutral imagery trials were treated as levels of replication (1,2,3,4) nested within the neutral imagery condition. A subsequent GLM was run using subject identity as a random factor (iterative REML) and testing for effects of imagery replication (four levels for neutral imagery, two levels each for personal and nonpersonal imagery, nested within condition), emotion (anger vs. fear), and imagery type (neutral, nonpersonal, personal). The dependent variables of interests were the PET ratings of emotional intensity (0–4), heart rate in beats per minute (BPM) and skin conductance on the left hand (area under the curve). Significant effects were followed by Tukey's honestly significant difference tests and Student's t-tests; these were used to interpret main effects. All GLMs were run using JMP-IN version 5.1.2 (Cary, NC, USA). The alpha level for all comparisons was set at .05.

PET contrasts. All PET data were analyzed with a pixelwise general linear model. Regression analyses were performed to test for differences between the personal and nonpersonal imagery conditions, as compared to the neutral condition. Prior to the regression analysis, data from the four neutral conditions were averaged and then subtracted from the personal and nonpersonal imagery conditions, and these difference images were used as the dependent measure. The following co-variables were included in the regression model: subject nested within emotion (fear or anger), a within-subjects factor of imagery type (personal vs. nonpersonal), a between-subjects factor of emotion (fear vs. anger), and an interaction term between condition and emotion.

Four contrasts were tested: the difference between personal and nonpersonal imagery (1), the difference between anger and fear (2), and the difference between personal and nonpersonal imagery nested within anger (3) and fear (4).

An ROI search was performed with a search volume that included regions known to be involved in emotion-feeling processes: orbitofrontal cortices, insula cortices, anterior cingulate cortices, somatosensory (SII) cortices, mesial parietal cortices, brainstem tegmentum and cerebellum (Damasio, 2003). The ROI was delineated by manual tracing on the averaged MR and measured 121.9cm^3, calculated to comprise 21.3 resels. A whole brain search was also performed to look for areas of activation outside of the ROI with a higher threshold. Critical t-values for the planned contrasts were determined using Gaussian random field theory (Worsley, 1994). The spatial extent statistic was used to look for especially large clusters in which the peak is less active than the critical t-value dictated by a magnitude threshold; the statistical images were converted to Z-score images and thresholded using clusters where $Z > \pm 3.0$ (Friston, Worsley, Frackowiak, Mazziotta, & Evans, 1994).

For this experiment, we expected a great amount of spatial overlap in activation across the personal and nonpersonal imagery conditions. Such overlap could produce a null result when contrasting the two conditions, thereby making interpretation of a null finding ambiguous with respect to activation of the individual conditions. We therefore performed a conjunction analysis for the personal and nonpersonal imagery conditions relative to the neutral condition using the minimum statistic compared to the conjunction null (MS/CN; Nichols, Brett, Andersson, Wager, & Poline, 2005). This conjunction approach tests for activation in both conditions as opposed to testing for activation in at least one condition (Nichols et al., 2005). The conjunction analysis was performed by individually thresholding the t-statistics from the personal and nonpersonal conditions (using random field theory, $p < .05$) and based on the individually significant results, determining overlapping areas of significant activation.

All imaging analyses were performed using custom software and visualized with Brainvox (Damasio & Frank, 1992; Frank, Damasio, & Grabowski, 1997). The alpha level for all comparisons was set at .05.

Results

Ancillary data. Intensity ratings were equivalent between the first two neutral imagery trials and the last two neutral imagery trials, $F(1, 41) = 0.075$, *ns*, justifying their equal treatment in a subsequent GLM testing effects across all three imagery types. In this GLM, there were no differences across replicates of the imagery types, $F(3, 92) = 0.813$, *ns*, or between fear and anger, $F(1, 92) = 0.063$, *ns*, but there was a strong main effect of imagery type, $F(2, 92) = 118.617$, $p < .0001$, due mainly to the neutral imagery trials being rated as less intense ($M = 0.670$) than the nonpersonal ($M = 2.429$) and personal ($M = 2.750$) trials; there was a trend for personal imagery to be more intense than the nonpersonal, $F(1, 92) = 3.280$, $p = .073$.

Heart rates were equivalent between the first and last two neutral imagery trials, $F(1, 33) = 0.529$, *ns*, justifying their equal treatment in a subsequent GLM testing effects across all three imagery types. In this GLM, there were no differences across replicates of the imagery types, $F(3, 80) = 0.203$, *ns*, or between fear and anger, $F(1, 80) = 2.519$, *ns*, but there were differences among the three imagery types, $F(2, 80) = 42.883$, $p < .0001$, due to lower heart rate during neutral imagery trials ($M = 66.67$ BPM) than on either nonpersonal ($M = 69.88$ BPM) or personal imagery trials ($M = 70.97$ BPM), with the personal showing a trend to being higher than the nonpersonal, $F(1, 80) = 3.623$, $p = .061$.

Skin conductance was equivalent for the first and last two neutral imagery trials, $F(1, 33) = 0.043$, *ns*, justifying their equal treatment in a subsequent GLM testing effects across all three imagery types. In this GLM, there were no differences between the first and second replicate within each imagery type, $F(3, 80) = 0.180$, *ns*, and no main effect of imagery type, $F(2, 80) = 1.322$, *ns*, but there was a significant difference between anger and fear, $F(1, 80) = 6.087$, $p = .0158$, due to higher skin conductance levels for anger ($M = 11.42$) than fear ($M = 3.57$).

PET contrasts

Anger > fear. Information about significant effects including peak locations, number of voxels and local maxima and mimina are reported in Table 3. In the ROI there was more activation for anger than fear in the right anterior cingulate ($t = 6.23$, $p < .05$) and more activation for fear than anger in the right posterior cingulate ($t = -7.25$ $p < .05$).

Using a whole-brain analysis with the spatial extent statistic, there were large clusters of activation that were greater for anger than fear in the fronto-temporal region, from the left temporal pole extending forward to the frontal operculum including peaks in the superior and middle temporal gyrus, temporal pole, and lateral orbital gyrus ($Z > 3.23$, $p = .005$). Using the whole brain analysis with the spatial extent statistic, there was more activation for fear than anger in the right basal ganglia and posterior cingulate and supramarginal gyrus ($Z < -3.69$, $p < .05$).

Personal > nonpersonal. There were no brain areas in which the activation was statistically different between the personal and nonpersonal imagery, either when the fear and anger groups were combined to increase the sample size, or when they were tested separately to reduce variance; this was true using the ROI, whole brain, and spatial extent statistic analyses ($-3.08 < t < 3.14$, *ns*). A perception–action model of empathy would predict great overlap between imagining a personal and a nonpersonal scenario, especially since participants were instructed to take the perspective of the other in the nonpersonal imagery, which greatly reduces any suspected cognitive differences in the process; however, as a null result, this could have occurred simply because of insufficient power. In order to rule out this latter hypothesis, a conjunction analysis was performed to determine whether the personal and nonpersonal imagery produced activation that was significantly different from the neutral imagery.

PET conjunction analysis—ROI. Information about significant effects including peak locations, number of voxels and local maxima and mimina are reported in Table 4A and visualized in Figures 1 and 2. Within the ROI, there was significantly more activation for both personal and nonpersonal imagery compared to neutral imagery in the midbrain. There were also effects in the insula; personal imagery activated the right ventral and anterior portion of the insula more than neutral imagery, nonpersonal imagery deactivated the left, more posterior and dorsal portion of the insula. SII on the left and the posterior cingulate on the right were less active during personal imagery than neutral imagery. The orbital frontal cortex (OFC) on the left was more active during nonpersonal imagery than neutral while the amygdala was less active during nonpersonal imagery than neutral imagery.

PET conjunction analysis—whole brain. Information about significant effects including peak locations, number of voxels and local maxima and mimina are reported in Table 4B and visualized in Figures 1 and 2. There were significant differences between both types of emotional imagery and the neutral imagery and there was almost complete overlap in the patterns of activation between the two, but the magnitude of the differences appeared to be greater between personal imagery and

TABLE 3

Significant effects in Experiment 1. There were no significant differences between the personal and nonpersonal conditions; therefore, results are reported for the contrast between imagery of fear and anger experiences using a Region of Interest (ROI) mask with a reduced threshold of $t = 5.29$ and a whole brain mask using the spatial extent statistic with a threshold of $Z = 3.00$. Between-subjects comparisons of fear and anger used error $df = 15$, resel size $= 17.9$. Peak coordinates and local maxima are presented in Talairach coordinates

			Peak locations								
Comparison	*Mask*	*Test (alpha)*	*x*	*y*	*z*	*t*	*Voxels*	*p*	*Local maxima x,y,z (t)*	*Side*	*Region*
Anger > Fear	ROI	*t* (5.29)	2	12	32	5.63	74		2, 13, 33 (6.23)	R	Anterior cingulate
			3	−35	37	−5.91	131		4, −36, 37 (−7.25)	R	Posterior cingulate
	Whole	*Z* (3.00)	−41	7	−20	3.18	7648	0.005	−47, −2, −29 (3.49)	L	Superior/middle temporal gyrus
									−38, 22, −22 (3.50)	L	Temporal pole
									−33, 12, −21 (3.44)	L	Superior temporal/parahippocampal gyrus
									−40, 22, −19 (3.52)	L	Temporal pole
									−43, 22, −12 (3.28)	L	Lateral orbital gyrus
									−45, 22, −11 (3.23)	L	Lateral orbital gyrus
									−53, 13, −9 (3.36)	L	Superior temporal gyrus
			24	−32	26	−3.32	4334	0.032	26, −30, 31 (−3.69)	R	Basal ganglia, posterior cingulate**
									22, −22, 30 (−3.84)	R	Basal ganglia, posterior cingulate**
									28, −38, 26 (−3.8)	R	Supra-marginal gyrus**

**Peak coordinates fall in white matter between three significant areas of deactivation.

TABLE 4A

Areas where there were significant differences between the neutral imagery and the emotional imagery types analyzed separately for the personal and nonpersonal imagery within the ROI (critical *t* value ±3.96); fear and anger participants are combined in these analyses. Talairach coordinates listed in the same row for personal and nonpersonal imagery refer to activation in the same region in both contrasts

		Personal					Nonpersonal				
Side	*Region*	*x*	*y*	*z*	*t*	*Voxels*	*x*	*y*	*z*	*t*	*Voxels*
L	Orbital frontal cortex						−16	23	−15	4.89	1192
R	Anterior insula	34	9	−5	4.63	835					
L	Posterior insula						−39	−15	7	−4.29	235
L	Amygdala						−22	−2	−18	−4.05	143
L	SII	−54	−22	18	−4.37	611					
R	Posterior cingulate	4	−37	36	−4.52	628					
L	Midbrain	3	−30	−20	5.16	5065	1	−33	−20	5.16	3823

TABLE 4B

Areas with significant differences between the neutral and emotional imagery analyzed separately for personal and nonpersonal imagery in the whole brain analysis (critical t value ± 4.84); fear and anger participants are combined in these analyses. Talairach coordinates listed in the same row for personal and nonpersonal imagery refer to activation in the same region in both contrasts

			Personal					Impersonal				
Lobe	*Side*	*Region*	*x*	*y*	*z*	*t*	*Voxels*	*x*	*y*	*z*	*t*	*Voxels*
Frontal	R	Dorsolateral prefrontal cortex	44	48	2	−4.96	213					
	L	Orbital frontal cortex						−17	24	−16	5.51	733
	L	Frontal operculum						−47	24	−5	5.17	451
	L	Inferior frontal gyrus	−53	−23	17	−4.87	18					
Cingulate	L	Anterior cingulate/superior frontal gyrus	−7	48	20	5.26	1290					
	R	Posterior cingulate	4	−38	38	−5.04	178					
	L	Retrosplenial	−20	−53	5	−4.97	136					
Temporal	R	Heschel's gyrus						48	−21	9	−5.24	692
	R	Heschel's gyrus						52	−9	0	−4.92	60
	R	Medial temporal gyrus/temporal pole	42	11	−22	5.8	3708	44	14	−27	4.97	96
	L	Medial temporal gyrus/temporal pole	−47*	17	−8	5.55	5402	−47	11	−24	5.09	636
	R	Hippocampus	28	−9	−22	−5.27	834					
	R	Parahippocampal gyrus						27	−10	−25	−5.57	3236
	R	Inferior temporal gyrus	55	−44	−13	−4.86	3	55	−40	−16	−5.1	413
	L	Inferior temporal gyrus						−51	−44	−18	−5.2	1103
Parietal	R	Supra-marginal gyrus	56	−29	38	−5.26	751	55	−28	40	−5.64	732
	L	Supra-marginal gyrus						−46	−39	42	−4.97	33
	R	Precuneus	9	−63	43	−5.21	820					
	R	Superior parietal lobule						21	−62	38	−5.25	3069
	L	Superior parietal lobule						−21	−72	42	−5.05	981
Occipital	R	Lateral occipital cortex	40	−67	−1	−5.29	2436					
	R	Middle occipital gyrus	21	−77	15	−5.18	6206					
	L	Middle occipital gyrus						−29	−70	14	−5.85	26405
	L	Middle occipital gyrus						−31	−82	5	−5.22	3663
Basal ganglia	R	Globus pallidus	12	−2	3	4.94	175	17	−7	5	5.27	1036
	L	Putamen	−22	0	4	5.02	401					
Midbrain	L	Hypothalamus/colliculi	−3	−23	0	5.23	605	−3	−28	−1	4.95	87
Cerebellum	R	Cerebellum	12	−59	−26	5.93	22599	10	−54	−27	5.66	21595

*Coordinate outside brain connecting activation in temporal pole and FOp.

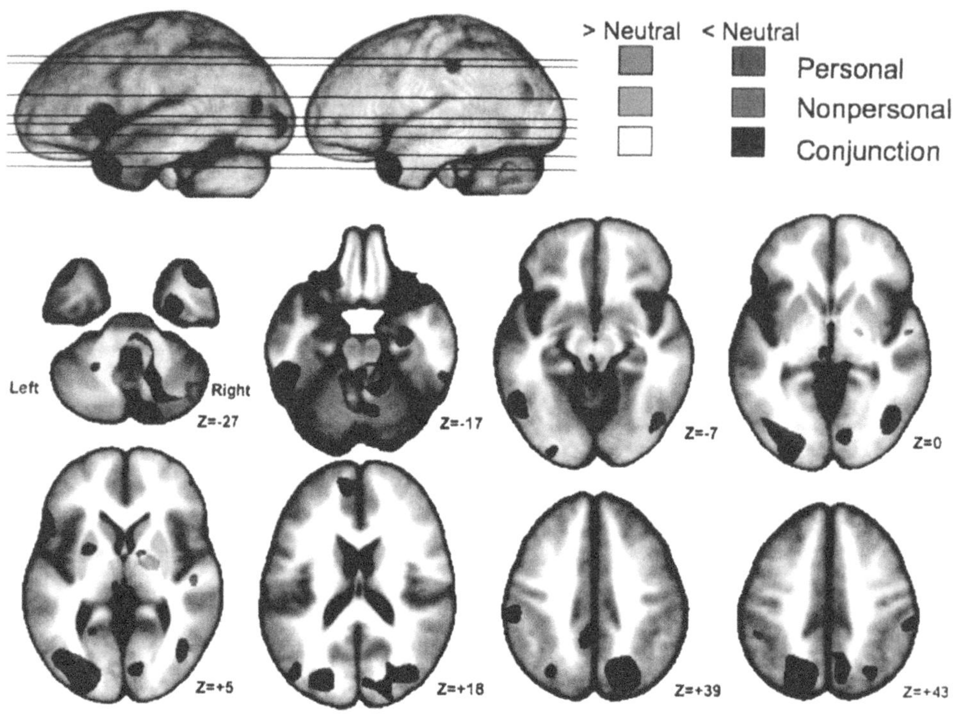

Figure 1. Experiment 1 brain activation data from conjunction analysis using a color-coded key. Areas that are significantly different in personal imagery from neutral imagery are depicted in red (>neutral) and green (<neutral); areas that are significantly different in nonpersonal imagery from neutral imagery are depicted in orange (>neutral) and blue (<neutral); areas where both personal and nonpersonal significantly overlap are depicted in yellow (>neutral) and purple (<neutral). Slices are labeled with respect to their *Z*-coordinate in the Talairach system.

neutral imagery than between nonpersonal and neutral imagery (Figure 1). There were large bilateral clusters that were more active for both types of emotional imagery than for neutral imagery in the cerebellum, brainstem, basal ganglia, and temporal poles. There were also large bilateral clusters that were less active during emotional imagery than neutral imagery in the left inferotemporal cortex (including the fusiform face area), right supra-marginal gyrus (SMG), right parahippocampal gyrus and left retrosplenial cortex. Personal imagery revealed an increase in activation compared to neutral imagery in the left anterior cingulate and decreased activation compared to neutral imagery in the dorsolateral prefrontal cortex (DLPFC), posterior cingulate, retrosplenial cortex, precuneus, and occipital cortex on the right. Nonpersonal imagery revealed an increase in activation compared to neutral imagery in the left orbital frontal cortex (OFC) and decreased activation compared to neutral imagery in right Heschl's gyrus, the right parahippocampal gyrus, left inferotemporal gyrus, bilateral superior parietal lobule, and the occipital cortex on the left.

Discussion

When participants were imagining, as if it were happening to them, an emotional experience from someone else's life selected to be the one they could relate to *the most*, they reported experiencing a similar level of emotional intensity as when they imagined an emotional experience from their own life. In either case, these emotional experiences were significantly more intense than those related to neutral scenarios. These findings were supported by self-report ratings of emotional intensity as well as heart rate, both of which revealed higher rates during emotional imagery than neutral imagery, without any distinction between personal imagery and nonpersonal, but "highly relatable" imagery scenarios. Skin conductance values did not differ across imagery types but were higher for anger than for fear scenarios, suggesting that anger imagery produces more sympathetic activation than fear imagery.

Anger produced more activation than fear in the anterior cingulate cortex on the right, and in a large sector of the superior temporal gyrus,

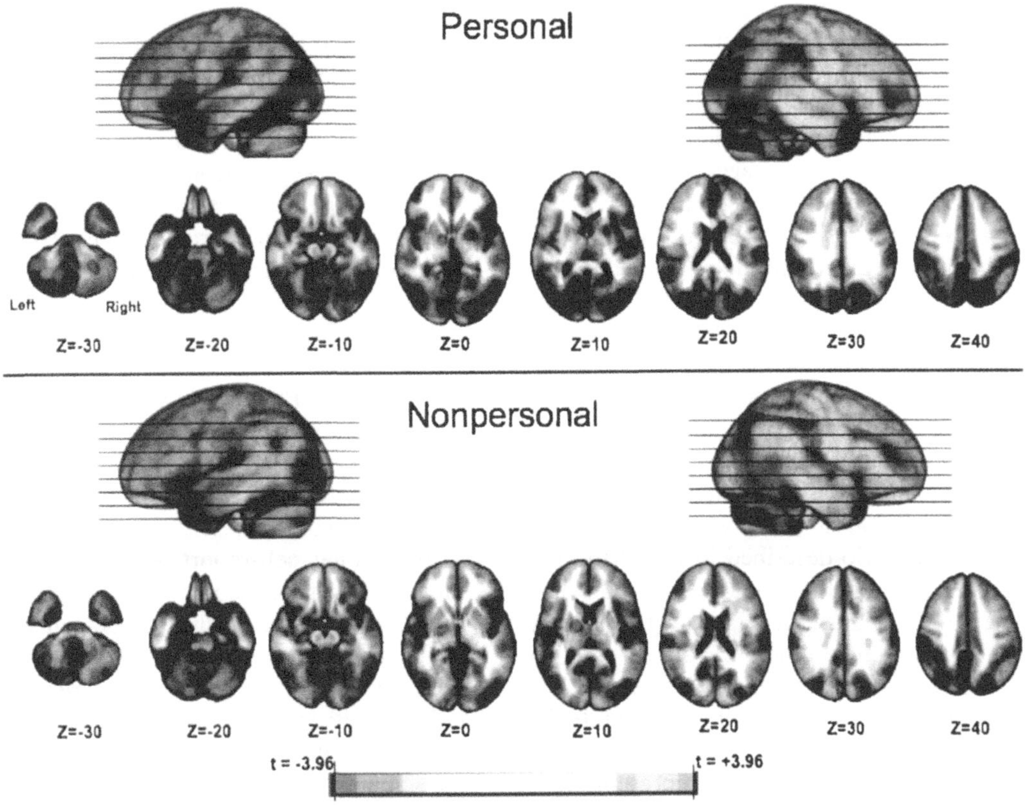

Figure 2. Experiment 1 brain activation data from the conjunction analysis showing personal imagery minus neutral imagery at the top and nonpersonal imagery minus neutral imagery at the bottom for comparison. For both conditions, significant increases in activation compared to neutral imagery are represented in red and all significant decreases in activation compared to neutral imagery are represented by royal blue as depicted in the color bar at the bottom ($t > \pm 3.96$). All images are presented in neurological convention with slices labeled with respect to their Z-coordinate in the Talairach system.

extending forward to the temporal pole, and in the lateral orbital cortex, which is consistent with activation related to personal imagery during anger in a prior study (Damasio et al., 2000). The predominance of anger activation on the left, especially in the frontal operculum (FOp) could reflect the nature of the task, as most of the scenarios involved heated arguments with another person, which would probably activate language-processing areas. Fear, in contrast, had had more activation compared to anger in the right hemisphere, in the posterior cingulate cortex, basal ganglia and supra-marginal gyrus, which is in keeping with the nature of fear scenarios, which usually involved physical events such as running, falling, or being attacked.

There were no significant differences in brain activation between personal and nonpersonal emotional imagery, using either the t-statistic at the level of the ROI or the whole brain, or using cluster analysis to identify large clusters of activation. Based on the results of the conjunction analysis, this null effect is not due to a lack of activation in emotional imagery trials compared to neutral imagery, as both personal and nonpersonal imagery showed many areas that were significantly different from neutral imagery. These patterns of activation were strikingly similar with large areas of increased activation compared to neutral imagery in multiple regions including bilateral areas of the temporal pole, basal ganglia, and cerebellum and decreased activation compared to neutral imagery in bilateral areas of the inferotemporal cortex, occipital cortex and parietal lobes. There was also activation in other areas associated with the representation of feeling states, but these areas were more lateralized, with greater orbital frontal cortex (OFC), anterior cingulate, and insular activation compared to neutral imagery in the left hemisphere, all of which were more pronounced in the personal than the nonpersonal imagery contrasts with neutral imagery.

Because increases in activation were mainly found in areas that represent feeling states while decreases in activation were mainly found in areas that represent visuo-spatial processing, it seems reasonable to suggest that emotional imagery relies importantly on reliving the feelings associated with the event and relies less on a concrete, visual image of the event. The fact that the personal and nonpersonal imagery produced such similar patterns of activation suggests that, indeed, participants do use their own substrates of processing to imagine being in the shoes of another, but this is likely to be due to the fact that they selected stories to which they could relate the most, in which case it is unclear the extent to which participants were actually imagining a related, personal scenario, akin to projection, or truly generating a new one based on the script. Based on subject comments after the nonpersonal imagery, it was clear that subjects were really trying to imagine the scenario described in the script, not simply imagining a highly similar scenario that had happened to them in the past. However, it is likely that when instructed: "Picture yourself in that situation and feel as scared/angry as you can," the subjects put great effort into trying to generate the feeling state associated with that emotion, which would necessarily activate their representations for that state based on personal experience and learning from others.

Although we found almost complete overlap between personal and nonpersonal imagery, it is still possible that there can be processing differences between personal and nonpersonal imagery in cases where subjects are less able to draw on a personal past experience in place of the other's experience. To test this, an additional experiment was conducted that was identical in design, but asked participants to select a nonpersonal experience to which they could *not* relate, or could relate to *the least*. Perhaps a more generative, creative, top-down cognitive process is needed to create an image of a novel situation or feeling, which would manifest as activation that is increased for nonpersonal imagery compared to personal. Additional self-report and psychophysiology measures were also added to aid interpretation of the data.

EXPERIMENT 2

Method

Participants. During the screening session, one participant was excluded due to ambidexterity and four participants were excluded because they did not have a significant personal memory of fear or anger. Five participants dropped out of the study between the screening session and the PET scan. Seventeen participants were tested in total, eight male, nine female, mean age 24.65 ($SD = 5.76$) ranging from 18 to 41. Eight participants imagined experiences of anger (4 males and 4 females) and nine participants imagined experiences of fear (4 males and 5 females; see Table 1). All participant assignment was randomly determined in advance.

Procedures. Procedures were identical to those described above for Experiment 1 except that participants in this experiment were specifically instructed to select the nonpersonal story to which they could relate the *least*.

Additional self-report and psychophysiology measures were added to help address different hypotheses. Specifically, participants selected which basic emotion they felt the most during their previous imagery, in order to verify that they indeed felt the assigned emotion during the emotional but not the neutral imagery trials, and participants rated the vividness of the imagery from 0 to 4 in order to determine if differences in the effectiveness of the personal and nonpersonal imagery were related to how easy it was to imagine the scenarios or specifically to an inability to become aroused by the scenario despite equivalent quality of the imagery. Trait scales of empathy were administered at the end of the prescreening session to use as a covariate for the PET imaging data because it would be interesting to know if people who self-report as experiencing more empathy, personal distress, or perspective taking were more likely to engage, or to engage to a greater degree, a particular neural region.

Additional psychophysiology measures were collected in order to specify the reactions of the participants during the emotional imagery periods. Respiration was measured with a belt attached around the torso and generalized motor activity was measured using an EMG electrode attached to the dorsum of the right hand.

Statistical analysis

Ancillary data. All measures were analyzed using the same series of two GLMs, the first to test whether the first two neutral imagery trials differed from the last two, modeling subject identity as random factor (reduced by iterative REML) and using a dummy-code for neutral imagery trials (neutral 1 and 2 = 1, neutral 3 and 4 = 2), the second testing for differences among the imagery types using subject identity as a random factor (iterative REML) and testing for effects of imagery replication (four levels for neutral, two levels each for personal and nonpersonal imagery, nested within condition), emotion (anger vs. fear), and imagery type (neutral, nonpersonal, personal). The dependent variables of interests were the ratings of emotional intensity (0–4), vividness, heart rate in beats per minute (BPM), and skin conductance on the left hand (area under the curve), respiration (number of peaks per unit time) and EMG (microvolts per second). Significant effects were followed by Tukey's honestly significant difference tests and Student's t-tests; these were used to interpret main effects. The alpha level for all comparisons was set at .05.

Descriptive statistics were calculated for the Interpersonal Reactivity Index (IRI) subscales (Empathic Concern, Fantasy, Perspective Taking, Personal Distress; Davis, 1994) and the Mehrabian and Epstein Emotional Empathic Concern Scale (EECS; Mehrabian & Epstein, 1972), which were used a covariates in the final PET analysis.

PET contrasts. Identical analyses were performed on the PET imaging data as in Experiment 1, using the regression model that included a variable for subject nested within emotion (fear or anger), a within-subjects factor of imagery type (personal vs. nonpersonal), a between-subjects factor of emotion (fear vs. anger), and an interaction term between condition and emotion. Four contrasts were again tested: personal versus nonpersonal (1), anger versus fear (2), and personal versus nonpersonal nested within anger (3) and fear (4). Again searches were performed using the ROI and the whole brain and the spatial extent statistic was calculated with statistical images converted to Z-images and thresholded using clusters where $Z > 3.0$ ($-3.0 < Z > 3.0$) and a corrected, cluster significance of $p < .05$ (Friston et al., 1994).

In addition to the analyses from Experiment 1, ANCOVAs were performed to determine if blood flow in the first, nonpersonal injection varied as a function of the trait empathy covariables. Due to the correlation among the trait covariables, each trait covariable was used in a separate regression analysis. In addition to the appropriate mean-centered trait covariate, each regression model contained the between-subjects factor of emotion (fear or anger). Each trait covariable was analyzed using a t-statistic with thresholds determined by random field theory using $p < .05$, corrected (Worsley, 1994).

Results

Ancillary data. Overall emotional intensity ratings were equivalent between the first two neutral imagery trials and last two neutral imagery trials, $F(1, 41) = 0.782$, *ns*, justifying their equal treatment in a subsequent GLM testing effects across all three imagery types. In this GLM, there were no differences in replicates within each imagery type, $F(3, 92) = 0.906$, *ns*, or between fear and anger, $F(1, 92) = 0.462$, *ns*, but there was a main effect of imagery type, $F(2, 92) = 58.686$, $p < .0001$, due to a linear scaling of ratings with personal imagery most intense ($M = 3.220$), nonpersonal imagery significantly less intense than personal ($M = 1.881$) and neutral imagery significantly less intense than nonpersonal ($M = 1.372$). The results were identical using subject's ratings for the intensity of the specific emotion they were imagining (fear or anger), $F(2, 92) = 181.009$, $p < .0001$.

Vividness ratings were equivalent between the first and last two neutral imagery trials, $F(1, 41) = 2.545$, *ns*, justifying their equal treatment in a subsequent GLM testing effects across all three imagery types. In this GLM, there were no differences in replicates within each imagery type, $F(3, 92) = 0.333$, *ns*, or between fear and anger, $F(1, 92) = 0.494$, *ns*, but there was a main effect of imagery type, $F(2, 92) = 15.005$, $p < .0001$, due to greater reported vividness for personal ($M = 3.020$) and neutral imagery ($M = 3.020$) than nonpersonal imagery ($M = 2.270$); personal and neutral were equivalent.

Heart rate was equivalent between the first and last two neutral imagery trials, $F(1, 48) = 1.75$, *ns*, justifying their equal treatment in a subsequent GLM testing effects across all three imagery types. In this GLM, there were no differences between the first and second replicate within each imagery type, $F(3, 110) = 0.222$, *ns*, but there was a main effect of condition, $F(2, 110) = 13.055$, $p < .0001$, due to higher heart rates in personal trials ($M = 65.684$) than in nonpersonal ($M = 62.634$) and neutral imagery trials ($M = 61.055$), which were equivalent. There was also a main effect of emotion, $F(1, 110) = 6.763$, $p = .011$, due to a higher heart rate in fear participants ($M = 69.075$) than anger ($M = 57.173$).

Skin conductance was lower on the first two ($M = 4.517$) than the last two ($M = 10.367$) neutral imagery trials, $F(1, 48) = 5.803$, $p = .0199$. Thus a subsequent GLM was used that kept the first two neutral imagery trials separate from the last two, yielding four conditions: neutral before, personal, nonpersonal, and neutral after. This GLM found no differences in replicates within each imagery type, $F(4, 108) = 0.079$, *ns*, or between fear and anger, $F(1, 108) = 0.251$, *ns*, but there was a main effect of condition, $F(3, 108) = 4.408$, $p = .0058$, due to lower skin conductance in the first two neutral imagery trials ($M = 4.358$) than in personal trials ($M = 16.667$).

Respiration was equivalent between the first and last two neutral imagery trials, $F(1, 45) = 0.015$, *ns*, justifying their equal treatment in a subsequent GLM testing effects across all three imagery types. In this GLM, there were no differences in replicates within each imagery type, $F(3, 103) = 0.094$, *ns*, or between fear and anger, $F(1, 103) = 0.056$, *ns*, but there was a main effect of imagery type, $F(2, 103) = 202.859$, $p < .0001$, due to higher respiration during personal ($M = 9.09$) than nonpersonal ($M = 6.66$) and neutral imagery ($M = 6.02$), which were equivalent.

The EMG data from the hand found less motor activity in the first two neutral imagery trials ($M = 62.127$) than in the last two ($M = 191.223$), suggesting some residual change in state after the emotional scenarios or an order effect due to fatigue, as was the case with SCR. Thus, the GLM was used that kept the first two neutral imagery trials separate from the last two, yielding four conditions: neutral before, personal, nonpersonal, and neutral after. There were no main effects of replicate, $F(4, 101) = 0.374$, *ns*, emotion, $F(1, 101) = 0.139$, *ns*, or imagery type, $F(3, 101) = 1.367$, *ns*.

Among the IRI subscales, the Empathic Concern scale had a mean of 20.82 ($SD = 4.83$, range 6–28), the Fantasy Scale had a mean of 18.06 ($SD = 4.38$, range 11–28) the Perspective Taking scale had a mean of 16.53 ($SD = 5.44$, range 9–26) and the Personal Distress scale had a mean of 9.19 ($SD = 4.92$), consistent with the published means (Davis, 1980). The Mehrabian and Epstein Scale of Emotional Empathy had a mean of 36.77 ($SD = 25.53$, range −18–90), consistent with the published mean of 33 ($SD = 24$).

PET contrasts

Anger > fear. Comparing the basic emotions of anger and fear, the results were similar to those of Experiment 1, with greater activation for fear in the left inferior frontal gyrus, anterior cingulate, and the posterior sector of the superior temporal gyrus, but these results were not significant in Experiment 2 using the ROI, whole brain, or spatial extent statistic analyses ($t > -5.44$, *ns*).

Personal > nonpersonal. Information about significant effects including peak locations, number of voxels and local maxima and mimina are reported in Table 5. There were no differences between personal and nonpersonal imagery in the ROI ($t < 3.22$, *ns*), but using the spatial extent statistic there were large clusters that were more active in nonpersonal than personal imagery (neutral intermediate), in the left inferotemporal/fusiform gyrus ($Z = -4.28$, $p < .05$) and lateral cerebellum ($Z = -3.63$, $p < .05$; Figure 3). Within the two emotions, fear participants did not show any significant differences between personal and nonpersonal imagery ($-3.39 < t < 3.03$, *ns*); anger participants did not have any significant differences within the ROI ($t < \pm 3.94$) but in the whole brain analysis there was increased activation in the personal imagery, compared to the nonpersonal, in the left superior temporal gyrus ($t = 5.02$, $p < .05$).

Discussion

In Experiment 1, the nonpersonal emotional scenario was selected to permit participants to relate to them easily. This produced equivalent levels of emotionality to the personal emotional scenario across multiple measures. In contrast, in Experiment 2, in which the nonpersonal emotional scenarios were selected to be hard to relate to, personal imagery produced greater emotionality than nonpersonal imagery. Participants reported higher levels of emotional intensity (overall and for the specific emotion), vividness of the imagery, had higher heart rates and greater respiration when imagining the personal than the nonpersonal scenarios.

Skin conductance no longer showed a difference between fear and anger. One possibility is that this effect was not reliable. Another possibility is that the strengths of the emotional experiences

TABLE 5

Significant effects in Experiment 2. Peak areas of activation and local maxima are presented using Talairach coordinates with inferential statistics. Results are reported for the contrast of personal and nonpersonal imagery at a whole brain level using cluster analyses with a critical *Z* of 3.00 and for the contrast of personal and nonpersonal imagery specifically in the emotion condition of anger using a whole brain analysis with a critical *t* of 4.29

Comparison	*Mask*	*Test (alpha)*	*Peak*				*Voxels*	*p*	*Local maximum (t)*	*Side*	*Region*
			x	*y*	*z*	*t*					
Personal > Impersonal	Whole	*Z* (3.00)	−45	−34	−18	−3.42	4511	0.029	−43, −28, −17 (−4.28)	L	ITG/fusiform
									−50, −41, −31 (−3.63)	L	Lateral cerebellum
Anger Personal > Impersonal	Whole	*T* (4.79)	−34	−34	13	4.89	36		−35 −35 14 (5.02)	L	STG/transverse

of anger and fear evoked during imagery in this sample were almost equivalent.

Both the skin conductance and EMG data showed differences between early and late neutral imagery trials, suggesting either that subjects are more aroused after experiencing emotional imagery, or simply move more towards the end of a long experiment (in the final two neutral imagery trials). However, this order effect does not impact the comparison of interest between the personal and nonpersonal imagery (which were counterbalanced) and only served to add noise to the imaging subtractions that averaged the neutral imagery trials together before subtracting them from the emotional imagery trials.

Interestingly, the self-reports of vividness followed a unique pattern, with neutral imagery and personal imagery being rated as significantly more vivid than the nonpersonal. Neutral imagery should not be intense or emotional, but because it was drawn from the participant's own life and had occurred recently, it was remembered as well as a highly salient memory from the subject's past. This is in contrast to a nonpersonal memory that could not be vivid because it had not been directly experienced by the subject and was, thus, harder to visualize.

In conclusion, across multiple measures, there was a difference between personal and nonpersonal imagery when participants could not relate easily to the other person's scenario, a difference that was not present when the participants could relate to the scenario, even though it was drawn from another person's life. This was also true of the brain activation data, which revealed significant differences between the personal and nonpersonal imagery under these conditions. With both emotions combined there was a large region in the inferior part of the left hemisphere including the inferior temporal and the fusiform gyri and the lateral cerebellum that was less active than neutral imagery during personal imagery (as it was in Experiment 1 for both the personal and the nonpersonal conditions) and more active than neutral imagery during nonpersonal imagery, something that did not occur in Experiment 1. It is possible that the left, lateral cerebellar activation is a spillover from the left IT/Fusiform activation, especially since the cerebellar activation is no longer present when the data are analyzed at a smaller smoothing kernel (6 FWHM). There were no significant differences between personal and nonpersonal imagery

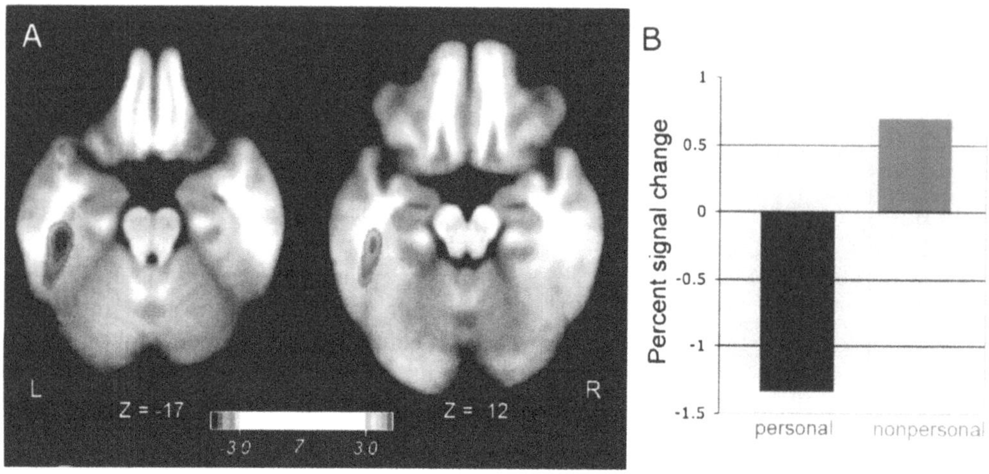

Figure 3. Experiment 2 brain activation data from the contrast of personal minus nonpersonal imagery, without respect to emotion. Part (a) depicts the extent of the cluster in the inferotemporal cortex, which was significantly more active for nonpersonal than personal imagery in two slices. The images have been masked to only show regions that were significantly different ($Z > -3.00$). All images are presented in neurological convention with slices labeled with respect to their Z-coordinate in the Talairach system. Part (b) depicts the mean activation levels for personal minus neutral imagery compared to nonpersonal minus neutral imagery.

within fear participants alone, which may be due in part to the low sample size within emotion conditions and to the manifest difficulty of inducing fear in laboratory conditions, which is consistent with lower accuracy ratings for fear in emotion recognition studies and with decreased SCR for fear in Experiment 1. However, subjective ratings were equivalent and heart rate was actually higher for fear in Experiment 2. Within anger participants, there was a significantly greater recruitment of the left superior temporal and transverse gyri for personal than for nonpersonal imagery.

GENERAL DISCUSSION

This study investigated the neural substrates of cognitive empathy, defined as the deliberate attempt to effortfully imagine what it is like to be in the emotional situation of another person as if it is happening to you. The results suggest that the substrates of cognitive empathy depend in large part on how well the participant can relate to the situation of the other. When participants selected the nonpersonal story that they could relate to the most, there were no statistically significant differences between areas recruited for personal and nonpersonal imagery. This is in keeping with all theories that propose that people activate their own emotion-producing substrates when observing the emotional state of another, including mirror neuron theories (Gallese, 2001; Iacoboni & Lenzi, 2002), perception–action theories (Preston & de Waal, 2002), neural versions of simulation theory (Adolphs, 1999; Gallese & Goldman, 1998) and Damasio's conceptualization of emotion (Damasio, 2003). This overlap also confirms multiple previous studies that found overlap in areas recruited for experiencing and observing basic emotions and pain in another (Carr et al., 2003; Decety & Chaminade, 2003; Singer et al., 2004; Wicker et al., 2003). However, when participants selected the nonpersonal story that they could relate to *the least*, there were differences between personal and nonpersonal imagery, suggesting that cognitive perspective-taking theories of empathy may be especially applicable when participants do not already possess representations that can encapsulate the experience of the other.

When the participants could relate less to the other person's scenario, in both fear and anger conditions, there was more activation for the nonpersonal than personal imagery in left inferior temporal and fusiform gyri and higher-order visual association areas, including regions often associated with face processing (Kanwisher, Mcdermott, & Chun, 1997). As for the greater activation in the left hemisphere, this has been noted with actual and imagined self-action as well as imitation (Decety et al., 2002; Goldenberg, 1999; Grafton, Arbib, Fadiga, & Rizzolatti, 1996) and it is, thus, possible that participants recruited the left hemisphere because they were actively imagining themselves in the situation. In contrast, in Experiment 1 there was a significant decrease compared to neutral imagery in the nonpersonal condition in Heschel's gyrus on the right, further suggesting that the nonpersonal imagery was less dependent on auditory than visual information.

There were strong differences in activation between personal and nonpersonal imagery in the cerebellum. Both fear and anger participants had greater activation for nonpersonal imagery in the ventro-lateral part of the cerebellum. This region may correspond to the posterior, lateral portion of the cerebellum, which has been associated with imagined movement (Allen, Buxton, Wong, & Courchesne, 1997; Grafton et al., 1996; Hanakawa et al., 2003).

Mental imagery is usually associated with activation in the hippocampus, amygdala, entorhinal cortex, and parahippocampal gyrus (Kreiman, Koch, & Fried, 2000). These areas were not differentially activated by the conditions in our studies, because participants were engaged in imagery in all three conditions (neutral, personal, and nonpersonal).

In conclusion, empathy depends on the degree to which we can relate to the experience of another. When we can relate to the experience of the other, we activate our own records of a comparable situation and produce an emotional experience, characteristic of "true empathy" (cf. Ungerer, 1990). It should be noted that when we activate personal representations of emotional situations in relation to the states of others, we may be predisposed to "project" our own feelings onto the other rather than truly "trying on" the state of the other, as presupposed by more cognitive accounts of empathy such as philosophical versions of simulation theory (Goldman, 1992) and perspective taking (Goldie, 1999; Smith, 1989). Thus, the fact that we access our own representations means that the difference between projection and empathy depends on the extent to which our representations are applicable to the experience of the other (Preston & de Waal, 2002). When we cannot relate to the experience of the other, presumably because we do not have records of comparable feelings and situations from our own past experience, we can achieve empathy by literally trying to imagine what it is like to be the other person, in keeping with traditional perspective-taking theories of empathy, which describe empathy as "putting oneself in the place of another" or "imaginatively projecting oneself into the situation of another" (Allport, 1937; Buchheimer, 1963; Demos, 1984; Goldie, 1999; Smith, 1989).

Our model is not in contradistinction to mirror neuron or simulation theories of empathy, which also propose that subjects use their own representations to internally simulate the states of others (Adolphs, 1999; Gallese & Goldman, 1998; Iacoboni & Lenzi, 2002), but an important difference of emphasis and practical application is apparent from our results. Perhaps because mirror neurons were discovered with respect to grasping actions that are virtually ubiquitous across individuals and even primate species, other models implicitly assume that subjects possess the representations needed to simulate or "try on" any conceivable object experience. In contrast, nonscientific use of the term "empathy" reserves its use for a minority of cases where there is serious emotional distress and the observer has had an almost identical experience. Thus, the child cannot be said to *empathize* with the parent nor the husband with the laboring wife. This distinction likely exists because, in an important way, subjects can only really feel states that they have experienced, creating an increasingly widening intersubjective gap for states of increasing distress, less and less likely to have been experienced by others. Our model puts a high explicit demand on the need for representations developed from past experience to successfully simulate how another feels. As our data show, without similar past experience, one *can* imagine being in another's shoes, but the result is much less vivid, less emotional and even of a different quality, which likely diminishes the accuracy of the empathy, the motivation to help and the likelihood that the object will receive comfort. From a phenomenological perspective, these differences in the quality of the empathic experience are of critical importance and should not be overlooked or underestimated.

REFERENCES

Adolphs, R. (1999). Social cognition and the human brain. *Trends in Cognitive Sciences, 3*(12), 469–479.

Allen, G., Buxton, R. B., Wong, E. C., & Courchesne, E. (1997). Attentional activation of the cerebellum independent of motor involvement. *Science, 275*(5308), 1940–1943.

Allport, G. W. (1937). *Personality: A psychological interpretation*. New York: Henry Holt.

Blakemore, S.-J., & Decety, J. (2001). From the perception of action to the understanding of intention. *Nature Reviews, 2*, 561–567.

Buchheimer, A. (1963). The development of ideas about empathy. *Journal of Counseling Psychology, 10*(1), 61–70.

Carr, L., Iacoboni, M., Dubeau, M. C., Mazziotta, J. C., & Lenzi, G. L. (2003). Neural mechanisms of empathy in humans: A relay from neural systems for imitation to limbic areas. *Proceedings from the National Academy of Sciences, 100*(9), 5497–5502.

Damasio, A. (1994). *Descarte's error: Emotion, reason, and the human brain*. New York: G. P. Putman's Sons.

Damasio, A. R. (2003). *Looking for Spinoza: Joy, sorrow, and the feeling brain* (1st ed). Orlando, FL: Harcourt Books.

Damasio, A. R., Grabowski, T. J., Bechara, A., Damasio, H., Ponto, L. L. B., Parvizi, J., et al. (2000). Subcortical and cortical brain activity during the feeling of self-generated emotions. *Nature Neuroscience, 3*(10), 1049–1056.

Damasio, H., & Frank, R. (1992). Three-dimensional in vivo mapping of brain lesions in humans. *Archives of Neurology, 49*(2), 137–143.

Damasio, H., Grabowski, T. J., Frank, R. J., Knosp, B., Hichwa, R. D., Watkins, G. L., et al. (1993). PET-Brainvox, a technique for neuroanatomical analysis of positron emission tomography images. In K. Uemura, N. A. Lassen, T. Jones, & I. Kanno (Eds.), *Quantification of brain function. Tracer kinetics and image analysis in brain PET* (pp. 465–473). Amsterdam: Elsevier.

Damasio, H., Tranel, D., Grabowski, T., Adolphs, R., & Damasio, A. (2004). Neural systems behind word and concept retrieval. *Cognition, 92*(1–2), 179–229.

Davis, M. H. (1980). Individual differences in empathy: A multidimensional approach. *Dissertation Abstracts International, 40*(7-B), 3480.

Davis, M. H. (1994). *Empathy: A social psychological approach*. Boulder, CO: Westview Press.

Decety, J., & Chaminade, T. (2003). Neural correlates of feeling sympathy. *Neuropsychologia, 41*(2), 127–138.

Decety, J., Chaminade, T., Grèzes, J., & Meltzoff, A. N. (2002). A PET exploration of the neural mechanisms involved in reciprocal imitation. *NeuroImage, 15*, 265–272.

Decety, J., & Grèzes, J. (2006). The power of simulation: Imagining one's own and other's behavior. *Brain Research, 1079*(1), 4–14.

Decety, J., & Sommerville, J. A. (2003). Shared representations between self and other: A social cognitive neuroscience view. *Trends in Cognitive Science, 7*(12), 527–533.

Demos, V. (1984). Empathy and affect: Reflections on infant experience. In J. Lichtenberg, M. Bornstein, & D. Silver (Eds.), *Empathy II* (Vol. 3, pp. 9–34). Hillsdale, NJ: Analytic Press.

di Pellegrino, G., Fadiga, L., Fogassi, L., Gallese, V., & Rizzolatti, G. (1992). Understanding motor events: A neurophysiological study. *Experimental Brain Research, 91*, 176–180.

Eisenberg, N. (1986). *Altruistic emotion, cognition, and behavior*. Hillsdale, NJ: Lawrence Erlbaum Associates, Inc.

Eisenberg, N., Fabes, R. A., Miller, P. A., Fultz, J., Shell, R., Mathy, R. M., et al. (1989). Relation of sympathy and personal distress to prosocial behavior: A multimethod study. *Journal of Personality and Social Psychology, 57*(1), 55–66.

Feshbach, N. D. (1975). Empathy in children: Some theoretical and empirical considerations. *The Counseling Psychologist, 5*, 25–30.

Frank, R. J., Damasio, H., & Grabowski, T. J. (1997). Brainvox: An interactive, multimodal visualization and analysis system for neuroanatomical imaging. *NeuroImage, 5*(1), 13–30.

Friston, K. J., Worsley, K. J., Frackowiak, R. S. J., Mazziotta, J. C., & Evans, A. C. (1994). Assessing the significance of focal activations using their spatial extent. *Human Brain Mapping, 1*, 214–220.

Gallese, V. (2001). The "shared manifold" hypothesis: From mirror neurons to empathy. In E. Thompson (Ed.), *Between ourselves: Second-person issues in the study of consciousness* (pp. 33–50). Thorverton, UK: Imprint Academic.

Gallese, V., & Goldman, A. (1998). Mirror neurons and the simulation theory of mind-reading. *Trends in Cognitive Sciences, 2*(12), 493–501.

Goldenberg, G. (1999). Matching and imitation of hand and finger postures in patients with damage in the left or right hemispheres. *Neuropsychologia, 37*(5), 559–566.

Goldie, P. (1999). How we think of others' emotions. *Mind & Language, 14*(4), 394–423.

Goldman, A. I. (1992). In defense of the simulation theory. *Mind & Language, 7*(1–2), 104–119.

Grabowski, T. J., Damasio, H., Frank, R., Hichwa, R. D., Boles Ponto, L. L., & Watkins, G. L. (1995). A new technique for PET slice orientation and MRI-PET coregistration. *Human Brain Mapping, 2*, 123–133.

Grafton, S. T., Arbib, M. A., Fadiga, L., & Rizzolatti, G. (1996). Localization of grasp representations in humans by positron emission tomography: Observation compared with imagination. *Experimental Brain Research, 112*(1), 103–111.

Hanakawa, T., Immisch, I., Toma, K., Dimyan, M. A., Gelderen, P. V., & Hallett, M. (2003). Functional properties of brain areas associated with motor execution and imagery. *Journal of Neurophysiology, 89*, 989–1002.

Hichwa, R. D., Ponto, L. B., & Watkins, G. L. (1995). *Chemists views of imaging centers*. New York: Plenum Press.

Hoffman, M. L. (1978). Empathy: Its development and prosocial implications. In J. H. E. Howe & C. B. Keasey (Eds.), *Nebraska symposium on motivation: Social cognitive development* (Vol. 25, pp. 169–217). Lincoln, NE: University of Nebraska Press.

Hoffman, M. L. (1982). Affect and moral development. *New Directions for Child Development, 16*, 83–103.

Hoffman, M. L. (2000). Empathy and moral development: Implications for caring and justice. Cambridge, UK: Cambridge University press.

Holmes, C. J., Hoge, R., Collins, L., Woods, R. P., Evans, A. C., & Toga, A. W. (1998). Enhancement of MR images using registration for signal averaging. *Journal of Computer Assisted Tomography, 22*, 324–333.

Iacoboni, M., & Lenzi, G. L. (2002). Mirror neurons, the insula, and empathy (commentary). *Behavioral and Brain Sciences, 25*(1), 39–40.

Jackson, P. L., Brunet, E., Meltzoff, A. N., & Decety, J. (2006). Empathy examined through the neural mechanisms involved in imagining how I feel versus how you feel pain. *Neuropsychologia, 44*(5), 752–761.

Jackson, P. L., Meltzoff, A. N., & Decety, J. (2005). How do we perceive the pain of others? A window into the neural processes involved in empathy. *NeuroImage, 24*(3), 771–779.

Kanwisher, N., Mcdermott, J., & Chun, M. M. (1997). The fusiform face area: A module in human extrastriate cortex specialized for face perception. *Journal of Neuroscience, 17*(11), 4302–4311.

Kreiman, G., Koch, C., & Fried, I. (2000). Imagery neurons in the human brain. *Nature, 408*, 357–361.

Lamm, C., Batson, C. D., & Decety, J. (2007). The neural substrate of human empathy: Effects of perspective-taking and cognitive appraisal. *Journal of Cognitive Neuroscience, 19*(1), 42–58.

Lipps, T. (1903). Einfühlung, innere Nachahmung und Organempfindung. *Archiv für die gesamte Psychologie, 1*, 465–519.

Mehrabian, A., & Epstein, N. (1972). A measure of emotional empathy. *Journal of Personality, 40*(4), 525–543.

Merleau-Ponty, M. (1970). *Phenomenology of perception* (5th ed.). London: Routledge & Kegan Paul. (Originally published 1962)

Nichols, T., Brett, M., Andersson, J., Wager, T., & Poline, J. B. (2005). Valid conjunction inference with the minimum statistic. *NeuroImage, 25*(3), 653–660.

Preston, S. D., & de Waal, F. B. M. (2002). Empathy: Its ultimate and proximate bases. *Behavioral and Brain Sciences, 25*(1), 1–71.

Prinz, W. (1987). Ideo-motor action. In H. Heuer & A. F. Sanders (Eds.), *Perspectives on perception and action* (pp. 47–76). Hillsdale, NJ: Lawrence Erlbaum Associates, Inc.

Ruby, P., & Decety, J. (2001). Effect of subjective perspective taking during simulation of action: A PET investigation of agency. *Nature Neuroscience, 4*(5), 546–550.

Ruby, P., & Decety, J. (2004). How would you feel versus how do you think she would feel? A neuroimaging study of perspective-taking with social emotions. *Journal of Cognitive Neuroscience, 16*(6), 988–999.

Singer, T., Seymour, B., O'Doherty, J., Kaube, H., Dolan, R. J., & Frith, C. D. (2004). Empathy for pain involves the affective but not sensory components of pain. *Science, 303*, 1157–1162.

Smith, D. W. (1989). *The circle of acquaintance: Perception, consciousness and empathy.* Dordrecht, The Netherlands: Kluwer.

Ungerer, J. A. (1990). The early development of empathy: Self-regulation and individual differences in the first year. *Motivation and Emotion, 14*(2), 93–106.

Wicker, B., Perrett, D. I., Baron-Cohen, S., & Decety, J. (2003). Being the target of another's emotion: A PET study. *Neuropsychologia, 41*(2), 139–146.

Wispé, L. (1987). History of the concept of empathy. In N. Eisenberg & J. Strayer (Eds.), *Empathy and its development* (pp. 17–37). New York: Cambridge University Press.

Woods, R. P., Cherry, S. R., & Mazziotta, J. C. (1992). A rapid automated algorithm for accurately aligning and reslicing PET images. *Journal of Computer Assisted Tomography, 16*, 620–633.

Woods, R. P., Dapretto, M., Sicotte, N. L., Toga, A. W., & Mazziotta, J. C. (1999). Creation and use of a Talairach-compatible atlas for accurate, automated, nonlinear intersubject registration, and analysis of functional imaging data. *Human Brain Mapping, 8*, 73–79.

Woods, R. P., Grafton, S. T., Watson, J. D., Sicotte, N. L., & Mazziotta, J. C. (1998). Automated image registration: II. Intersubject validation of linear and nonlinear models. *Journal of Computer Assisted Tomography, 22*, 153–156.

Woods, R. P., Mazziotta, J. C., & Cherry, S. R. (1993). PET-MRI registration with automated algorithm. *Journal of Computer Assisted Tomography, 17*, 536–546.

Worsley, K. J. (1994). Local maxima and the expected Euler characteristic of excursion sets of Chi(2), F, and t fields. *Advances in Applied Probability, 26*, 13–42.

SOCIAL NEUROSCIENCE, 2007, 2 (3–4), 276–291

Psychology Press
Taylor & Francis Group

Different circuits for different pain: Patterns of functional connectivity reveal distinct networks for processing pain in self and others

Jamil Zaki, Kevin N. Ochsner, Josh Hanelin, and Tor D. Wager
Columbia University, New York, USA

Sean C. Mackey
Stanford University, Stanford, California, USA

The ability to empathize with the suffering of others is critical for maintaining relationships and engaging in prosocial behavior. Recently, a series of studies have demonstrated that while watching other people experience pain (other pain), participants engage the anterior insula (AI) and anterior cingulate cortex (ACC), brain regions involved in the direct experience of pain (self pain). Here we test the hypothesis that common activity in ACC and AI may reflect the operation of distinct but overlapping networks of regions that support perception of self or other pain. To address this possibility, we scanned participants using fMRI while they received noxious thermal stimulation (self pain) or watched short videos of other people sustaining painful injuries (other pain). We isolated overlapping regions for self and other pain in the ACC and AI and then used them as seed regions for two kinds of functional connectivity analyses. These analyses identified areas whose activity co-varied with ACC and AI activity during self or other pain either across time (intra-individual connectivity) or across participants (inter-individual connectivity). Both connectivity analyses identified clusters in the midbrain and periaqueductal gray with greater connectivity to the AI during self pain as opposed to other pain. The opposite pattern was found in the dorsal medial prefrontal cortex, that showed greater connectivity to the ACC and AI during other pain than during self pain using both types of analysis. Intra-individual connectivity analyses also revealed regions in the superior temporal sulcus, posterior cingulate, and precuneus that became more connected to ACC during other pain as compared to self pain. Together, these data demonstrated that regions showing similar activity during self and other pain may nonetheless be part of distinct functional networks. These networks could not have been detected in prior work that examined overlap between self and other pain in terms of average activity, but not connectivity.

INTRODUCTION

The ability to empathize with the suffering of other people is a central part of human social behavior. Empathy promotes prosocial behavior towards close others and strangers (Batson et al., 2003; Eisenberg & Miller, 1987), and a lack of emotional empathy is an important symptom of psychopathy and sociopathy (Blair, 2005).

One important conceptual issue in the study of empathy is the extent to which it relies on "shared representations" of pain in the self and in others, enabling perceivers to understand the sensory and affective state of a social target. While some theories hold shared representations as the single most central component of empathy (Gallese, 2003; Gallese, Keysers, & Rizzolatti, 2004), others claim that "empathy is not a simple resonance of

Correspondence should be addressed to: Jamil Zaki, Department of Psychology, Columbia University, Schermerhorn Hall, 1190 Amsterdam Ave., New York, NY 10027, USA. E-mail: jamil@psych.columbia.edu

www.psypress.com/socialneuroscience DOI:10.1080/17470910701401973

affect between the self and other," and instead requires more complex perspective taking and distinct representations of pain-related responses for self and other (Decety & Jackson, 2004).

In the past few years, these issues have been examined in several fMRI studies of pain empathy. These studies generally compared the experience of painful physical stimulation (self pain) with the experience of seeing other people experience pain (other pain). All studies employing this kind of design have reported that two key regions, the anterior cingulate cortex (ACC) and the anterior insula (AI) become engaged both during self and other pain (Botvinick et al., 2005; Morrison, Lloyd, di Pellegrino, & Roberts, 2004; Singer et al., 2004, 2006). Further studies have also shown that activity in ACC tracks with the amount of pain perceivers believe someone else is experiencing (Jackson, Meltzoff, & Decety, 2005; Saarela et al., 2007).

The role of ACC and AI in empathic pain has been interpreted as reflecting recruitment of systems for computing the emotional or affective value of pain experienced directly by a perceiver or by another person.

Findings of self/other overlap in pain processing have been influential, and have often been used as evidence for the centrality of shared representations in empathy. However, both conceptually and empirically, there are reasons to believe that there may be differences between pain processing for self and other. On one hand, primary sensory cortex involved in coding the location and intensity, rather than the affective value of pain (Peyron et al., 1999; Rainville, 2002) has only been engaged during perception of pain in the self, and not when pain is given to others (Singer et al., 2004). Furthermore, the direct experience of self pain may uniquely engage forms of coping or regulation not engaged during other pain. For example, descending projections from the ACC, AI, and thalamus innervate pain-related brainstem nuclei such as the periaqueductal gray (PAG) and the midbrain nucleus cuneiformis, which can modulate the output of pain-related afferents, changing the experience of pain according to changes in expectancy (Fields, 2004; Keltner et al., 2006; Wager, 2005a; Zambreanu, Wise, Brooks, Iannetti, & Tracey, 2005).

On the other hand, registering the suffering of another may require taking that person's perspective in order to draw inferences about what he or she might be feeling. These mental state attributions may affect a perceiver's empathic reaction to a target's pain, including the perceiver's subsequent behavior towards that target (Batson et al., 2003). Candidate brain regions for making mental state attributions during the perception of others' pain include the medial prefrontal cortex (MPFC), which is involved in both cognitive and affective perspective taking (Mitchell, Heatherton, & Macrae, 2002; Ochsner et al., 2004a; Vollm et al., 2006), and superior temporal sulcus (STS), which is involved in assessing the social significance of nonverbal cues (Pelphrey, Morris, & McCarthy, 2004; Saxe, Xiao, Kovacs, Perrett, & Kanwisher, 2004).

Previous functional imaging studies have provided initial support for dissociations in activity between self and other pain. For example, some studies have found areas in PFC and STS to be preferentially active during other pain, and cerebellum and brainstem to be active preferentially during self pain (Botvinick et al., 2005; Morrison et al., 2004; Ochsner et al., under review; Singer et al., 2004). However, no study has found the MPFC to be active during empathic pain, a striking absence in light of the MPFC's centrality to social cognition. Furthermore, it is unclear whether cerebellar, frontal, and other activations found uniquely during self or other pain represent processing steps unrelated to the perception of pain, or whether they are part of unique, overlapping circuits used to process self and other pain, respectively.

One way to test this is through analyses of *functional connectivity* during self and other pain. Whereas standard contrast analyses create a "snapshot" of regional brain activity in response to a task or condition, functional connectivity analyses can identify patterns of communication between regions that contrast analyses may not detect. Connectivity analyses can identify regions whose activity co-varies with activity in ACC and AI during self and other pain, helping to create a dynamic model of circuits underlying each type of pain perception. In the context of empathy, this type of analysis provides a means for testing the extent to which overlapping representations underlie empathy for pain.

In order to explore patterns of connectivity during self and other pain, we scanned participants using fMRI while they either received painful heat stimulation, or watched short videos of other people in pain. We identified regions of overlap between self and other pain in AI and ACC that were similar to those reported by previous studies (the main effects of this study are reported in Ochsner et al., under review). We then used two methods to assess differences in functional connectivity with the commonly recruited regions of AI and ACC.

One analysis examined *inter-individual* connectivity. This analysis searched for regions showing stronger task-related activation in participants who also showed stronger activation in ACC or AI. Significant correlations indicate that individuals who show high ACC (or AI) activity also show high activity in "connected" regions; thus, there is a coherent network involving the ACC (or AI) that is more engaged by some individuals than others. The interpretation of task-state effects is that coherent individual differences in correlated activity across regions are greater in some tasks (i.e., self pain) than others (i.e., other pain). Variants of inter-individual connectivity analyses have previously been used to asses several types of network activity, including prefrontal regulation of pain-related activity in the thalamus and midbrain (Lorenz, Minoshima, & Casey, 2003), and prefrontal modulation of emotion-related activity in the amygdala (Ochsner et al., 2004b).

A second analysis examined *intra-individual* connectivity using psychophysiological interaction analyses (PPI; Friston et al., 1997). This analysis searched for brain regions whose activation across time co-varied with that of ACC and AI more in one state than in another in each subject, and then assessed the reliability of these connectivity patterns across subjects. Within subject connectivity analyses have been used previously to assess functional circuits engaged by emotion regulation and executive control of attention (Egner & Hirsch, 2005; Etkin, Egner, Peraza, Kandel, & Hirsch, 2006).

We chose these two analyses because they complement each other methodologically, and consistent findings across the two would strengthen our ability to draw inferences about the brain regions interacting with our seed regions. Task-evoked changes in functional connectivity might be expected to be detected by both methods, but they measure somewhat different quantities and so have complementary advantages. The main advantage of the PPI analysis is that it assesses co-variance between regions across time, and so provides a test of task effects on connectivity within subjects. Advantages of the inter-individual connectivity analysis are: (a) it is less susceptible to timeseries artifacts than PPI; (b) it is more robust to variations in the shape of the hemodynamic response across regions; and (c) it directly compares inter-regional correlations across task conditions, whereas PPI tests a difference in the slope of the relationship among regions, which may be influenced by changes in activation magnitudes in one region.[1] It is important to note, however, that neither of these analyses can establish causal links between regions (i.e., activity in region A causes activity in region B). Instead, these analyses allow inferences that regions are coactive across participants (inter-individual) and across time (PPI), that they are doing so in a way that is functionally relevant, and that the strength of co-activation is modulated by task state.

We expected several differences in connectivity patterns to emerge via these analyses. On one hand, we hypothesized that regions involved in perspective taking, such as medial PFC and the STS, would be preferentially connected to ACC and AI during other pain. On the other hand, we hypothesized that regions involved in processing the sensory and autonomic aspects of pain, such as somatosensory cortex, and the periaqueductal gray (PAG) would be more connected to ACC and AI seed regions during self pain. The goal was to directly test the hypothesis that pain affect is associated with overlapping, but qualitatively different, processing networks depending on whether pain is experienced by participants themselves or by another person.

METHODS

Participants

Thirteen participants (*M* age = 29.5 years; 6 male) were recruited in compliance with the human subjects regulations of Stanford University Medical School.

Behavioral paradigm

In a single experimental session described in detail elsewhere (Ochsner et al., under review), participants completed both self pain and other pain tasks in counterbalanced order. In the self

[1] That is, if Regions A and B are functionally coupled, and task state S increases activity in A but not B, then apparent connectivity in a PPI analysis may increase, even if no true change in functional coupling has occurred.

pain task, noxious (painful) thermal and non-noxious (neutral) thermal stimulation was delivered to the right distal lateral forearm by a computer-controlled thermal stimulator with an MRI compatible 30 mm^2 Peltier probe (TSA-2001, Medoc, Chapel Hill, NC). Twenty-second blocks of noxious thermal stimuli (45–50°C) alternated with 30 s blocks of neutral stimuli (38°C) five times. Temperatures used for the noxious thermal blocks were determined on a participant-by-participant basis in a pre-scanning session. Noxious temperatures elicited the maximum level of pain without causing movement, which roughly corresponded to a participant-defined 7 out of 10 on a verbal rating scale (0 = *no pain*, 10 = *worst pain imaginable*). Immediately upon exiting the scanner, participants used a verbal rating scale (0 = *not unpleasant*, 10 = *most unpleasant experience imaginable*) to rate the unpleasantness of the noxious stimulation they had received (i.e., pain affect).

In the other pain task participants viewed a 2-minute video clip depicting 17 events in which individuals suffered injuries in sporting events (e.g., a leg break in a soccer match), stunts (e.g., injuring an arm or leg while skateboarding), or vehicular accidents (e.g., a young girl being struck by train). Participants were instructed to attend to and watch all events presented during the course of the video.

As described in detail in Ochsner et al. (under review), a separate behavioral study was conducted to characterize and verify pain perception during the novel other pain task. Thirteen participants (age and gender matched to those in the imaging study) completed the same self and other pain paradigms as in the scanner study while also using a 10-point scale to provide continuous ratings of either self or other pain affect. Paired sample *t*-tests comparing self report painful and non-painful video segments verified that injuries included in the video were effective in creating the perception of pain, $t(12) = 3.20$, $p = .008$. Similar ratings during the self pain condition verified the efficacy of the thermal stimulus in generating the experience of pain during the application of noxious heat, $t(12) > 20$, $p < .001$.

MRI data acquisition

During completion of both tasks a T2*-sensitive gradient echo spiral-out pulse sequence (30 ms TE, 2000 ms TR, 2 interleaves, 60° flip angle, 24 cm field of view, 64 × 64 data acquisition matrix) was used to collect whole brain fMRI data (32 axial slices, 3.5 mm thick) at 3T (GE Signa LX Horizon Echospeed scanner). High-order shimming was performed before functional scans (Glover, 1999). For anatomical reference, T2-weighted flow-compensated spin-echo scans were acquired using the same slice prescription (2000 ms TR; 85 ms TE).

Data analysis

Preprocessing and statistical analyses were carried out using SPM2 (Wellcome Department of Cognitive Neurology). Functional images were slice-time and motion corrected, normalized using parameters derived from the normalization of coregistered anatomical images to a the Montreal Neurological Institute (MNI) standard template brain, interpolated to 2 × 2 × 2 mm voxels, and smoothed with a Gaussian filter (6 mm full width–half maximum). First-level analyses for the self pain task modeled noxious and neutral blocks with boxcar regressors convolved with the canonical hemodynamic response. First-level analyses for the other pain task modeled observed physical injuries as events (whose onset was the moment a physical injury occurred) convolved with the canonical hemodynamic response. All other portions of the video depicting the same actors engaged in non-painful activities served as the no-pain baseline against which activation related to other pain events was determined. These regressors were fitted to fMRI data from each task using the general linear model. Contrast images for each participant summarized differences between (1) noxious and non-noxious blocks for the self pain task, and (2) differences between observed painful events and all other portions of the video depicting non-painful activities for the other pain task. These contrast images were used to create second-level group average SPM maps of regions more active either for the experience of noxious heat as compared to non-noxious warmth, or during the observation of painful as compared to non-painful events experienced by others.

To identify differential patterns of functional connectivity during self and other pain, we followed a two-step analysis procedure. First, we identified commonly activated regions for self and other pain by using the *t*-map for the self pain contrast as an inclusive mask for the other pain contrast. Each contrast was voxel-level thresholded at $p < .005$, which yields maps of active overlap regions at a height threshold of $p < .000025$ across both tasks using the Fisher method for combining probabilities (Kampe, Frith, & Frith, 2003; Ochsner et al., 2004b). As described in detail elsewhere (Ochsner et al., under review), this analysis identified regions of ACC and AI active during both self and other pain similar to those identified in other studies of pain empathy (Botvinick et al., 2005; Jackson et al., 2005; Morrison et al., 2004).

To assess intra-individual connectivity, we employed a psychophysiologic interaction (PPI) analysis implemented in SPM to identify regions whose timecourse co-varied with the timecourse of activity in either the ACC or AI. A PPI analysis convolves the timecourse of activity in a seed region with a psychological variable of interest and uses this vector as a regressor in a subsequent whole-brain analysis. In this way, PPI analyses identify regions that are functionally connected to the seed region *more strongly* in one condition than in another (Friston et al., 1997). We performed PPI analyses separately for self and other pain. For each condition, we identified the activation peak in each individual that was closest—in Euclidian distance—to the group overlap peaks in ACC and AI, extracted timecourses from each peak, and computed PPI models for each subject based on their own ACC and AI peaks. These individual PPI analyses were then used in group-level paired-sample *t*-tests, comparing connectivity during self pain with connectivity during other pain. Contrasts on those *t*-tests were then estimated to identify: (1) regions that were more connected to AI or ACC during self pain than during other pain; (2) regions that were more connected to AI or ACC during other pain than during self pain; and (3) regions that were connected to AI or ACC during both self and other pain. These contrasts were thresholded at $p < .005$, uncorrected, with an extent threshold of 10 voxels.

To verify that each region from PPI analyses was consistently active in conditions of interest, the Brain Imaging Toolbox (http://web.mit.edu/swg/www/software.htm) was used to extract parameter estimates for each participant from the peak voxel of each overlap cluster identified at the group level. Mean parameter estimates for each condition were then computed for each condition and were compared using paired-sample *t*-tests.

The second, *inter-individual* method for computing functional connectivity examined how individual differences in task-evoked activation of ACC and AI co-varied with task-evoked activation in other brain regions, and whether this connectivity was modulated by the self versus other task condition. These analyses were performed using custom Matlab scripts written by one of the authors (TDW). Data consisted of contrast magnitude maps for self and other pain for each participant. Mean contrast values in the AI and ACC were used as "seeds," and we calculated maps of correlations between each voxel and the seed for both self and other pain contrasts. We searched for voxels with higher correlations during self versus other pain by comparing the two correlation values using methods developed by Steiger (1980). This search was conducted in regions of interest defined by PPI analyses, namely: MPFC; superior temporal sulcus (STS); ACC; PCC; insula; supplementary motor area; and midbrain, and thresholded at $p < .005$, uncorrected, with an extent threshold of 10 contiguous voxels. As noted in the introduction, task-evoked changes in functional connectivity might be expected to be detected by both methods, but they measure somewhat different quantities and so have complementary advantages.

RESULTS

Imaging results

Regions common to self and other pain perception. As reported elsewhere (Ochsner et al., under review), "standard" contrast analyses revealed common recruitment of anterior mid cingulate cortex (MNI coordinates: 4, 10, 40) and right ventral anterior insula (MNI coordinates: 42, 22, −12), as well as middle frontal gyrus, premotor cortex, and dorsal thalamus during both self and other pain as compared to self and other non-pain conditions. Additional interaction contrasts revealed a set of premotor, parietal, and lateral prefrontal regions that responded preferentially to other pain, and clusters in the anterior and posterior insula responding more strongly to self pain. As these data have been commented upon in detail elsewhere, we will not elaborate them here. For our present purposes, the results of connectivity analyses with the ACC and AI regions commonly recruited for both self and other pain are of primary interest.

Connectivity with ACC. Regions showing increased intra-individual connectivity in PPI analyses with the ACC seed region during other pain as compared to self pain included the right STS, MPFC and rostrolateral prefrontal cortex, posterior cingulate cortex, parahippocampal gyrus, premotor cortex, and fusiform gyrus (Table 1). Interestingly, some of these regions (i.e., premotor cortex and RLPFC) had already been shown by contrast analyses to be more *active* during other pain than during self pain. However, the PPI analysis identified two additional regions—STS and MPFC—that, while not more active during the perception of other pain, did show increased connectivity with ACC during this condition.

Regions showing increased connectivity to the ACC seed during self pain than during other pain included an additional region in ACC anterior and superior to the seed region, as well as an insula region posterior and superior to the overlap AI region and clusters in the parietal lobe and cerebellum. Analysis of parameter estimates for the ACC region demonstrated that differential connectivity for self and other pain was driven by a strong *decoupling* of this region's activity from that of the seed region during other pain, as opposed to a significant positive correlation between these regions during self pain (see Figure 1 and Table 2). This was the only region whose differential connectivity was driven by such an effect.

The inter-individual connectivity analyses revealed patterns consistent with PPI connectivity analyses in MPFC and STS but did not reproduce the differential connectivity pattern observed in the ACC.

Connectivity with AI. Regions showing increased intra-individual connectivity with the AI during other as compared to self pain again included a priori regions of interest in the MPFC (which was found in inter-individual analyses also), as well as clusters in the posterior cingulate and precuneus, lingual gyrus, and precentral gyrus (see Figure 2 and Table 2). Of these, only the precentral and lingual gyri were found by simple contrasts to become more active during other pain.

Regions showing increased connectivity with the AI during self as compared to other pain included the periaqueductal gray (PAG), right mid insula and right middle temporal gyrus, as well as a cluster in the midbrain adjacent to the nucleus cuneiformis (Keltner et al., 2006). Inter-individual connectivity analyses showed results consistent with much of this pattern, mirroring timecourse connectivity results in the PAG and mid insula.

Common patterns of connectivity with ACC or AI. To determine whether any regions showed functional connectivity with ACC or AI during both self and other pain, we calculated overlap images of the results of each of the four primary analyses (i.e., intra- and inter-individual connectivity analyses for the ACC and AI). The only analysis yielding a pattern of functional connectivity common to both self and other pain was the intra-individual connectivity analysis for the AI,

TABLE 1
Regions showing common activation for self pain and other pain, as assessed by masked contrast analyses

		Coordinates				
Region of activation	*Lat*	*x*	*y*	*z*	*T-score*	*Volume (vox)*
Middle frontal gyrus	R	46	28	20	4.4	59
Anterior cingulate	M	4	10	40	4.13	11
Premotor gyrus	R	48	8	40	3.99	23
Anterior insula (AI)	R	42	22	−12	7.76	56
	R	28	28	2	5.2	57
	R	28	16	6	5.18	(L)
	R	30	20	−2	4.75	(L)
Dorsal thalamus	R	12	−2	10	5.2	56
	R	20	−6	14	4.03	(L)
	R	16	−10	20	3.78	(L)

Notes: Local maxima for clusters are designated with (L). Hemisphere is designated as midline (M) for maxima within 6 mm of the midline. Coordinates are in MNI space. Vox = voxels.

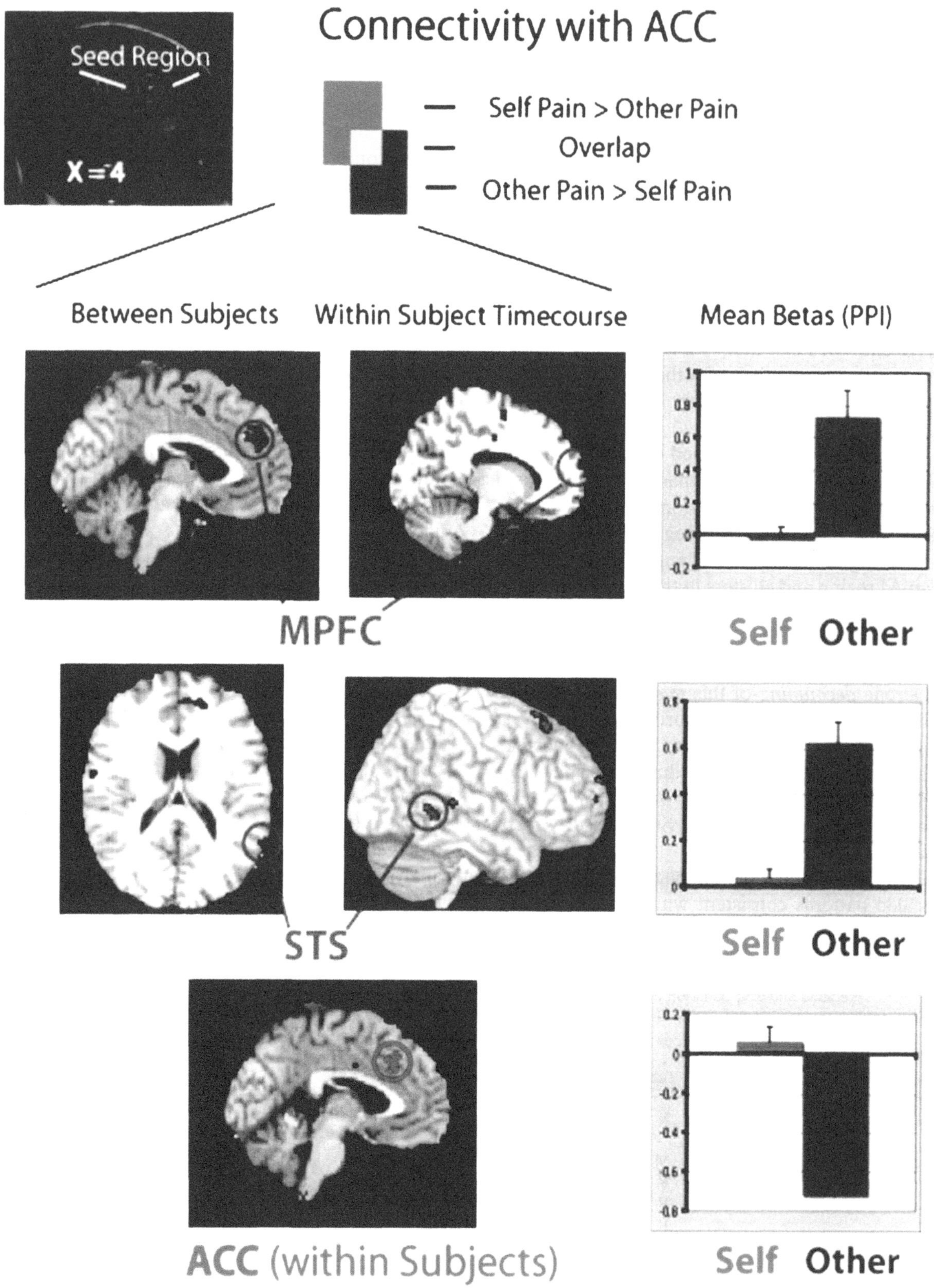

Figure 1. Regions showing increased connectivity with the ACC during other and self pain, respectively. Parameter estimates are drawn from within subject timecourse connectivity analyses. MPFC = medial prefrontal cortex; STS = superior temporal sulcus; STG = superior temporal gyrus; ACC = anterior cingulate cortex.

TABLE 2

Comparison of regions found in intra-individual (within subjects) and inter-individual (between subjects) connectivity analyses

		Intra-Individual Analysis					*Inter-Individual Analysis*				
		Coordinates					*Coordinates*				
Region of activation	*Lat*	*x*	*y*	*z*	*T-score*	*Nvox*	*x*	*y*	*z*	*T-score*	*Nvox*
ACC Self > Other											
Anterior cingulate cortex	M	2	26	32	4.19	22					
Inferior parietal lobe	L	−60	−44	30	4.53	32					
	−60	−42	22	4.47	(L)						
ACC Other > Self											
Dorsomedial frontal gyrus	R	12	64	18	5.36	16	12	48	12	3.98	32
							−4	60	18	4.56	26
Rostrolateral PFC	R	24	60	4	4.99	32					
Rostrolateral PFC	R	24	52	10	3.93	(L)					
Precentral gyrus	L	−30	−18	66	4.33	40					
Precentral gyrus	L	−30	−20	58	4.05	(L)					
Superior temporal sulcus	R	60	−46	−4	7.29	26	50	−42	−12	5.92	24
Superior temporal gyrus	R	52	−18	−12	5.29	24					
Superior temporal gyrus	R	56	−12	−4	3.51	(L)					
Superior temporal gyrus	R	42	−20	16	5.2	47					
Superior parietal lobe	L	−22	−50	68	4.2	25					
Superior parietal lobe	L	−18	−50	52	3.91	(L)					
Inferior parietal lobe	L	−50	−54	34	5.47	30					
Posterior cingulate cortex	M	8	−10	38	4.63	20	6	6	44	4.32	16
AI Self > Other											
Mid insula	R	46	−2	−4	4.37	21	44	−2	8	5.12	40
Middle temporal	R	56	−68	12	4.74	37					
PAG	M	−6	−32	−10	4.6	20	−6	−34	−6	4.27	14
Midbrain	M	6	−24	−26	4.92	29	−4	−20	−24	4.19	32
AI Other > Self											
Medial frontal gyrus	R	10	56	34	5.07	26	6	40	24	4.75	18
Posterior cingulate cortex	R	12	−14	48	4.86	15	8	−6	50	5.34	38
Precentral gyrus	R	18	−30	74	3.88	17					
Lingual gyrus	R	20	−48	−2	5.27	35					
Precuneus	L	−20	−78	42	5.53	39					
Overlap											
Inferior temporal gyrus	R	50	−60	−8	5.88	36					

Notes: Local maxima for clusters are designated with (L). Hemisphere is designated as midline (M) for maxima within 6 mm of the midline. Coordinates are in MNI space. Nvox = cluster size (in number of voxels).

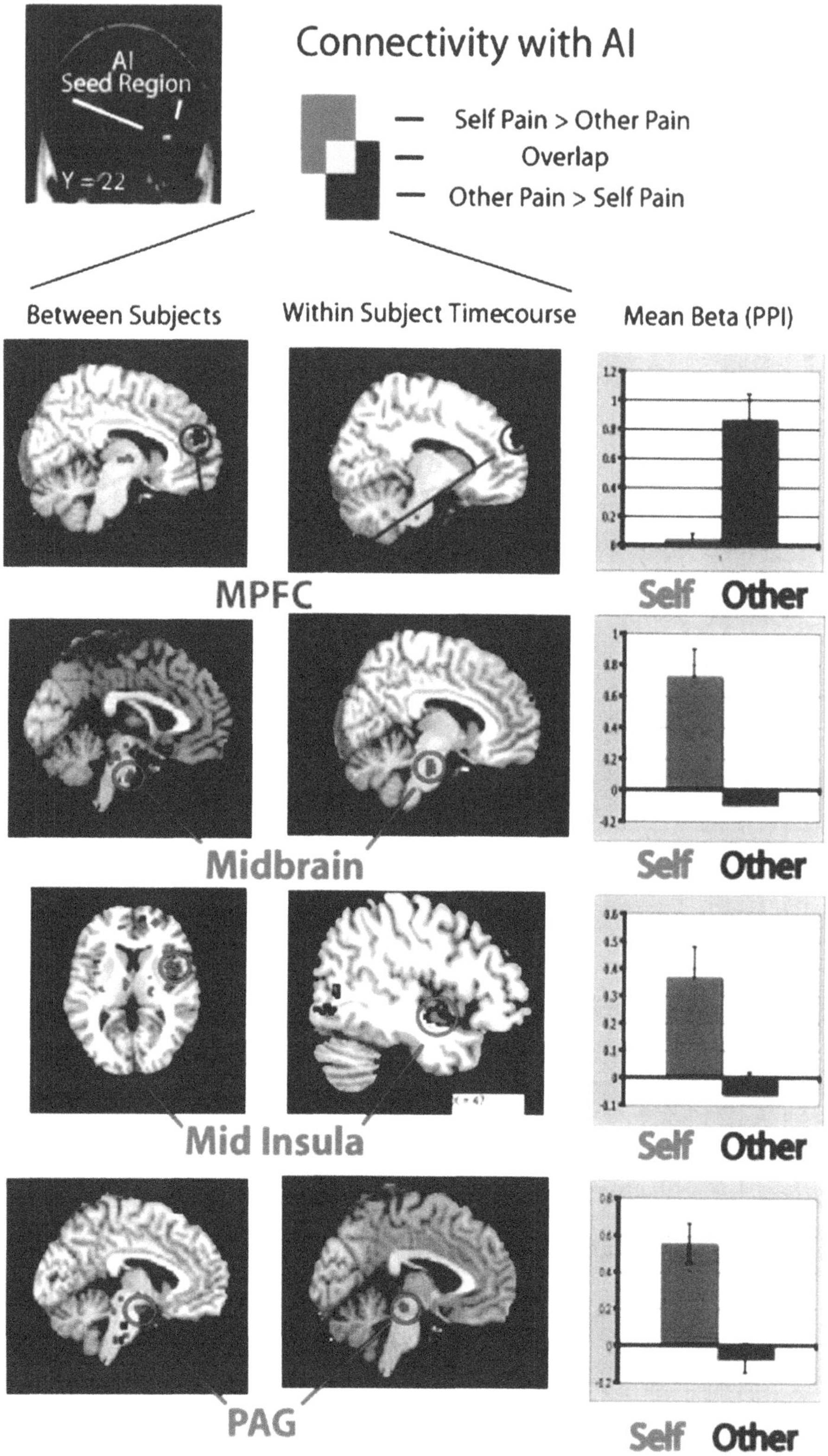

Figure 2. Regions showing increased connectivity with the AI during other and self pain, respectively. Parameter estimates are drawn from within subject timecourse connectivity analyses. The inset shows orthogonal views of the cluster in the PAG, overlaid on a mean anatomical image drawn from all of our subjects. MPFC = medial prefrontal cortex; PAG = periaqueductal gray; MI = mid insula.

which identified a region of the right inferior temporal gyrus connected to the seed region during both self and other pain.

DISCUSSION

Both experiencing pain directly and observing others in pain have been shown to recruit regions of mid ACC and AI implicated previously in the emotional distress accompanying physical pain (Botvinick et al., 2005; Hutchison, Davis, Lozano, Tasker, & Dostrovsky, 1999; Jackson, Brunet, Meltzoff, & Decety, 2006; Morrison et al., 2004; Singer et al., 2004, 2006). This is the first neuroimaging study in which functional connectivity analyses have been used to address questions about how networks involved in pain perception may vary depending on whether the person in pain is the subject themselves or another person. Consistent with our expectations, analyses indicated that overlapping, but qualitatively different, functional networks are involved in perceiving either one's own pain or the pain of another person.

To appreciate the functional significance of these different patterns of connectivity one must first consider the functional roles played by the commonly recruited regions of ACC and AI. Consider, for example, that the mid portion of the ACC activated in our self/other overlap contrast receives ascending nociceptive inputs (Craig, 2003; Devinsky, Morrell, & Vogt, 1995; Vogt, 2005) and in imaging studies is activated by perception of pain applied either externally or internally to the viscera (Farrell, Laird, & Egan, 2005; Hebben, Corkin, Eichenbaum, & Shedlack, 1985; Jackson et al., 2006; Morrison et al., 2004; Peyron et al., 2000). ACC activity in response to pain is modulated by attention and anxiety (Seminowicz & Davis, 2006; Valet et al., 2004) and expectation (Wager, 2005b; Wager et al., 2004). Additionally, this region projects to motor and premotor cortex and has been associated with the initiation of actions, as well as monitoring of the affective consequences of these actions (Paus, 2001; Rushworth, Walton, Kennerley, & Bannerman, 2004). Finally, the dorsal ACC has been associated with monitoring and responding to conflict, as for example when subjects make an error on a cognitive control task (Carter et al., 1998).

In this context, engagement of the ACC during both self and other pain conditions could reflect several different types of processing. ACC activity could play a role in motivational aspects of pain, including urges or desires to stop painful events. It could also represent participants' inhibition of urges to pull away from painful stimulation, or to look away from gruesome depictions of others in pain. Importantly, the fact that ACC activity may not only represent a single process of "pain affect," but instead may be related to several types of motivation, leaves open the possibility that ACC activation indexes different processes across self and other pain conditions. If anything, our connectivity results suggest that this may be the case, as ACC activity co-varies with quite different brain regions in each condition.

The overlap region of insula found in our study and others has also been implicated in affective processing. This region corresponds most closely to the anterior and ventral agranular portion of the insula (Wager & Feldman Barrett, 2004), which is heavily interconnected with the orbitofrontal cortex and temporal operculum (Mesulam & Mufson, 1982), and is considered part of a set of "paralimbic" regions responsible for assessing the hedonic value of internal and external cues. This region may also be involved in interoceptive processing, or gathering information about and actively regulating internal bodily states (Craig, 2002; Critchley, 2004; Critchley, Melmed, Featherstone, Mathias, & Dolan, 2002). Taken together, these data suggest that, during empathic pain, the ACC may be involved in directing attention and mobilizing responses to painful stimuli, whereas the AI may be involved in associating pain cues with negative bodily and affective states.

Against this backdrop, we can now consider how affective representations in ACC and AI can be generated through the concerted actions of processing networks whose constituent regions may differ during the perception of self or other pain. This experiment tested two hypotheses about such networks. The first hypothesis was that empathic pain would involve affective perspective taking. Our second hypothesis was that the perception of self pain would differentially depend upon regions involved in coding somatic and autonomic responses to stimuli delivered directly to one's body, and that the network of brain regions involved in perceiving pain in the self, but not in others, would include midbrain and somatosensory cortex. Consistent with our predictions, we found that during other pain, the ACC and AI were functionally connected with prefrontal and posterior cortical regions related to ToM and interpretation of socially relevant nonverbal cues. By contrast, during self pain, the ACC and AI were more functionally connected with regions involved in the ascending transmission of pain.

Functional networks supporting the perception of pain in self and other

Based on these data, as well as knowledge about anatomical connections between regions involved in nociceptive processing, we can begin to construct a model of the common and distinct networks involved in the empathic perception of pain in others as well as in the direct perception of one's own pain. Figure 3 represents an initial step towards such a model, though it is by no means a complete picture of the anatomical or functional connections between regions supporting direct or empathic pain processing. Here, red and blue lines indicate functional connectivity found in our data to be greater during self and other pain, respectively. Yellow lines highlight relevant inter-regional anatomical connections we can infer are involved in perspective taking or nociception, based on prior anatomical studies.

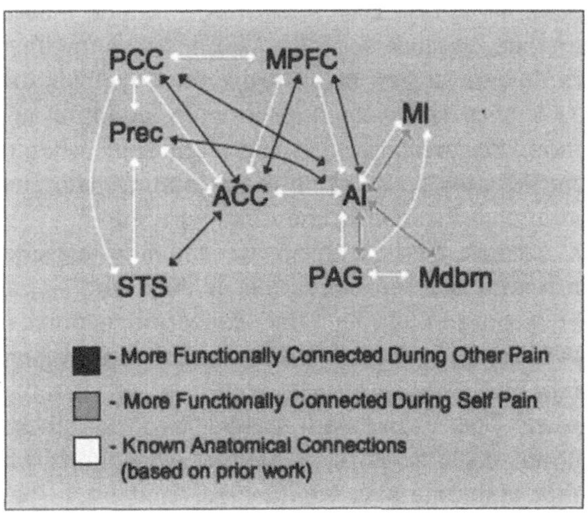

Figure 3. A circuit model diagramming the interaction of brain areas during self and other pain only, as well as interactions occurring during both types of pain. Connections in the model are based both on data from this study, and on existing information about intrinsic physical connections between these regions. Note that data from the current study imply *functional* connectivity, but do not prove that these connections occur through monosynaptic or direct connections. MPFC = medial prefrontal cortex; STS = superior temporal sulcus; MI = mid insula; ACC = anterior cingulate cortex; AI = anterior insula; PAG = periaqueductal gray; Prec = precuneus; PCC = posterior cingulate cortex; MdBrn = midbrain.

As highlighted by the blue connections in Figure 3, the empathic perception of pain in others depends upon regions of ACC and AI that register the affective value of painful events as well as a network of regions implicated in various aspects of social cognition. Perhaps foremost among the social-cognition-related regions are portions of the rostral and dorsal medial prefrontal cortex, which here were identified in both the intra- and inter-individual connectivity analyses as functionally connected with both ACC and AI during other pain. These MPFC subregions have been implicated in tasks involving mental state attribution, including inferring false beliefs and judgments of preferences, emotions, and traits in others (Amodio & Frith, 2006; Mitchell, Macrae, & Banaji, 2004, 2005; Ochsner et al., 2004a). Examination of parameter estimates demonstrated that connectivity between the MPFC and seed regions in the ACC and AI during other pain is not only robust, but is virtually absent during self pain.

Another social-cognition-related region, the superior temporal sulcus (STS), showed increased connectivity during other pain with only the ACC. The STS is involved in interpreting the intentional and emotional meaning of socially significant movements (Grosbras & Paus, 2006; Pelphrey et al., 2004) and speech (Sander et al., 2004). Previously, empathic pain studies using static pain-related images did not report STS activity during other pain (Jackson et al., 2006; Saarela et al., 2007), whereas those using videos—including ours—have found STS activity (Botvinick et al., 2005). The fact that the STS became selectively connected with the ACC but not the AI during other pain is consistent with the interpretation that the social salience of someone else's pain influences allocation of attentional resources via connectivity between the STS and the ACC. At least in rhesus monkeys, the posterior STS and supergenual ACC do not share many direct connections (Vogt & Pandya, 1987), and, as such, functional connectivity between these regions should be explained via a third region. Candidate regions for transmitting socially significant information between the STS and ACC are the precuneus and adjacent posterior cingulate cortex (PCC). These two regions share connections with both STS and ACC, and are thought to be involved in helping guide attention towards salient environmental cues (Cavanna & Trimble, 2006; Vogt, Vogt, & Laureys, 2006). Our analyses showed that the PCC became more functionally connected with AI and ACC during other pain than during self pain, consistent with this interpretation.

In comparison to the results for the empathic perception of pain in others, heat pain administered directly to participants caused quite disparate regions to become functionally connected with the AI and ACC. As highlighted by the red connections in Figure 3, the direct perception of one's own pain depends upon regions of ACC and AI that register the affective value of painful events as well as a network of regions implicated in somatic and autonomic responses to pain. Most notably, we found that during self, but not other, pain the AI became functionally connected with the periaqueductal gray (PAG), a midbrain region dorsal to the pons, adjacent to the nucleus cuneiformis (NCOs), and the mid insula, posterior to the AI seed cluster. The PAG is a key region in the top-down control of pain transmission in the spinal cord through influences on the medulla (Fields, 2004; Wager, 2005a). While the PAG is often associated with inhibiting pain transmission via top-down mechanisms such as the placebo effect, it can also become activated during anticipatory anxiety preceding pain. Furthermore, this region shows dense interconnectivity with the midbrain NCOs, which is also involved in pain anticipation. Similarly, the mid insula is thought to be involved in processing information about the location and intensity of pain, as opposed to the affective and motivational information encoded in more ventral and anterior portions of the insula (Wager & Feldman Barrett, 2004).

Increased connectivity of mid insula, PAG, and midbrain with the AI during self as opposed to other pain suggests that the direct experience of pain is necessary to engage these ascending nociceptive pathways. While the PAG showed self pain related connectivity with the AI through both analyses, it showed significant self-pain-related connectivity with the ACC only when examined through inter-individual analyses. This discrepancy could be caused by our inter-individual connectivity analysis being more sensitive to the coactivity of regions that show sustained activation throughout the entire duration of a task.

Overall, our data suggest that partially dissociable networks drive empathic pain-related activity in the ACC and AI. During self pain, brainstem afferents, as well as the mid insula, interact with affective pain processing in the AI. Conversely, during other pain medial prefrontal and posterior cingulate activity interact with both the ACC and AI, whereas posterior cortical areas, such as the superior temporal sulcus and extra striate visual areas, become connected selectively with the ACC via the precuneus.

Caveats

It is important to acknowledge two limitations to the current research. First, the self and other pain conditions in this study differed along multiple dimensions. While self pain was administered through heat in periods of 20 seconds, other pain was presented in events that occurred throughout a short video. This opens our data to the possibility that differences in pain-related activity and connectivity across conditions may have been related to differences between the predictability, timecourse, or complexity of their respective stimuli. That being said, it is important to consider these potential limitations in light of principled reasons for choosing these particular stimuli.

All studies of empathic pain necessarily involve dissociations between the experiences participants have across conditions. Self pain always involves direct stimulation with the threat of tissue damage, whereas other pain involves (usually visual) cues about injury or discomfort in someone else. As such, some difference in computations between conditions is inevitable. Previous studies of empathic pain have dealt with this fact in differing ways. Some (Morrison et al., 2004; Singer & Frith, 2005) have attempted to equate their stimuli as much as possible by using similar pain-delivery techniques to subjects and their social targets. Others (Botvinick et al., 2005) have used stimuli that differ (heat vs. facial expressions of pain) in ways similar to ours. Studies employing both strategies have been consistent both in finding overlapping activation for self and other pain in the ACC and AI, and in finding dissociations in activation in the prefrontal cortex, which responds more to other pain, and sensory cortex, which responds preferentially to self pain. Our findings dovetail with these prior results, and encourages our belief that these findings occur not based on stimulus non-comparability, but rather as a function of self and other pain processing *per se*.

We do, however, acknowledge the fact that the differences in processing self and other pain (and in the resulting connectivity of neural circuitry involved in each process) found in this study may partially result from the differences in stimulus type across conditions. We believe this to be a limitation to the study of empathic pain and empathy more generally

A second limitation in this study is inherent in the connectivity analyses used. Because we were interested in exploring many possible regions for interconnections with ACC and AI during pain processing, we used exploratory functional connectivity analyses rather than model-based approaches such as structural equation modeling or dynamic causal modeling. Consequently, the results of our connectivity analyses, unlike model-based approaches, do not allow us to make inferences about the *direction* of influence between connected regions. In other words, we cannot be sure whether regions such as the medial prefrontal cortex drive ACC activity in the context of other pain, ACC activity drives MPFC activity, or if this correlation represents processes in the ACC and MPFC *both* being driven by a third locus of activation not identified in our analyses. Future research should address the directionality of influence in these circuits.

Implications and future directions

The present findings have a number of implications for current and future work on pain empathy and social cognition. First, they suggest new interpretations for findings of overlapping activation in ACC and AI during self and other pain. Until now, these activations have been taken as supporting shared representations in empathy. Shared representations, in turn, have been cited as the fundamental feature underlying empathic and social cognitive ability in humans. One problem with positing shared representations as the sole basis for social cognition, however, is that regions involved in shared representations (such as the ACC, AI, and premotor mirror neuron regions) are not typically activated in studies that require participants to report on the mental states of another person. Instead, tasks requiring judgments about mental states show activity in a network of distinct regions including the STS, PCC, and MPFC. Here, we report connectivity between brain regions underlying social cognition and shared representations only in the context of perceiving others in pain.

These findings build a bridge between pain empathy studies that until now have emphasized the functional roles played by "shared representations" of pain affect in ACC and AI, and studies of mental state attribution that until now have emphasized the role of MPFC and other midline structures in social cognition. More broadly, they show the way that brain structures can support the relationship between different types of empathic subprocesses. An outstanding problem in empathy research has been the use of the broad term "empathy" to refer to several subprocesses that may in fact differ extensively from each other. This problem was noted by Davis (1994) who wrote that a "central, recurrent, and intractable problem" in pain empathy research is that "the term empathy is routinely used to refer to two distinctly separate phenomena, cognitive role taking and affective reactivity to others" (p. 9). The current findings suggest mechanisms through which these subprocesses of empathy may work together when a perceiver sees someone else in pain. Negative pain affect may be supported by the ACC and AI for both self and other pain, whereas only when observing others in painful situations do these systems interact with MPFC systems supporting perspective taking. The present findings can not tell us whether one of these two types of processing precedes or drives the other, of course. It is possible either that taking the perspective of another facilitates affective responses, or that automatically evoked affective reactions motivate perspective taking and higher-level cognition about another's plight.

A second implication of the present work concerns the use of functional connectivity analyses. Previous work on pain empathy has not shown activity in MPFC regions. The fact that we observed MPFC using connectivity analyses, but not in simple contrasts (see Ochsner et al., under review), suggests that empathy-related computations performed by MPFC may have gone unnoticed in prior work. More generally, our findings highlight the kinds of functional inferences that can be drawn on the basis of connectivity as compared to contrast analyses. Whereas contrasts can identify regions that are functionally co-active but whose activity may be uncorrelated, connectivity analyses can identify regions whose activity co-varies but may not appear as active in group-averaged contrasts. Future work employing both kinds of analyses will refine our understanding of the regions showing tonic as opposed to phasic network activity during empathic interactions.

Third, the present findings may suggest a route by which our understanding of another's intentions may impact our feelings of empathy for them. Prior work has shown that social context can importantly alter our reactions to the suffering of others (de Quervain et al., 2004; Lanzetta & Englis, 1989; Singer et al., 2006). For example, competition with someone can weaken a perceiver's autonomic and neural reaction to that opponent's pain, whereas instructing participants to take the perspective of another can increase these responses (Jackson et al., 2006; Lamm, Batson, & Decety, 2007). These manipulations can have important behavioral consequences, changing perceivers' willingness to engage in helping behavior (Batson et al., 1997, 2003). Given the role of MPFC in regulating emotion via strategies that involve changing one's interpretation of another's affective states (Ochsner et al., 2004a), our results suggest that MPFC may in some cases support the role of perspective taking in modulating responses to another's pain and distress.

REFERENCES

Amodio, D. M., & Frith, C. D. (2006). Meeting of minds: the medial frontal cortex and social cognition. *Nature Reviews Neuroscience*, *7*(4), 268–277.

Batson, C. D., Lishner, D. A., Carpenter, A., Dulin, L., Harjusola-Webb, S., Stocks, E. L., et al. (2003). "...As you would have them do unto you": Does imagining yourself in the other's place stimulate moral action? *Personality & Social Psychology Bulletin*, *29*(9), 1190–1201.

Batson, C. D., Polycarpou, M. P., Harmon-Jones, E., Imhoff, H. J., Mitchener, E. C., Bednar, L. L., et al. (1997). Empathy and attitudes: can feeling for a member of a stigmatized group improve feelings toward the group? *Journal of Personality and Social Psychol*, 72(1), 105–118.

Blair, R. J. (2005). Responding to the emotions of others: dissociating forms of empathy through the study of typical and psychiatric populations. *Conscious Cognition*, *14*(4), 698–718.

Botvinick, M., Jha, A. P., Bylsma, L. M., Fabian, S. A., Solomon, P. E., & Prkachin, K. M. (2005). Viewing facial expressions of pain engages cortical areas involved in the direct experience of pain. *NeuroImage*, *25*(1), 312–319.

Botvinick, M. M., Braver, T. S., Barch, D. M., Carter, C. S., & Cohen, J. D. (2001). Conflict monitoring and cognitive control. *Psychological Review*, *108*(3), 624–652.

Carter, C. S., Braver, T. S., Barch, D. M., Botvinick, M. M., Noll, D., & Cohen, J. D. (1998). Anterior cingulate cortex, error detection, and the online monitoring of performance. *Science*, *280*(5364), 747–749.

Cavanna, A. E., & Trimble, M. R. (2006). The precuneus: a review of its functional anatomy and behavioural correlates. *Brain*, *129*(3), 564–583

Craig, A. D. (2002). How do you feel? Interoception: the sense of the physiological condition of the body. *Nature Reviews Neuroscience*, *3*(8), 655–666

Craig, A. D. (2003). Interoception: the sense of the physiological condition of the body. *Current Opinion in Neurobiology*, *13*(4), 500–505

Critchley, H. D. (2004). The human cortex responds to an interoceptive challenge. *Proceedings of the National Academy of Sciences of the USA*, *101*(17), 6333–6334.

Critchley, H. D., Melmed, R. N., Featherstone, E., Mathias, C. J., & Dolan, R. J. (2002). Volitional control of autonomic arousal: a functional magnetic resonance study. *NeuroImage*, *16*(4), 909–919.

Davis, M. (1994). *Empathy: a social psychological approach*. Madison, WI: Brown & Benchmark.

de Quervain, D. J., Fischbacher, U., Treyer, V., Schellhammer, M., Schnyder, U., Buck, A., et al. (2004). The neural basis of altruistic punishment. *Science*, *305*(5688), 1254–1258.

Decety, J., & Jackson, P. L. (2004). The functional architecture of human empathy. *Behavioral Cognitive Neuroscience Review*, *3*(2), 71–100.

Devinsky, O., Morrell, M. J., & Vogt, B. A. (1995). Contributions of anterior cingulate cortex to behaviour. *Brain*, *118*(1), 279–306.

Egner, T., & Hirsch, J. (2005). Cognitive control mechanisms resolve conflict through cortical amplification of task-relevant information. *Nature Neuroscience*, *8*(12), 1784–1790.

Eisenberg, N., & Miller, P. A. (1987). The relation of empathy to prosocial and related behaviors. *Psychological Bulletin*, *101*(1), 91–119.

Etkin, A., Egner, T., Peraza, D. M., Kandel, E. R., & Hirsch, J. (2006). Resolving emotional conflict: a role for the rostral anterior cingulate cortex in modulating activity in the amygdala. *Neuron*, *51*(6), 871–882.

Farrell, M. J., Laird, A. R., & Egan, G. F. (2005). Brain activity associated with painfully hot stimuli applied to the upper limb: a meta-analysis. *Human Brain Mapping*, *25*(1), 129–139.

Fields, H. (2004). State-dependent opioid control of pain. *Nature Reviews Neuroscience*, *5*(7), 565–575.

Friston, K. J., Buechel, C., Fink, G. R., Morris, J., Rolls, E., & Dolan, R. J. (1997). Psychophysiological and modulatory interactions in neuroimaging. *NeuroImage*, *6*(3), 218–229.

Gallese, V. (2003). The roots of empathy: the shared manifold hypothesis and the neural basis of intersubjectivity. *Psychopathology*, *36*(4), 171–180.

Gallese, V., Keysers, C., & Rizzolatti, G. (2004). A unifying view of the basis of social cognition. *Trends in Cognitive Science*, *8*(9), 396–403.

Glover, G. H. (1999). Simple analytic spiral K-space algorithm. *Magnetic Resonance Medicine*, *42*(2), 412–415.

Grosbras, M. H., & Paus, T. (2006). Brain networks involved in viewing angry hands or faces. *Cerebral Cortex, 16*(8), 1087–1096.

Hebben, N., Corkin, S., Eichenbaum, H., & Shedlack, K. (1985). Diminished ability to interpret and report internal states after bilateral medial temporal resection: case H. M. *Behavioral Neuroscience, 99*(6), 1031–1039.

Hutchison, W. D., Davis, K. D., Lozano, A. M., Tasker, R. R., & Dostrovsky, J. O. (1999). Pain-related neurons in the human cingulate cortex. *Nature Neuroscience, 2*(5), 403–405.

Jackson, P. L., Brunet, E., Meltzoff, A. N., & Decety, J. (2006). Empathy examined through the neural mechanisms involved in imagining how I feel versus how you feel pain. *Neuropsychologia, 44*(5), 752–761.

Jackson, P. L., Meltzoff, A. N., & Decety, J. (2005). How do we perceive the pain of others? A window into the neural processes involved in empathy. *NeuroImage, 24*(3), 771–779.

Kampe, K. K., Frith, C. D., & Frith, U. (2003). Hey John": signals conveying communicative intention toward the self activate brain regions associated with "mentalizing," regardless of modality. *Journal of Neuroscience, 23*(12), 5258–5263.

Keltner, J. R., Furst, A., Fan, C., Redfern, R., Inglis, B., & Fields, H. L. (2006). Isolating the modulatory effect of expectation on pain transmission: a functional magnetic resonance imaging study. *Journal of Neuroscience, 26*(16), 4437–4443.

Lamm, C., Batson, C. D., & Decety, J. (2007). The neural substrate of human empathy: Effects of perspective-taking and cognitive appraisal. *Journal of Cognitive Neuroscience, 19*(1), 42–58.

Lanzetta, J. T., & Englis, B. G. (1989). Expectations of cooperation and competition and their effects on observers vicarious emotional responses. *Journal of Personality and Social Psychology, 56*(4), 543–554.

Lorenz, J., Minoshima, S., & Casey, K. L. (2003). Keeping pain out of mind: the role of the dorsolateral prefrontal cortex in pain modulation. *Brain, 126*(5), 1079–1091.

Mesulam, M. M., & Mufson, E. J. (1982). Insula of the old world monkey. I. Architectonics in the insulo-orbito-temporal component of the paralimbic brain. *Journal of Comparative Neurology, 212*(1), 1–22.

Mitchell, J. P., Heatherton, T. F., & Macrae, C. N. (2002). Distinct neural systems subserve person and object knowledge. *Proceedings of the National Academy of Sciences of the USA, 99*(23), 1523815243.

Mitchell, J. P., Macrae, C. N., & Banaji, M. R. (2004). Encoding-specific effects of social cognition on the neural correlates of subsequent memory. *Journal of Neuroscience, 24*(21), 4912–4917.

Mitchell, J. P., Macrae, C. N., & Banaji, M. R. (2005). Forming impressions of people versus inanimate objects: social-cognitive processing in the medial prefrontal cortex. *NeuroImage, 26*(1), 251–257.

Morrison, I., Lloyd, D., di Pellegrino, G., & Roberts, N. (2004). Vicarious responses to pain in anterior cingulate cortex: is empathy a multisensory issue? *Cognitive Affective, & Behavioral Neuroscience, 4*(2), 270–278.

Ochsner, K. N., Knierim, K., Ludlow, D. H., Hanelin, J., Ramachandran, T., Glover, G., et al. (2004a). Reflecting upon feelings: an fMRI study of neural systems supporting the attribution of emotion to self and other. *Journal of Cognitive Neuroscience, 16*(10), 1746–1772.

Ochsner, K. N., Ray, R. D., Cooper, J. C., Robertson, E. R., Chopra, S., Gabrieli, J. D., et al. (2004b). For better or for worse: neural systems supporting the cognitive down- and up-regulation of negative emotion. *NeuroImage, 23*(2), 483–499.

Ochsner, K. N., Zaki, J., Hanelin, J., Ludlow, D. H., Knierim, K., Ramachandran, T., et al. (under review). *Your pain or mine? Common and distinct neural systems supporting the perception of pain in self and others.*

Paus, T. (2001). Primate anterior cingulate cortex: where motor control, drive and cognition interface. *Nature Reviews Neuroscience, 2*(6), 417–424.

Pelphrey, K. A., Morris, J. P., & McCarthy, G. (2004). Grasping the intentions of others: the perceived intentionality of an action influences activity in the superior temporal sulcus during social perception. *Journal of Cognitive Neuroscience, 16*(10), 1706–1716.

Peyron, R., Garcia-Larrea, L., Gregoire, M. C., Costes, N., Convers, P., Lavenne, F., et al. (1999). Haemodynamic brain responses to acute pain in humans: sensory and attentional networks. *Brain, 122*(9), 1765–1780.

Peyron, R., Laurent, B., & Garcia-Larrea, L. (2000). Functional imaging of brain responses to pain. A review and meta-analysis. *Neurophysiologic Clinique, 30*(5), 263–288.

Rainville, P. (2002). Brain mechanisms of pain affect and pain modulation. *Current Opinion in Neurobiology, 12*(2), 195–204.

Rainville, P., Duncan, G. H., Price, D. D., Carrier, B., & Bushnell, M. C. (1997). Pain affect encoded in human anterior cingulate but not somatosensory cortex. *Science, 277*(5328), 968–971.

Rushworth, M. F., Walton, M. E., Kennerley, S. W., & Bannerman, D. M. (2004). Action sets and decisions in the medial frontal cortex. *Trends in Cognitive Science, 8*(9), 410–417.

Saarela, M. V., Hlushchuk, Y., Williams, A. C., Schurmann, M., Kalso, E., & Hari, R. (2007). The compassionate brain: Humans detect intensity of pain from another's face. *Cerebral Cortex, 17*, 230–237.

Saxe, R., Xiao, D. K., Kovacs, G., Perrett, D. I., & Kanwisher, N. (2004). A region of right posterior superior temporal sulcus responds to observed intentional actions. *Neuropsychologia, 42*(11), 1435–1446.

Seminowicz, D. A., & Davis, K. D. (2007). Interactions of pain intensity and cognitive load: The brain stays on task. *Cerebral Cortex, 17*(6), 1412–1422.

Singer, T., & Frith, C. (2005). The painful side of empathy. *Nature Neuroscience, 8*(7), 845–846.

Singer, T., Seymour, B., O'Doherty, J., Kaube, H., Dolan, R. J., & Frith, C. D. (2004). Empathy for pain involves the affective but not sensory components of pain. *Science*, *303*(5661), 1157–1162.

Singer, T., Seymour, B., O'Doherty, J. P., Stephan, K. E., Dolan, R. J., & Frith, C. D. (2006). Empathic neural responses are modulated by the perceived fairness of others. *Nature*, *439*(7075), 466–469.

Steiger, J. H. (1980). Tests for comparing elements of a correlation matrix. *Psychological Bulletin*, *85*, 211–229.

Valet, M., Sprenger, T., Boecker, H., Willoch, F., Rummeny, E., Conrad, B., et al. (2004). Distraction modulates connectivity of the cingulo-frontal cortex and the midbrain during pain—An fMRI analysis. *Pain*, *109*(3), 399–408.

Vogt, B. A. (2005). Pain and emotion interactions in subregions of the cingulate gyrus. *Nature Reviews Neuroscience*, *6*(7), 533–544.

Vogt, B. A., & Pandya, D. N. (1987). Cingulate cortex of the rhesus monkey: II. Cortical afferents. *Journal of Comparative Neurology*, *262*(2), 271–289.

Vogt, B. A., Vogt, L., & Laureys, S. (2006). Cytology and functionally correlated circuits of human posterior cingulate areas. *NeuroImage*, *29*(2), 452–466.

Vollm, B. A., Taylor, A. N., Richardson, P., Corcoran, R., Stirling, J., McKie, S., et al. (2006). Neuronal correlates of theory of mind and empathy: a functional magnetic resonance imaging study in a nonverbal task. *NeuroImage*, *29*(1), 90–98.

Wager, T. (2005a). Neural bases of placebo effects in pain. *Current Directions in Psychological Science*, *14*(4), 175–179.

Wager, T., & Feldman Barrett, L. (2004). From affect to control: Functional specialization of the insula in motivation and regulation. *PsychExtra*.

Wager, T. D. (2005b). Expectations and anxiety as mediators of placebo effects in pain. *Pain*, *115*(3), 225–226.

Wager, T. D., Rilling, J. K., Smith, E. E., Sokolik, A., Casey, K. L., Davidson, R. J., et al. (2004). Placebo-induced changes in fMRI in the anticipation and experience of pain. *Science*, *303*(5661), 1162–1167.

Wispe, L. (1986). The distinction between sympathy and empathy: To call forth a concept, a word is needed. *Journal of Personality & Social Psychology*, *50*(2), 314–321.

Zambreanu, L., Wise, R. G., Brooks, J. C., Iannetti, G. D., & Tracey, I. (2005). A role for the brainstem in central sensitization in humans. Evidence from functional magnetic resonance imaging. *Pain*, *114*(3), 397–407.

SOCIAL NEUROSCIENCE, 2007, 2 (3–4), 292–306

Psychology Press
Taylor & Francis Group

Empathy and enduring depersonalization: The role of self-related processes

Emma J. Lawrence
Institute of Psychiatry, King's College London, London, UK

Philip Shaw
National Institute of Mental Health, Bethesda, Maryland, USA

Dawn Baker, Maxine Patel, Mauricio Sierra-Siegert, Nicholas Medford, and Anthony S. David
Institute of Psychiatry, King's College London, London, UK

Empathy has two key components: affective and cognitive. It relies on "embodied" processes such as the generation, representation and perception of feeling states. People diagnosed with Depersonalization Disorder (DPD) report disturbances in affective experience, such as emotional numbing, alongside aberrations in "body image" such as increased self-focus and feelings of "disembodiment". DPD therefore provides a test bed for the role of such self-related processes in empathy. We tested 16 participants diagnosed with DPD and 48 control volunteers on measures of cognitive and affective empathy. We used self-report measures (EQ; Baron-Cohen & Wheelwright, 2004), an objective measure of cognitive empathy—the "Eyes" task (Baron-Cohen, Wheelwright, Hill, Raste, & Plumb, 2001), and a novel task tapping affective empathy, utilizing speech rate as an implicit measure of physiological arousal. We also measured participants' tendency to use mental representations that relate to the self during the affective empathy task. The DPD group showed intact performance on the cognitive empathy task. However, there was a disruption in the physiological component of affective empathy alongside a more pronounced reliance on mental representations of the self. These findings suggest affective empathy to be reliant on intact emotional experience in the observer. In addition, excessive self-focus may be detrimental to an empathic response.

INTRODUCTION

Empathy is a multi-dimensional construct (Davis, 1980; Decety & Jackson, 2004; Lawrence, Shaw, Baker, Baron-Cohen, & David, 2004) encompassing: (1) cognitive empathy—"understanding and predicting someone else's mental state"; and (2) affective empathy—"*experiencing* an [appropriate] emotion as the result of someone else's mental state" (Baron-Cohen et al., 2004, emphasis added). However, affective empathy can be further fractionated into parallel and reactive empathy. At its simplest, a person may be in a state of *parallel* affective empathy (Davis, 1994) when they experience an emotion that is congruent to that of a protagonist. However, it is important to distinguish empathic states from simple emotion contagion, which may be present

Correspondence should be addressed to: E. J. Lawrence, Section of Cognitive Neuropsychiatry, Department of Psychological Medicine, Institute of Psychiatry, King's College London, London SE5 8AF, UK. E-mail: e.lawrence@iop.kcl.ac.uk

EL was supported by an MRC studentship during this study.
We would like to thank for the volunteers for giving up their time to take part in this project.

www.psypress.com/socialneuroscience DOI:10.1080/17470910701391794

in non-empathic responses to other people's mental states. For instance, a person could feel distress that is personally oriented or egocentric, i.e., anxiety (Batson, Early, & Salvarani, 1997; Davis, 1980, 1994; Lamm, Batson, & Decety, 2007) upon perceiving a negative state in another. In this instance, such emotional contagion, even if congruent, would not constitute an empathic response. Reactive empathy, on the other hand, refers to forms of affective empathy such as sympathy and compassion, which go beyond the simple matching of affect (Davis, 1994).

Depersonalization

In common usage, the term depersonalization refers to a subjective sense of unreality in one's self (Hunter, Phillips, Chalder, Sierra, & David, 2003). This may encompass the following phenomena: (1) "emotional numbing," which involves a disruption to the qualia of affective states; (2) "disembodiment" or disturbances in "body image"[1] (Gallagher, 2005) characterized by a "non-specific feeling of not being in the body"; and (3) derealization—the sense that the world and objects around are unreal (Sierra, Baker, Medford, & David, 2005). It can occur in healthy individuals often as a transitory state during times of extreme stress (Noyes & Kletti, 1977). There may also be a higher incidence of depersonalization in particularly individualistic cultures (Sierra et al., 2006b). Depersonalization as a disorder (DPD) is diagnosed when these feelings are severe and enduring, cannot be attributed to any underlying condition, and occur alongside functional impairment (Baker et al., 2003).

Experimental data are consistent with subjective reports of "emotional numbing" in DPD. Functional magnetic resonance imaging (fMRI) has detected differential neural activation in response to aversive scenes compared with that displayed by both healthy and clinical controls (Phillips et al., 2001). People who report DPD did not show the increased blood flow to the anterior insula seen in other groups. The anterior insula has been implicated in interoceptive awareness and the representation of feeling states (Critchley, Wiens, Rotshtein, Öhman, & Dolan, 2004; Pollatos, Gramman, & Schandry, 2007). Lack of anterior insula response to emotional material has since been replicated in DPD during an emotion-processing paradigm (Medford, Brierley, Brammer, Bullmore, & David, 2006). This was alongside hyperactivity in areas of the ventrolateral prefrontal cortex, which is thought to be involved in the suppression of emotion in healthy volunteers (Ochsner & Gross, 2005). In addition, a factor analysis of a rating scale devised to measure DPD symptomatology, revealed items relating to the awareness of visceral states such as hunger and thirst to load onto "emotional numbing" in people with DPD (Sierra et al., 2005). This suggests there may be a disruption to interoceptive processing in this condition (Wiens, 2005).

In contrast to healthy and clinical controls, skin conductance responses (SCR) in people with DPD are attenuated when viewing aversive images, matching subjective arousability and intensity ratings (Sierra et al., 2002). This suggests that the "emotional numbing" reported may not be due to a failure in interoception alone (Mauss, Wilhelm, & Gross, 2004). This attenuated autonomic arousal also transfers to person-based stimuli of negative valance, i.e., facial expressions depicting disgust (Sierra, Senior, Phillips, & David, 2006b). Consistently, some studies have also found lowered noradrenalin and cortisol levels in DPD (Simeon et al., 1998; Stanton et al., 2001). These data fit with the idea that DPD is related to an ordinarily protective response to threat in the environment, which dampens the potentially overwhelming affect associated with trauma or stress, and in turn, allows a person to remain alert and vigilant (Sierra & Berrios, 1998). One study with healthy volunteers illustrates the potentially protective nature of such "detachment," which was found to reduce pain reactivity (Kalisch et al., 2005). However, it is thought that this mechanism somehow goes awry in DPD both in terms of intensity and duration.

Finally, there is also the suggestion of aberrations in aspects of body image in DPD, such as increased self-focused attention (Hunter et al., 2003; Michal, Kaufhold, Overbeck, & Grabhorn, 2006). In support of this being of etiological importance, promising results have been found for a treatment program that includes a strategy to reduce self-focus (Hunter et al., 2003, 2005). Indirect evidence for the role of excessive self-focus in DPD can also be found in the psychological literature (Silvia & Abele, 2002). In one study with healthy volunteers, increased self-awareness was implicitly induced using mirrors

[1] Defined as a "system of perceptions, attitudes and beliefs pertaining to one's own body" (Gallagher, 2005).

and self-narrative, and was found to reduce the reported intensity of emotional experiences during mood induction.

In terms of empathy, theoretical links can be drawn between the phenomenology of affective empathy and the disturbances reported in depersonalization. Buddhist texts have long advocated that the ability to fully engage with one's own pain increases empathic capacity. By definition, affective empathy includes an emotional component, so disruptions in the generation, representation or perception of one's own feeling states may directly impact upon empathy. In addition, recent models of this component of empathy implicate self-related processes as key (Decety & Grezes, 2006) and so disruptions in the way self-related information is processed may impact on the empathic response. In contrast, no problems are anticipated with the cognitive aspect of empathy such as mental state attribution in DPD. DPD may therefore provide a clinical model from which to explore the various components of empathy. Indeed, early data suggest that people with DPD report problems with empathy as it is "applied" in social situations reflecting social competence (Lawrence et al., 2004). Consistently, this group also report a higher incidence of social isolation (Michal et al., 2006).

Empathy and the role of affect

Affective empathy is embodied insofar as it relies on aspects of body image (Gallagher, 2005) such as the representation and perception of one's own feeling states. However, it less clear whether simply having a *mental* representation of another person's affective state, i.e., a form of cognitive empathy, is embodied in this way. Adolphs, Damasio, Tranel, Cooper, & Damasio (2000) found that lesions to the right somatosensory cortices lead to impaired emotion recognition indicative of the role of somatosensory representations in retrieving conceptual knowledge about affective states. However, they also point out that "there is no need for the individual to *consciously* feel the state of the other" [emphasis added] during this form of cognitive empathy. Similarly, people with pure autonomic failure can label emotional expressions in line with controls but have difficulty with more complex scripted emotion attribution tasks (Heims, Critchley, Dolan, Mathias, & Cipolotti, 2004). The suggestion here is that empathic bodily responses, whether overt or covert, may be necessary for emotion perception, particularly with more subtle affective states.

Empathy is a socially valued construct and so its assessment may be problematic when using explicit methods prone to social desirability effects. This can be avoided by using implicit measures of psychophysiology and/or measuring responses not amenable to conscious control, i.e., involuntary physiological responses such as pupil size (Harrison, Singer, Rotshtein, Dolan, & Critchley, 2006). An alternative implicit method of measuring physiological arousal is speech rate. Vocal expression is complex and varies in intensity, loudness, pitch, frequency and tempo (Cook, 2003). There are many studies examining the perception of affective states via prosody (Kucharska-Pietura, Phillips, Gernard, & David, 2003; Ross, Thompson, & Yenkosky, 1997) but less using vocal cues as an index of the emotional state of the speaker. Evidence suggests that we speak faster when scared or anxious and more slowly when sad or depressed (see Siegman & Boyle, 1993). In addition, Breitenstein, Van Lancker, & Daum (2001) systematically manipulated speech rate and pitch for five emotions and found a faster rate to be associated with fear and anger and a lower rate to be associated with sadness. A review of emotion in voice and music confirmed these patterns, and found that articulation rate also increased when the speaker was happy (Cook, 2003; Scherer, 1995).

Teasdale and Fogarty (1979) tested control volunteers using mood induction and a counting task to assess speech rate. They observed significant differences for mood on speech rate. McKenna and Lewis (1994) asked participants to count both at normal speech rate and as fast as they could, before and after a mood-induction procedure. They found that mood induction affected both maximal and self-paced speech rate. Lastly, Barrett and Paus (2002), using a well-controlled mood-induction technique, found participants to have slower, quieter and more monotonous speech when sad. The utility of the speech rate method is also supported by clinical studies that employed it as an indirect measure of depressive symptomatology (Teasdale, Fogarty et al., 1980) and responsiveness to treatment (Godfrey & Knight, 1984).

Empathy and mental representations of the self

The theory of simulation: the "attempt to replicate, mimic or impersonate the mental life of the target" (Gallese & Goldman, 1998) has gained support from the discovery of neurons that respond to both observing and executing an action (Gallese & Goldman, 1998). Much evidence has since accumulated to suggest that at a neuronal level such "shared representations" are also employed in decoding other people's mental and affective states (Decety & Sommerville, 2003). In addition, Davis, Conklin, Smith, and Luce (1996) found that this occurs at a cognitive level during perspective taking, with participants constructing mental representations of the target that closely resembled that of the self. This strategy is also thought to occur spontaneously, in naturalistic settings, in the absence of explicit perspective-taking instructions (Davis et al., 2004).

However, if we rely on similar neural and cognitive processes to experience, communicate and perceive mental and affective states, how we distinguish between self and other becomes important (Kilner, Paulignan, & Blakemore, 2003). This "sharedness" cannot be absolute (Decety & Jackson, 2004; Ramnani & Miall, 2004; Ruby & Decety, 2004), as an excess of overlap between mental representations of the self and other could lead to conceptual confusion (Decety & Grezes, 2006; Decety & Sommerville, 2003; Lawrence et al., 2006).

Davis (1994) points out that perspective taking can prime people for at least two different kinds of resultant affective state: (1) empathy, sympathy and compassion; and/or (2) feelings of unease and personal distress. This distinction is rarely considered within social neuroscience literature, but see (Lamm et al., 2007). Batson et al. (1997) found evidence to suggest that these outcomes stem from distinct cognitive representations. *Imaging how the other feels* was found to lead to feelings of empathy, while *imaging how you would feel* led to feelings of both empathy and personal distress. However, it is unclear whether this study truly overcomes the potential confound of experimenter demands inherent in a study using explicit perspective-taking instructions. Batson et al. (2003) found further evidence that imagining *oneself* in another's position was not as effective in inducing an empathic response, as imagining the *other*. As Decety and Grezes (2006) point out, it appears that "agency is a crucial aspect in promoting a *selfless* regard for the other rather than a *selfish* desire to escape aversive arousal."

A helpful model that further specifies this distinction can be found within the psychopathology literature. Langdon, Coltheart, Ward, & Catts (2001) define allocentric simulation as the ability to reconcile one's own first-person view of the world with other people's viewpoints and *on an equal footing*. Egocentric simulation, however, is described as being tied to one's own point of view. Similarly, Vogeley and Fink (2003) define an allocentric viewpoint as a framework that is independent from the agent's position (see Figure 1).

The idea that shared perspectives can be egocentric is not new. Feningstein and Abrams (1993) found that people are more likely to assume that "others" think the same way as them when there is an increase in self-focused attention. But this has to be balanced with the known increase in empathy than can occur when the target is in a similar situation to the self (Houston, 1990). Nevertheless, Lawrence et al. (2006), using a version of a previously validated trait-overlap task (Davis et al., 1996), found that less overlap between the mental representations of self and other was associated with increased accuracy in a dynamic social-perception task. A further test of this may be afforded by the study of people with DPD who show an increased focus on self-qualia alongside perceived disruptions in empathic responses to others.

Aims

To explore whether disturbances in affect, as reported in DPD, influence the capacity to

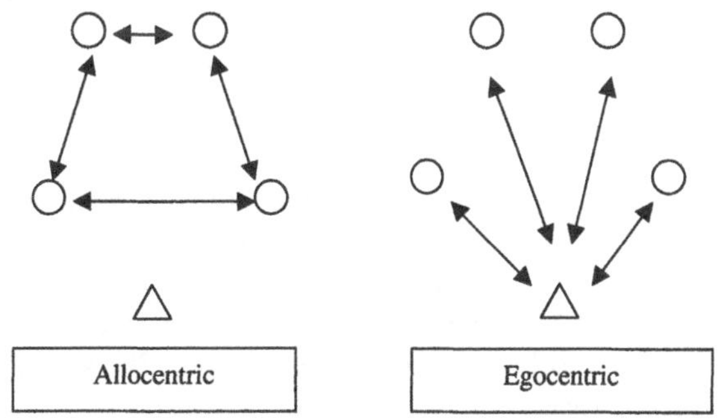

Figure 1. Allocentric and egocentric frames of reference (adapted from Vogeley & Fink, 2003).

experience empathy. If the participant is in a state of parallel affective empathy, it is expected that their speech rate will accelerate after processing "fearful" and "happy" situations and decelerate after "sad" material, reflecting physiological arousal congruent with the target. In addition, the reliance upon self-knowledge when attributing affective states will be measured using an adapted version of the trait paradigm devised by Davis et al. (1996). Self-report measures of empathy and personal distress will also be administered along with an objective measure of mental state labeling, i.e., the Reading the Eyes in the Mind task (Baron-Cohen et al., 2001), which purports to tap cognitive empathy. Consistent with previous studies no mental state recognition deficit is expected.

In terms of group differences it is predicted that:

1. People with DPD will show comparable scores to controls on the 'Reading the Mind in the Eyes' task.
2. People with DPD will show reduced physiological responses to affective material indicative of a disturbance in affective empathy.
3. People with DPD will show inappropriate use of self-knowledge, consistent with increased self-focus, when attributing affective states to others.

METHOD

Participants

Controls. Forty-eight control volunteers were recruited via a circular e-mail at the Institute of Psychiatry (IOP), a database of volunteers maintained by staff at the Institute and postcards in local newsagents, local library and job centre. Twenty-six (54%) were recruited from academic or clinical staff and the remaining 22 (46%) people were recruited from the local area and non-mental-health workers. The occupational status of the sample was therefore varied.

The sample consisted of 23 (48%) men and 25 (52%) women. The mean age was 34.4 years ($\pm$11.2). Thirty-six (75%) reported their ethnicity as White European, with the remainder describing themselves as Afro-Caribbean, Black, and Mixed race.

Of the group, 6 (13%) were educated to secondary-school standard, 4 (8%) to further education, 13 (28%) to higher and 24 (50%) to postgraduate level; 21 (44%) people were single, 20 (42%) co-habiting or married and the remainder divorced, separated, widowed or with a non-cohabiting partner.

Depersonalization group. All participants in this group were diagnosed with DPD following a full psychiatric interview and assessment at the Depersonalization Research Unit at the Maudsley Hospital, London. A final clinical diagnosis was made according to DSM-IV (American Psychiatric Association, 1994) and ICD-10 using the Present State Exam or PSE (Wing, Cooper, & Sartorius, 1974). The PSE includes items for depersonalization and derealization: 0 = *not present*, 1 = *moderately intense or transient*, and 2 = *intense and persistent*. Our case definition required a total score of 2 or above without an obvious additional clinical diagnosis or prominent non-dissociative symptomatology.

Demographic details. The DPD group consisted of 4 (25%) women and 12 men (75%). The mean age was 32 years ($\pm$7.8). Of the participants, 11 (68.8%) people were single, 2 (12.5%) were co-habiting, 2 (12.5%) married and 1 (6.3%) was separated; 2 (12.5%) participants were educated to secondary level, 2 (12.5%) had completed further education, 9 (56.3%) had been educated to degree level and 3 (18.8%) had postgraduate qualifications. There were 3 (18.8%) students, 3 (18.8%) unemployed and the rest were employed in a variety of occupations.

Clinical characteristics. All participants in the DPD group completed the Dissociative Experiences Scale version II (DES; Bernstein & Putnam, 1986; Carlson & Putnam, 1993), the Cambridge State and Trait Depersonalization scales (Carlson & Putnam, 1993), the Beck Anxiety Inventory (BAI: Beck, Epstein, & Brown, 1988a) and Beck Depression Inventory (BDI: Beck, Steer, & Garbin, 1988b). These were sent to participants prior to referral to the clinic, and the information was entered into a comprehensive database (Baker et al., 2003). There were two instances of missing data due to participants' failure to return the questionnaires.

The DES is a 28-item self-report questionnaire with a cut-off score of 30 for severe dissociative disorders (Carlson & Putnam, 1993). Factor analysis has revealed three main components of the scale: "depersonalization/derealization

(DPD/DR)", "amnesia" for dissociative experiences, and "absorption" and imaginative involvement (Carlson et al., 1991). Eight items make up the DES-Taxon, which is sensitive to the detection of DPD with a cut-off score of 13 (Simeon et al., 1998). A score above 8 on the DP/DR scale being suggestive of a dissociative disorder. The BAI and BDI were also given to participants, due to the comorbidity observed between depersonalization disorder, depression and anxiety (Simeon et al., 1998). A score below 11 on either scale is considered within the "normal" range, whereas a score above 30 is classed as "severe."

The mean DES score for the DPD group was 24.1 ($\pm$12.6), with the mean DES taxon score being 23.9 ($\pm$15.4) and the DES DP/DR factor being the highest at 37.8 ($\pm$21.4). This suggest the participants in the DPD group scored in the clinical range for the DES taxon and the DP/DR factor, but not on the DES overall. This is consistent with previous data reported for a larger DPD group, which included some of the current sample (Lawrence et al., 2004) and other studies (Lambert, Senior, Fewtrell, Phillips, & David, 2001).

The Cambridge Depersonalization "trait" and "state" self-report questionnaires measure depersonalization independently from other dissociative disorders. The trait scale measures the frequency and duration of DPD symptoms during the last six months, while the state scale simply measures the frequency of symptoms over the last week. The median trait score in the original paper was 113 and the cut-off point for depersonalization is >70. The median score in this sample was 130 (range 0–270).

The mean score on the BAI was 22 ($\pm$9) and for the BDI it was 22.8 ($\pm$10.7). This therefore suggests that this group suffered mild to moderate depression and anxiety, mirroring the characteristics of the larger cohort reported elsewhere (Lambert et al., 2001).

Materials

Empathy measures. The EQ is a self-report measure purporting to tap empathy. There are 60 item including 20 filler questions. Responses are given on a 4-point scale ranging from "*strongly agree*" to "*strongly disagree*". Participants received 0 for a "non-empathic" response, whatever the magnitude, and 1 or 2 for an "empathic response" depending on the strength of the reply and so the total score is out of 80. Principal components analysis of the EQ has derived the existence of 3 factors (Lawrence et al., 2004):

1. *Cognitive empathy*: I am good at predicting how someone will feel.
2. *Emotional reactivity*: seeing people cry doesn't really upset me [*reversed scored*].
3. *Social skills*: I find it hard to know what to do in a social situation.

This factor structure has since been replicated in a different laboratory (Muncer & Ling, 2006). The EQ does not, however, distinguish between affective empathy and non-empathic states such as personal distress, hence the use of more accurate label of emotional reactivity as opposed to affective empathy *per se*. For this reason, a selection of items from the "personal distress" scale of the Interpersonal Reactivity Index (Davis, 1980) was also administered, for example:

1. Do you feel apprehensive and ill at ease in emergency situations?
2. When you see someone get hurt do you tend to remain calm?
3. Do you feel helpless when you are in the middle of a very emotional situation?

These items were scored categorically as yes/no responses.

Counting task. A novel task was developed to tap physiological arousal after reading an emotive diary extract. In an attempt to increase ecological validity (Parkinson & Manstead, 1993), 4 vignettes were adapted from real-life diary extracts displayed on a website called www.opendiary.com (see Appendix 1). There are 3 vignettes with affective content and one control vignette. The protagonists were two men and two women and the age range was varied. To control for practice and other order effects, the final vignettes were counterbalanced using a 4×4 Latin square.

Speech rate was measured using a desk-mounted microphone and a portable recording device. Participants were asked to count from 1 to 10 immediately after reading the vignette. Participants were next required to answer questions about the vignette before moving onto the next one (data not reported). This lessened the likelihood of carry-over effects. Post-experimental questioning revealed that many participants thought the counting task was intended as a distraction and none had guessed its real purpose.

This suggests social desirability effects were minimal.

As speech rate is known to vary between individuals this variable was calculated within participants. The time taken to count from 1 to 10 was converted to a percentage of the baseline speech rate for each participant, namely the control vignette speech rate. Piloting revealed that people count more quickly as they move though the task, which was controlled by counter-balancing.

Trait task. The paradigm used by Davis et al. (1996) was adapted for use in the current study. Rather than relying on freely generated trait lists, the "Trait Adjective Checklist" (Loehlin & Nichols, 1976), consisting of 159 traits presented in alphabetical order, was used in the pilot study.

Piloting. The main change resulting from piloting was a reduction in the number of traits used. The list was pared down to 75 adjectives by using the traits that were endorsed most frequently while ensuring an even balance of positively and negatively valenced traits.

At least 24 hours before the experimental session (although in most cases the interval was longer) participants were presented with a list of traits and asked to tick those that best described themselves. During the experimental session, participants were then asked to endorse traits that best describe the protagonist depicted in the vignette (see Appendix 2).

Coding. The number of traits that the participant had previously endorsed for themselves and later attributed to the target was calculated and a mean was then calculated for all 3 affective state vignettes. The proportion overlap was calculated according to Davis et al. (1996): (1) percentage of the "self" traits that were later endorsed for the target (self-overlap); and (2) percentage of a "target" traits that were earlier endorsed for the self (target-overlap).

Eyes test. In the "Reading the Mind from the Eyes Test" (Baron-Cohen et al., 2001) participants are required to look at a photograph of an actor's eyes and choose one of four words that best describes what the actor is thinking or feeling. The task is one of mental state inference from socio-perceptual cues. One point was allocated for a correct answer yielding a total score out of 36.

RESULTS

There was a trend towards a significantly different gender distribution between the groups ($\chi^2 = 3.55$, $df = 1$, $p = .059$) with more men then women in the DPD group (3:1) and a near equal distribution of men and women in the control group (23:25). However, there was no difference in ages between the groups ($t = 0.91$, $df = 62$, $p > .05$). Educational achievement also differed with more controls having entered postgraduate education ($\chi^2 = 4.8$, $df = 1$, $p = .028$). In addition, controls had higher verbal IQs ($t = 2.2$, $df = 61$, $p = .031$) as estimated from the National Adult Reading Test (Nelson, 1982) although the DPD were in the above average range (117.8 ± 5.2). These observed demographic differences, i.e., gender and verbal IQ were adjusted for statistically. Verbal IQ was chosen over educational status, a continuous variable with more specificity. There was one case of missing verbal IQ in the DPD group, which was replaced with the group mean.

Self-report measures. Total scores on all factors of the empathy self-report measures are shown in Table 1. A 2×3 repeated measures ANOVA on EQ scores revealed a significant main effect for Group, $F(1, 62) = 5.8$, $p = .019$, and a trend towards an interaction between Group × EQ factor, $F(1, 62) = 2.5$, $p = .084$. Exploration of the means suggests these findings were in the main driven by the DPD group's lower score on the "social skills" factor.

A $2 \times 3 \times 2$ repeated measures ANCOVA was then conducted controlling for the gender and IQ

TABLE 1
Mean score on self-report measures

	Gender			
	Male		*Female*	
Group	*Mean*	*SD*	*Mean*	*SD*
CONTROL				
Cognitive empathy	4.9	1.8	6.8	1.8
Emotional reactivity	4.9	2.9	6.5	1.9
Social skills	6	2.5	6.4	2.5
Personal distress	2	1	2	1
DPD				
Cognitive empathy	5.2	3.5	6.5	2.6
Emotional reactivity	5	2.3	4.7	1.5
Social skills	4.2	2.4	3	2.6
Personal distress	2	1	1	1

differences between groups. The previously observed findings held (see Figure 2). A significant main effect for group was observed, $F(1, 59) = 4.2$, $p = .045$, alongside a trend towards an interaction between Group × EQ factor, $F(1, 118) = 2.5$, $p = .066$. There was no main effect for Gender, $F(1, 59) = 1.4$, $p > .05$, or Group × Gender, $F(1, 59) = 1.9$, $p > .05$, nor EQ factor × Gender, $F(2, 118) = 2.1$, $p > .05$, or a 3-way interaction between Group × EQ factor × Gender, $F(2, 118) = 1.3$, $p > .05$. A 2 × 3 ANCOVA confirmed the near significant interaction effect to be driven by the DPD group's lower score on the "social skills" factor, $F(1, 64) = 8.3$, $p = .006$. It is of note that no association was found between BAI scores, i.e., general anxiety, and this factor ($r = -.058$, $p = .85$). In addition, there was a trend for people with DPD to report *less* personal distress than controls, $F(1, 62) = 3.45$, $p = .069$. However, within this group such distress was positively related to BAI scores ($r = .56$, $p = .058$).

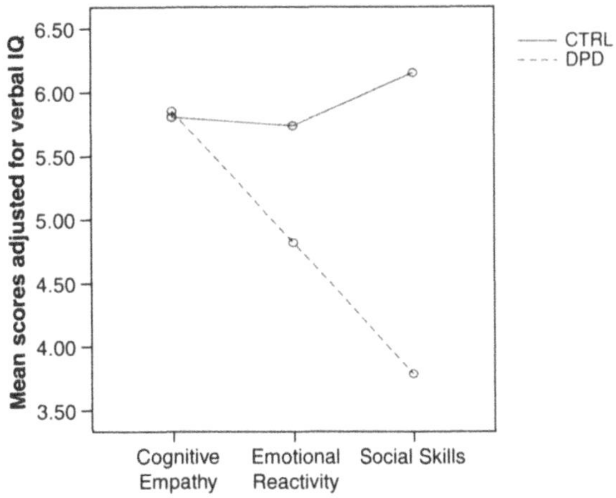

Figure 2. Mean scores on subscales of the EQ.

Mental state labeling. In the Eyes Task there was one instance of missing data in the DPD group.[2] As predicted, simple *t*-tests revealed no significant differences on this task ($t = -0.51$, $df = 61$, $p > .05$). However, when demographic variables were entered into the analysis a main effect of group was observed, $F(1, 62) = 4.4$, $p = .040$, suggesting slightly improved performance in the DPD group (mean DPD: 27.47 ± 3.2; control: 27.06 ± 3.5).

Speech rate. Due to technical issues there were two instances of missing data in the DPD group and one in the control group. One participant in the DPD group also had missing data for the fear vignette. The data were first examined for outliers and 2 extreme cases in the control dataset were dropped and the data log transformed. Simple analysis of the control data alone confirmed the observed speech-rate patterns to be in line with predictions. Paired *t*-tests revealed a significant difference between the speech rate in the control group following the "sad" and "happy" vignettes ($t = 2.325$, $df = 44$, $p = .025$), with the difference between the "sad" and "fear" vignette approaching significance ($t = -1.942$, $df = 44$, $p = .059$) but no difference between the "happy" and "fear" vignettes ($t = 0.370$, $df = 44$, $p > .05$). As predicted, speech rate decelerated after the sad vignette (despite a tendency to speed up over the course of the testing session) and accelerated with the fearful and happy vignettes.

A 3 × 2 repeated measures ANOVA with speech rate as the within-participant variable with 3 levels (sad/happy/fear) was conducted. The was no main effect for Speech Rate, $F(2, 112) = 2.2$, $p > .05$, or group, $F(2, 112) = 0.9$, $p > .05$, but there was a significant interaction between Speech × Group, $F(2, 112) = 6.5$, $p = .002$. Simple *t*-tests confirmed the difference to be on the "sad" vignette with the DPD group "speeding up" ($t = 2.1$, $df = 57$, $p = .045$) after this vignette (see Figure 3). A trend indicative of group differences was also observed on the "happy" vignette ($t = -1.9$, $df = 57$, $p = .072$).

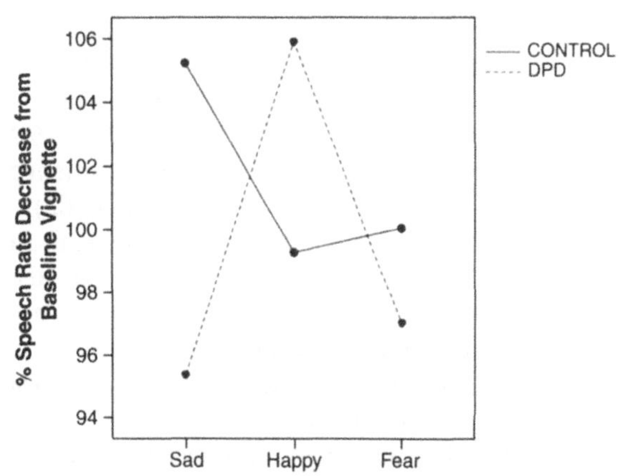

Figure 3. Mean speech rate adjusted for verbal IQ. *Note*: An increase on the *Y*-axis represents a decrease in speech rate from participant's individual baseline speech rate on the neutral vignette, which is represented by 100%.

[2] A participant in the control group scored in the range usually observed in those with autistic spectrum disorders on the Eyes task and so was removed as an outlier.

Next a 3 × 2 × 2 repeated measures ANCOVA was performed to adjust for gender and verbal IQ differences. The was no main effect for Speech Rate, $F(2, 112) = 0.8$, $p > .05$, nor an interaction between Speech Rate × Verbal IQ, $F(2, 112) = 0.8$, $p > .05$. The Speech Rate × Group interaction still held, $F(2, 112) = 4.5$, $p = .012$ (see Figure 2). There were, however, no main effects or interactions with gender. Univariate ANCOVA revealed the main difference between groups to be on the "sad" vignette, whereby it appears the DPD participants' speech rate accelerated.

Trait task. These data were analyzed using the same procedure as that adopted by Davis et al. (1996). Both the actual "*number of traits attributed to the self*" and the "*number of traits attributed to the targets*" provide an index as to the sheer number of traits a participant tends to endorse during the tasks. This variable is subject to individual variation as participants are free to endorse as many or as few as they like. These two indices were intercorrelated, so only the index with the strongest correlation with the dependant variable, i.e., overlap index, was co-varied in each analysis. This was "number of traits attributed to self" in the target-overlap (% of "target" traits that were earlier endorsed for the self) and "mean number of traits attributed to the target" in the self-overlap (% of "self" traits that were later endorsed for the target). Valence of vignette (neutral/sad/happy/fear) was also entered as a repeated-measures variable.

A 2 × 2 × 4 ANCOVA was conducted for "target-overlap" including gender and verbal IQ. There was a significant interaction between Group × Trait, $F(3, 168) = 5.1$, $p = .002$ (see Figure 4). Univariate ANCOVAs confirmed these differences to be on the neutral vignette with the DPD group displaying less overlap, $F(1, 62) = 6.7$, $p = .012$, and both vignettes of negative valence, sad: $F(1, 62) = 4.1$, $p = .047$ and fear: $F(1, 62) = 5.5$, $p = .022$, with the DPD group displaying more overlap. An interaction with gender was also observed on the sad vignette, $F(1, 62) = 4.1$, $p = .048$, with the main difference being driven by males. No differences were observed on the vignette of positive valence.

As an additional test of construct validity, univariate ANCOVAs were also conducted without traits that could feasibly stem from both comorbid anxiety and depression in the DPD group, and the emotion displayed by the target in

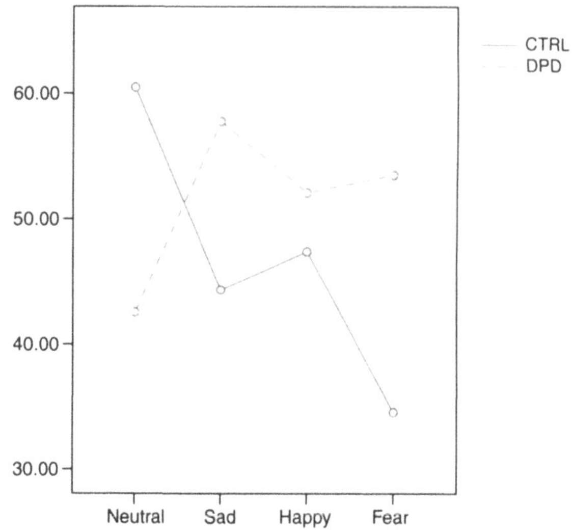

Figure 4. Percentage of "target" traits that were also endorsed for the self (adjusted for verbal IQ and number of traits attributed to the self).

the sad and fearful vignettes.[3] The pattern of findings remained for both the fear, $F(1, 62) = 3.74$ $p = .058$, and sad vignette, $F(1, 62) = 5.45$, $p = .023$, which again was influenced by gender, $F(1, 62) = 4.59$, $p = .036$. The magnitude of the effect for fear was slightly reduced reflecting the fact that the negative traits available for endorsement relate more to fearful than sad emotional states.

The same analysis was followed for the "self-overlap," i.e., the percentage of the "self" traits that were later endorsed for the target. However, the association between this variable and the "mean number of traits" and "number of self traits", necessary covariates to control for variations in the number of traits endorsed in each condition, differed between groups preventing full analysis.

DISCUSSION

As predicted, people who describe feeling continually depersonalized self-report less empathic ability overall, and less egocentric feelings of personal distress, than non-depersonalized volunteers. Despite this, they were as accurate as controls at labeling mental states from photographs of the eye region. However, speech-rate patterns in this group were indicative of a lack of

[3] There were seven such traits being: anxious, fearful, moody, nervous, tense, timid and worrying, which were removed from this analysis.

congruent physiological arousal in response to emotive narratives. In addition, the DPD group displayed more overlap between mental representations of the self and target than controls, during an emotion attribution task. This may indicate an increased self-focus when attributing affective states to others.

The DPD group self-reported less empathic ability overall as evident from their total EQ scores. As shown previously, the difference appeared to be most clear on the "social skills" factor of the EQ, which measures social competence (Lawrence et al., 2004). It is possible that this relates to social anxiety in this group, however, no association was observed with general anxiety ratings on the Beck Anxiety Inventory (Beck et al., 1988a). The DPD group also reported less "personal distress" in response to others' emotional states, which is consistent with described emotional numbing. This also fits with the idea that, in this group, the emotional numbing reported may to some extent be specific to external or social stimuli of negative valence (Sierra et al., 2006). It is of note, however, that a positive association was also observed between personal distress and BAI ratings in this group, confirming the construct validity of these two measures.

Turning to the "Reading the Mind in the Eyes" task, the finding that the DPD group performed as well as controls, suggests this group can competently decode complex mental states. This is consistent with previous studies that suggest no deficit in emotion recognition (Phillips et al., 2001; Sierra et al., 2006). Given previously reported reductions in both objective and subjective measures of arousal in response to faces depicting mental states of negative valence (Sierra et al., 2006), these data are consistent in suggesting that arousal, whether overt or covert, is not a prerequisite for success in emotion labeling tasks. It may, however, be important for intensity ratings, which were not tested by the current paradigm (Adolphs et al., 2000). In addition, these data suggest superior performance by the DPD group on this task, once IQ and gender were taken into account. These data are also in line with the performance of people diagnosed with psychopathy, who have been shown to do well on this paradigm, despite attenuated physiological responses to the distress of others (Richell et al., 2003).

The main finding of interest in this study is that in contrast to controls, people with DPD display differential physiological reactions to other people's affective states. Whereby control participants displayed speech-rate changes in line with predictions, the DPD group showed arousal levels (as indexed by speech rate) contrary to the predicted patterns. In contrast to controls, the DPD group's speech rate tended to accelerate after the sad vignette, and slow down after the happy vignette. These patterns are indicative of a lack of congruent physiological arousal, and hence parallel empathy, and indicate that the DPD group was not emotionally "in tune" with the protagonist.

However, this group did display some changes in arousal as their speech rate did not remain constant from vignette to vignette, or simply attenuated, as predicted by pure emotional numbing. This suggests that the emotional numbing reported is more complex than a simple lack of arousal. For instance, that the DPD group accelerated after the sad vignette, a speech-rate pattern associated with increased physiological arousal, may instead reflect a state of anxiety (Michal et al., 2006) in response to the perception of a negative state in another person. However, the reduced self-reported personal distress in this group is inconsistent with this interpretation. One possible explanation is that this speech-rate pattern represents the physiological correlates of feelings of unease and distress, but that a failure in interoceptive processing leads to self-reports of a reduction in such states or "emotional numbing." Further studies are needed to ascertain the point at which the emotional experience becomes blunted. A slowing in speech rate, indicative of reduced arousal, after the happy vignette is also difficult to decipher, and highlights a limitation of this and other current measures of physiological arousal, i.e., that they fail to usefully discriminate valence. This also prevents the exploration of states of reactive empathy.

In terms of the validity of the speech-rate task, however, it is of note that the predicted speech-rate pattern was also observed in participants diagnosed with Asperger's syndrome who are thought not to have problems with the physiological components of affective empathy (Lawrence, 2004). However, further validation using explicit measures such as SCR would add weight to the utility of this task as an implicit gauge of physiological arousal.

Turning to the mental representation accompanying these arousal patterns, the personality traits endorsed by the DPD group suggest consistently more overlap between representations of the self, and of the target, during affective state attribution, than observed in controls. This pattern was specific to the vignettes of negative valence. These effects were also independent of verbal IQ, gender, and the number of traits a participant chose to endorse. This excess of overlap may be best explained with reference to the "egocentric/allocentric" distinction with additional overlap representing "egocentric" simulation (Langdon et al., 2001). That less overlap was shown in the DPD group on the neutral vignette, and no group difference was observed after reading the happy narrative, suggests that this strategy is a response to negative affective states.

These data indicate that for people reporting symptoms of DPD, perceiving the negative states of others may rely more extensively on self-knowledge than is the case for controls. It is interesting to note that such an egocentric mental representation is ordinarily thought to result in feelings of personal distress (Batson et al., 1997; Batson, Fultz, & Schoenrade, 1987). However, excessive focus on the self has also been found to be related to "emotional numbing" (Silvia & Abele, 2002). These data taken together seem to suggest that the reported lack of "personal distress" in the DPD group, alongside a self-focused mental representation of the target mental state, may be due to the effects of "emotional numbing."

These results are also in line with the increased self-focused attention reported in DPD (Hunter et al., 2003) and are consistent with therapeutic interventions geared towards defocusing the client from aspects of the self. The psychological literature from healthy controls (Silvia & Abele, 2002) suggests a defocusing strategy may have the indirect result of reinstating the intensity of affective experience, which may in turn impact on the empathic response. The prediction is that an allocentric viewpoint would be more likely to foster an empathic response. This is again also consistent with previous studies, which suggest imagining how the other feels is more effective at promoting an empathic response than imagining how the self would feel (Batson et al., 1987, 1997).

This study is subject to various limitations, for instance the group differences in demographic variables such as IQ and gender. Nevertheless, the data suggest that DPD provides a useful clinical test of models of empathy. The cognitive components of empathy such as mental state labelling appear to be intact, alongside deficits in experiencing the physiological aspects of empathy and other forms of emotional contagion, consistent with reported disruptions in self-affectivity. In addition, further disturbances such as increased self-focus appear to influence the empathic process.

REFERENCES

Adolphs, R., Damasio, H., Tranel, D., Cooper, G., & Damasio, A. R. (2000). A role for somatosensory cortices in the visual recognition of emotion as revealed by three-dimensional lesion mapping. *Journal of Neuroscience, 20*(7), 2683–2690.

American Psychiatric Association. (1994). *Diagnostic and statistical manual of mental disorders* (4th ed.). Washington, DC: Author.

Baker, D., Hunter, E., Lawrence, E., Medford, N. C., Patel, M., Senior, C., et al. (2003). Depersonalization disorder: Clinical features of 204 cases. *British Journal of Psychiatry, 182*, 428–433.

Baron-Cohen, S., & Wheelwright, S. (2004). The Empathy Quotient: An investigation of adults with Asperger's syndrome or high functioning autism, and normal sex differences. *Journal of Autism & Developmental Disorders, 34*(2), 163–175.

Baron-Cohen, S., Wheelwright, S., Hill, J., Raste, Y., & Plumb, I. (2001). The Reading the Mind in the Eyes Test revised version: A study with normal adults, and adults with Asperger syndrome or high-functioning autism. *Journal of Child Psychology & Psychiatry & Allied Disciplines, 42*(2), 241–251.

Barrett, J., & Paus, T. (2002). Affect-induced changes in speech production. *Experimental Brain Research, 146*, 531–537.

Batson, C., Early, S., & Salvarani, G. (1997). Perspective-taking: Imagining how another feels versus imagining how you would feel. *Personality and Social Psychology Bulletin, 23*(7), 751–758.

Batson, C., Lishner, D., Carpenter, A., Harjusola-Webb, S., Stocks, E., Hassan, O., et al. (2003). As you would have them do unto you: Does imagining yourself in the other's place stimulate moral action? *Personality and Social Psychology Bulletin, 29*(9), 1190–1201.

Batson, C. D., Fultz, J., & Schoenrade, P. A. (1987). Distress and empathy: Two qualitatively distinct vicarious emotions with different motivational consequences. *Journal of Personality, 55*(1), 19–39.

Beck, A. T., Epstein, N., & Brown, G. (1988a). An inventory for measuring clinical anxiety: Psychometric properties. *Journal of Consulting and Clinical Psychology, 56*(6), 893–897.

Beck, A. T., Steer, R. A., & Garbin, M. G. (1988b). Psychometric properties of the Beck Depression Inventory: Twenty-five years of evaluation. *Clinical Psychology Review, 8*, 77–100.

Bernstein, E. M., & Putnam, F. W. (1986). Development, reliability, and validity of a dissociation scale.

Journal of Nervous & Mental Disease, 174(12), 727–735.

Breitenstein, C., Van Lancker, D., & Daum, I. (2001). The contribution of speech rate and pitch variation to the perception of vocal emotions in a German and an American sample. *Cognition and Emotion, 15*(1), 57–79.

Carlson, E. B., & Putnam, F. W. (1993). An update on the dissociative experiences scale. *Dissociation: Progress in the Dissociative Disorders, 6*(1), 16–27.

Carlson, E. B., Putnam, F. W., Anderson, G., Clark, P., Torem, M., Coons, P., et al. (1991). *Factor analysis of the dissociative experiences scale: A multi-center study.* Chicago: Rush Presbyterian.

Cook, N. D. (2003). Tone of voice and mind: The connections between intonation, emotion, cognition and consciousness. *Advances in Consciousness Research, 47*, 97–99.

Critchley, H. D., Wiens, S., Rotshtein, P., Öhman, A., & Dolan, R. (2004). Neural systems supporting interoceptive awareness. *Nature Neuroscience, 7*(2), 189–195.

Davis, M. (1980). A multidimensional approach to individual differences in empathy. *Catalog of Selected Documents in Psychology, 10*(4), 85.

Davis, M., Conklin, L., Smith, A., & Luce, C. (1996). Effect of perspective-taking on the cognitive representation of persons: A merging of self and other. *Journal of Personality and Social Psychology, 70*(4), 713–726.

Davis, M., Soderlund, T., Cole, J., Gadol, E., Kute, M., Myers, M., et al. (2004). Cognitions associated with attempts to empathize: How do we imagine the perspective of another? *Personality and Social Psychology Bulletin, 30*(12), 1625–1635.

Davis, M. H. (1994). *Empathy: A social psychological approach.* Dubuque, IA: Brown & Benchmark.

Decety, J., & Grezes, J. (2006). The power of simulation: Imagining one's own and other's behaviour. *Brain Research, 1079*, 4–14.

Decety, J., & Jackson, P. (2004). The functional architecture of human empathy. *Behavioural and Cognitive Neuroscience Reviews, 3*(2), 71–100.

Decety, J., & Sommerville, J. (2003). Shared representations between self and other: a social cognitive neuroscience view. *Trends in Cognitive Sciences, 7*(12), 527–533.

Fenigstein, A., & Abrams, D. (1993). Self-attention and the egocentric assumption of shared perspectives. *Journal of Experimental Social Psychology, 29*(4), 287–303.

Gallagher, S. (2005). *How the body shapes the mind.* Oxford, UK: Oxford University Press.

Gallese, V., & Goldman, A. (1998). Mirror neurones and the simulation theory of mind-reading. *Trends in Cognitive Sciences, 2*(12), 493–501.

Godfrey, H. P., & Knight, R. G. (1984). The validity of actometer and speech activity measures in the assessment of depressed patients. *British Journal of Psychiatry, 145*, 159–163.

Harrison, N., Singer, T., Rotshtein, P., Dolan, R., & Critchley, H. (2006). Pupillary contagion: Central mechanisms engaged in sadness processing. *Social Cognitive and Affective Neuroscience, 1*, 5–17.

Heims, H. C., Critchley, H. D., Dolan, R., Mathias, L., & Cipolotti, L. (2004). Social and motivational functioning is not critically dependent on feedback of autonomic responses: neuropsychological evidence from patients with pure autonomic failure. *Neuropsychologia, 42*, 1979–1988.

Houston, D. (1990). Empathy and the self: Cognitive and emotional influences on the evaluation of negative affect in others. *Journal of Personality and Social Psychology, 59*(5), 859–868.

Hunter, E., Phillips, M., Chalder, T., Sierra, M., & David, A. S. (2003). Depersonalisation disorder: A cognitive-behavioural conceptualization. *Behavior Research and Therapy, 41*(12), 1451–1467.

Hunter, E. C., Baker, D., Phillips, M., Sierra, M., & David, A. S. (2005). Cognitive behaviour therapy for depersonalisation disorder: An open study. *Behaviour Research and Therapy, 43*(9), 1121–1130.

Kalisch, R., Wiech, K., Critchley, H., Seymour, B., O'Doherty, J., Oakley, D., et al. (2005). Anxiety reduction through detachment: Subjective, physiological, and neural effects. *Journal of Cognitive Neuroscience, 17*(6), 874–883.

Kilner, J., Paulignan, Y., & Blakemore, S. J. (2003). An interference effect of observed biological movement on action. *Current Biology, 13*, 522–525.

Kucharska-Pietura, K., Phillips, M. L., Gernand, W., & David, A. S. (2003). Perception of emotions from faces and voices following unilateral brain damage. *Neuropsychologia, 41*(8), 1082–1090.

Lambert, M. V., Senior, C., Fewtrell, W., Phillips, M., & David, A. S. (2001). Primary and secondary depersonalisation disorder: a psychometric study. *Journal of Affective Disorders, 63*, 249–256.

Lamm, C., Batson, C., & Decety, J. (2007). The neural substrate of human empathy: Effects of perspective-taking and cognitive appraisal. *Journal of Cognitive Neuroscience, 19*(1), 42–58.

Langdon, R., Coltheart, M., Ward, P., & Catts, S. (2001). Visual and cognitive perspective-taking impairments in schizophrenia: a failure of allocentric simulation? *Cognitive Neuropsychiatry, 6*(4), 241–269.

Lawrence, E. J. (2004). *Empathy: Cognitive and neural correlates.* Doctoral thesis, Institute of Psychiatry, King's College London.

Lawrence, E. J., Shaw, P., Baker, D., Baron-Cohen, S., & David, A. S. (2004). Measuring empathy—reliability and validity of the Empathy Quotient. *Psychological Medicine, 34*(5), 911–919.

Lawrence, E. J., Shaw, P., Giampietro, V., Surguladze, S., Brammer, M., & David, A. S. (2006). The role of "shared representations" in social perception and empathy: An fMRI study. *NeuroImage, 29*, 1173–1184.

Loehlin, J., & Nichols, R. (1976). *Heredity, environment and personality: A study of 850 sets of twins.* Austin, TX: University of Texas Press.

Mauss, I., Wilhelm, F., & Gross, J. (2004). Is there less to social anxiety than meets the eye? Emotion experience, expression and bodily responding. *Cognition and Emotion, 18*(5), 631–662.

McKenna, F. P., & Lewis, C. (1994). A speech rate measure of laboratory induced affect: The role of demand characteristics revisited. *British Journal of Clinical Psychology*, *33*(3), 345-351.

Medford, N., Brierley, B., Brammer, M., Bullmore, E. T., & David, A. S. (2006). Emotional memory in depersonalization disorder: a functional MRI study. *Psychiatry Research: Neuroimaging*, *148*, 93-102.

Michal, M., Kaufhold, J., Overbeck, G., & Grabhorn, R. (2006). Narcissistic regulation of the self and interpersonal problems in depersonalized patients. *Psychopathology*, *39*, 192-198.

Muncer, S., & Ling, J. (2006). Psychometric analysis of the empathy quotient (EQ) scale. *Personality & Individual Differences*, *40*, 1111-1119.

Nelson, H. E. (1982). *National Adult Reading Test (NART): Test manual*. Windsor, UK: NFER-Nelson.

Noyes, R., & Kletti, R. (1977). Depersonalization in response to life-threatening danger. *Comprehensive Psychiatry*, *18*(4), 375-384.

Ochsner, K., & Gross, J. (2005). The cognitive control of emotion. *Trends in Cognitive Sciences*, *9*, 242-249.

Parkinson, B., & Manstead, A. (1993). Making sense of emotion in stories and social life. *Cognition and Emotion*, *7*(3/4), 295-323.

Phillips, M. L., Medford, N., Senior, C., Bullmore, E. T., Suckling, J., Brammer, M. J., et al. (2001). Depersonalization disorder: Thinking without feeling. *Psychiatry Research: Neuroimaging*, *108*(3), 145-160.

Pollatos, O., Gramman, K., & Schandry, R. (2007). Neural systems connecting interoceptive awareness and feelings. *Human Brain Mapping*, *28*, 9-18.

Ramnani, N., & Miall, C. (2004). A system in the human brain for predicting the actions of others. *Nature Neuroscience*, *7*(1), 85-90.

Richell, R., Mitchell, D., Newman, C., Leonard, A., Baron-Cohen, S., & Blair, J. (2003). Theory of mind and psychopathy: Can psychopathic individuals read the "language of the eyes"? *Neuropsychologia*, *41*, 523-526.

Ross, E. D., Thompson, R. D., & Yenkosky, J. (1997). Lateralization of affective prosody in brain and the callosal integration of hemispheric language functions. *Brain and Language*, *56*, 27-54.

Ruby, P., & Decety, J. (2004). How would you feel versus how do you think she would feel? A neuroimaging study of perspective-taking with social emotions. *Journal of Cognitive Neuroscience*, *16*(6), 988-999.

Scherer, K. (1995). Expression of emotion in voice and music. *Journal of Voice*, *9*(3), 235-248.

Siegman, A. W., & Boyle, S. (1993). Voices of fear and anxiety and sadness and depression: The effects of speech rate and loudness on fear and anxiety and sadness and depression. *Journal of Abnormal Psychology*, *102*(3), 430-437.

Sierra, M., Baker, D., Medford, N., & David, A. S. (2005). Unpacking the depersonalization syndrome: An exploratory factor analysis on the Cambridge Depersonalization. *Psychological Medicine*, *35*, 1523-1532.

Sierra, M., & Berrios, G. (1998). Depersonalization: neurobiological perspectives. *Biological Psychiatry*, *44*, 898-908.

Sierra, M., Gomez, J., Molina, J., Luque, R., Muñoz, J., & David, A. (2006a). Depersonalization in psychiatric patients: A transcultural study. *Journal of Nervous and Mental Illness*, *194*(5), 356-361.

Sierra, M., Senior, C., Dalton, J., McDonough, M., Bond, A., Phillips, M. L., et al. (2002). Autonomic response in depersonalization disorder. *Archives of General Psychiatry*, *59*(9), 833-838.

Sierra, M., Senior, C., Phillips, M., & David, A. (2006b). Autonomic response in the perception of disgust and happiness in depersonalization disorder. *Psychiatry Research*, *145*, 225-231.

Silvia, P. J., & Abele, A. E. (2002). Can positive affect induce self-focused attention? Methodological and measurement issues. *Cognition and Emotion*, *16*(6), 845-853.

Simeon, D., Guralnik, O., Gross, S., Stein, D. J., Schmeidler, J., & Hollander, E. (1998). The detection and measurement of depersonalization disorder. *Journal of Nervous & Mental Disease*, *186*(9), 536-542.

Stanton, B. R., David, A. S., Cleary, A. J., Sierra, M., Lambert, M. V., Phillips, M. J., et al. (2001). Basal activity of the hypothalamic-pituitary-adrenal axis in patients with depersonalization disorder. *Psychiatry Research: neuroimaging*, *104*(1), 85-89.

Teasdale, J. D., & Fogarty, S. J. (1979). Differential effects of induced mood on retrieval of pleasant and unpleasant events from episodic memory. *Journal of Abnormal Psychology*, *88*(3), 248-257.

Teasdale, J. D., Fogarty, S. J., & Williams, M. (1980). Speech rate as a measure of short term variation in depression. *British Journal of Social and Clinical Psychology*, *19*, 271-278.

Vogeley, K., & Fink, G. R. (2003). Neural correlates of first person perspective. *Trends in Cognitive Sciences*, *7*(1), 38-42.

Wiens, S. (2005). Interoception in emotional experience. *Current Opinion in Neurology*, *18*, 442-447.

Wing, J. K., Cooper, J. E., & Sartorius, N. (1974). *The Measurement and Classification of Psychiatric Symptoms*. London: Cambridge University Press.

APPENDIX 1

Neutral

I was thinking today how the Internet has changed my life in so many ways. I can't think of any other invention as of late that has as many uses as the Internet. Ellie prefers to use the library and says the Internet makes people less sociable. She just sees how long I spend on the computer—she doesn't realize it promotes communication. For me, the biggest attribute of the Internet is to be able to access any and all types of information instantly. Since I do a lot of long-distance walking, being able to locate specific routes is brilliant. The same goes for classic car things, as well. I belong to five classic car clubs, and I met all of those groups through the Internet. Ellie thinks the Internet isn't any quicker in the long run but it is if you know how to use it properly.

Sad

Tomorrow is the funeral of my Nan, so as you can imagine I am really not looking forward to it. The hardest part about it is that when I see that coffin tomorrow I will know that this is the last time I will ever be that close to my Nan again. I hate the thought of that. Why did you have to die? I hate never being able to see you again. I hate that we will all be sitting in your house tomorrow and you won't be there. She loved that house. I just can't help how much I will miss her, I loved her so much. I know these are selfish thoughts, because I know how much she missed my Pop and she just wanted to be with him, so I should be happy for her. I can't believe this has happened. She was so young for her age, and fit and healthy.

Happy

I woke up this morning in the recovery room to a male nurse standing over me saying, "Mrs —, did they tell you that you have a son?" I couldn't believe it—a boy as I'd hoped—I wanted to ask how he was, but I couldn't even get my eyes open, let alone my mouth. Eventually, I croaked out, "Is he OK?" "Yes," the nurse said, "he was pretty big, too." Soon after that they took me to see him—he was in the Intensive Care Unit as a precaution because he was born with a fever. He had "TONS" of long, black hair. This surprised us both since we were both born with moderate amounts of medium colored hair. My baby boy was 8 pounds 3 ounces and 22 inches long. He looked just like my husband and I just wanted to hold him and not let go.

Fear

I usually run at the sight of needles—but its too late for that now—there's no getting out of it. I can see lots of needles. The guys in front of me are being injected one at a time alternating each arm—left then right. "Oh crap—I'm next". The technician swabs my left arm with a cotton ball dipped in alcohol—that sickening smell of alcohol. I looked at him, the needle, then looked away. It seemed like an eternity before I felt the intense sting of this contraption inject its contents into my arm. He removed it from my arm and some residual fluid ran down. It was cold. The process continued all over again. Then two more times after that. By the time I was finished I felt like a pin cushion full of burning holes. I can remember turning pale and having to sit down.

Note: Extracts adapted from www.opendiary.com, 2001.

APPENDIX 2

Trait task instructions—self

Which of the following adjectives do you consider to be descriptive of yourself? Place a tick beside any adjective that you might use in describing yourself to someone else. Your behavior will vary with the situation, of course, so place a tick beside the adjectives that might apply to you frequently, even though they are not appropriate all the time. Work rapidly putting down all your first thoughts.

Trait task instructions—target

Please place a tick next to the following adjectives that best describe James. We appreciate that you only have limited information but we are interested in your initial thoughts and there are no wrong or right answers. So please work rapidly putting down what first comes to mind.

Adaptable	Excitable	Moody
Alert	Fearful	Nervous
Ambitious	Feminine	Obliging
Anxious	Friendly	Opinionated
Assertive	Generous	Outgoing
Capable	Good-natured	Outspoken
Cautious	Helpful	Patient
Cheerful	Highly Strung	Pleasant
Confident	Honest	Quiet
Confused	Humorous	Realistic
Conscientious	Idealistic	Resourceful
Conventional	Impulsive	Responsible
Curious	Independent	Self-centered
Defensive	Industrious	Sensitive
Dependable	Inhibited	Serious
Dependent	Insightful	Shy
Determined	Intelligent	Sociable
Dissatisfied	Irresponsible	Suspicious
Dominant	Irritable	Tense
Easy Going	Kind	Thoughtful
Efficient	Lazy	Timid
Egotistical	Logical	Warm
Emotional	Loyal	Withdrawn
Energetic	Mature	Witty
Enthusiastic	Mischievous	Worrying

SOCIAL NEUROSCIENCE, 2007, 2 (3–4), 307–319

Psychology Press
Taylor & Francis Group

Infant genotype may moderate sensitivity to maternal affective communications: Attachment disorganization, quality of care, and the DRD4 polymorphism

Judit Gervai, Alexa Novak, Krisztina Lakatos, Ildiko Toth, and Ildiko Danis
Hungarian Academy of Sciences, Budapest, Hungary

Zsolt Ronai, Zsofia Nemoda, and Maria Sasvari-Szekely
Semmelweis University, Budapest, Hungary

Jean-Francois Bureau, Elisa Bronfman, and Karlen Lyons-Ruth
Harvard Medical School, Cambridge, Massachusetts, USA

Disorganized attachment is an early predictor of the development of psychopathology in childhood and adolescence. Lyons-Ruth, Bronfman, and Parsons (1999) developed the AMBIANCE coding scheme to assess disrupted communication between mother and infant, and reported the link between maternal behavior and disorganized attachment. The Hungarian group found an association between a polymorphism of the *DRD4* gene and disorganized attachment (Gervai et al., 2005; Lakatos et al., 2000, 2002). The present collaborative work investigated the interplay between genetic and caregiving contributions to disorganized attachment. Mother–infant dyads (138), from a Hungarian low-social-risk sample (96) and a US high-social-risk sample (42), were assessed for infant disorganized attachment behavior, for *DRD4* gene polymorphisms, and for disrupted forms of maternal affective communication with the infant. In accord with literature reports, we found a robust main effect of maternal AMBIANCE scores on infant disorganization. However, this relation held only for the majority of infants who carried the short form of the DRD4 allele. Among carriers of the 7-repeat DRD4 allele, there was no relation between quality of maternal communication and infant disorganization. This interaction effect was independent of degree of social risk and maternal DRD4 genotype.

INTRODUCTION

Advances in molecular genetic techniques and the description of the human genome have made it possible to focus on locating and identifying specific genes underlying genetic variance estimated in earlier twin and adoptive studies. Exploring interactions between genetic and environmental influences that were not easily grasped using quantitative genetic techniques have also become central to recent behavior genetic studies (Rutter, 2006). Such interaction effects may take a variety of forms. Genetic effects may be expressed more strongly under some environmental conditions than others, and environmental effects may be conditional on certain genotypes. Also, exposure to specific environments may be influenced by the individuals' genetic make-up (for a review, see Rutter, Moffitt, & Caspi, 2006). These mechanisms may

Correspondence should be addressed to: Judit Gervai, Institute for Psychology, Hungarian Academy of Sciences, PO Box 398, Budapest, Hungary, 1394. E-mail: gervju@mtapi.hu

This work was supported by grants to KL-R (NIMH RO1062030; Fogarty International Research Center Award RO306014), JG (Hungarian Science Fund—OTKA T 038407) and KL (OTKA D 45940).

www.psypress.com/socialneuroscience DOI:10.1080/17470910701391893

direct individual development onto different trajectories from early in life and affect the long-term development of mental health and disorder.

Animal and human studies have demonstrated relations between measured aspects of the early nurturing environment and a variety of infant outcomes, including physiological responsiveness to stressors (e.g., Fish et al., 2005; Gunnar, 2005; Liu et al., 1997; Nachmias, Gunnar, Mangelsdorf, Parritz, & Buss, 1996; Spangler, Schieche, Ilg, Maier, & Ackermann, 1994) and cognitive and social development (e.g., NICHD Early Child Care Research Network, 2001, 2005). In both animal models and twin studies, the quality of the early parent–child relationship has been shown to contribute to outcome independently of genetic factors (Caspi et al., 2004; Francis, Champagne, Liu, & Meaney, 1999), with some of these effects being transmitted across generations (Belsky, Jaffee, Sligo, Woodward, & Silva, 2005; Fleming et al., 2002; Maestripieri, 2005; Meaney, 2001). However, other studies using different samples and sensitive molecular genetic methods suggest that, besides these experience-induced effects of early parenting, behavioral responses to the early caregiving environment show variability, which is moderated by specific genetic factors in rhesus monkeys (Barr et al., 2003; Champoux et al., 2002), and also in humans (Caspi et al., 2002, 2003; Kim-Cohen et al., 2006).

Attachment is the enduring personal bond that develops between infant and caregiver, providing the primary social environment for physical and mental development (Thompson, 1999). Disorganization of infant attachment, i.e., difficulties in developing and maintaining consistent strategies for obtaining comfort from the caregiver at times of fearful arousal, is one of the rare early predictors of later psycho-social maladjustment. It is a major risk factor for aggressive/externalizing behavior in childhood (Lyons-Ruth, Easterbrooks, & Cibelli, 1997; Moss, Bureau, Cyr, Mongeau, & St-Laurent, 2004a; Moss, Cyr, & Dubois-Comtois, 2004b; van IJzendoorn, Schuengel, & Bakermans-Kranenburg, 1999) and psychopathology in adolescence (Carlson, 1998; Ogawa, Sroufe, Weinfield, Carlson, & Egeland, 1997; Weinfield, Whaley, & Egeland, 2004). About 15–25% of typically developing 1-year-olds and 50–80% of at-risk (e.g., maltreated) infants show disorganized attachment behavior toward the caregiver in situations when comfort and proximity is needed (van IJzendoorn et al., 1999).

Attachment disorganization has been consistently related to a disturbed caregiving environment. In an earlier meta-analysis, the incidence of infant disorganization was related to parental psychiatric impairment but not to serious infant physical disorders, such as cerebral palsy or heart defects (van IJzendoorn, Goldberg, Kroonenberg, & Frenkel, 1992). A more recent meta-analysis confirmed that infant disorganization is related to atypical forms of maternal behavior toward the infant as assessed from videotape and mediates a part of the established relation between other maternal risk factors (i.e., maternal unresolved loss or trauma on the Adult Attachment Interview) and infant disorganization (Madigan et al., 2006). Assessed in this meta-analysis were both a narrower coding system for frightened or frightening forms of parental behavior (FR; Main & Hesse, 1990) and a broader coding system for parental disrupted affective communication with the infant (AMBIANCE; Lyons-Ruth, Bronfman, & Parsons, 1999), which includes codes for frightened or frightening behavior. Meta-analysis of studies using either coding system confirmed equivalent effect sizes and external validity for the two systems in relation to measures of maternal and infant attachment (Madigan et al., 2006). The AMBIANCE coding system for parental disrupted communication was used in the work reported here. Examples of the various forms of disrupted maternal communication can be seen under Methods.

A central proposition of attachment theory is that individual differences in child–parent attachment security can be explained primarily by shared environmental effects. This hypothesis has been examined in a few, relatively small-size twin studies designed to estimate genetic and environmental variance components of child–parent attachment. In a study including 207 two-year-old twin pairs, Finkel and Matheny (2000) found a significant difference between concordances of organized forms of attachment (secure/avoidant/resistant) of monozygotic (MZ) and dizygotic (DZ) twin pairs, suggesting a genetic effect on attachment behavior. Finkel and Matheny (2000) had no data regarding disorganized attachment. In another study of 110 twin pairs aged 3.5 years (O'Connor & Croft, 2001), only shared and non-shared environmental effects were significant. (The distinction between shared and nonshared environmental effects is based on whether such environmental influences make siblings more or less alike.) Due to the small sample size, O'Connor and Croft (2001) declined to separately test genetic influences on disorganization. In a third study of 138 one-year-old twin pairs (Bokhorst et al., 2003), only non-shared environmental factors accounted for the variance in twin concordances of disorganized versus organized attachment, while both shared and nonshared environmental effects accounted for the variance in secure versus insecure attachment.

These twin studies aimed at partitioning the population variance due to genetic and environmental effects by comparing MZ and DZ twin correlations or concordances. Sensitivity to detect significant differences between twin correlations (i.e., finding a genetic effect) is dependent on effect size as well as on sample size, with a good deal of power needed to detect such differences between correlations. Also, twin studies often lack the power to test gene–environment correlations and interactions. While two of the above studies converge in finding primarily environmental effects on attachment behavior, the divergent conclusions of the Finkel and Matheny study regarding the detection of genetic effects on attachment could easily result from the relatively modest sample sizes across studies. With only one twin study reporting on disorganized forms of attachment, no firm conclusions on heritability estimates can yet be drawn.

The development of molecular genetic techniques has now made it possible to detect small effects of single genetic polymorphisms. The D4 dopamine receptor gene has been considered as a candidate gene for infant attachment behavior (Lakatos et al., 2000), because it is preferentially expressed in the brain regions of the mesocorticolimbic dopamine pathway mediating reward related to social interaction, including mother–infant attachment (Insel, 2003; Muller, Brunelli, Moore, Myers, & Shair, 2005). In the third exon of the *DRD4* gene, there is a repeating sequence of 48 base pairs ranging from 2 to 10 repeats (Van Tol et al., 1992). The 7-repeat allele was found to have a lower potency for dopamine-mediated coupling to adenylate cyclase than receptors encoded by the 2- or 4-repeat forms (Asghari et al., 1995). Although some studies did not confirm this difference (Kazmi, Snyder, Cypess, Graber, & Sakmar, 2000; Watts et al., 1999), more recent research (Schoots & Van Tol, 2003; Van Craenenbroeck et al., 2005) provided new evidence for the functional significance of the exon III repeat polymorphism. These new results showed that the 7-repeat (D4.7) variant of the gene was significantly less efficient at the levels of transcription, translation and second-messenger generation compared to the most frequent short (D4.2, D4.4) forms. The 48 bp repeat polymorphism of the *DRD4* gene was associated with normal variations of human neonatal, infant, adult temperament (see Ebstein, 2006, for a review) and attention processes in healthy individuals (Fan, Fossella, Sommer, Wu, & Posner, 2003; Fossella et al., 2002; Schmidt, Fox, Perez-Edgar, Hu, & Hamer, 2001), but has also been related to attention deficit/hyperactivity disorder (ADHD; Faraone et al., 2005).

In a population association study, Lakatos et al. (2000) reported an association between the infant's D4.7 genotype and disorganized attachment. Infants carrying the 7-repeat variant were four times more likely than others (36% vs. 9%) to be classified as disorganized with their mothers in the Strange Situation. Analysis of the −521 C/T regulatory (promoter) polymorphism of the same gene revealed that the association of the 7-repeat allele and disorganized attachment was observed only in the presence of the −521T allele (Lakatos et al., 2002). While a Dutch study failed to replicate this result in a small twin sample (Bakermans-Kranenburg & van IJzendoorn, 2004), further family-based transmission disequilibrium tests (TDT) in the Hungarian sample have found a highly significant non-transmission of the 7-repeat allele to securely attached infants, as well as a trend for preferential transmission to disorganized infants (Gervai et al., 2005).

The present collaborative study was concerned with testing the hypothesis of an interplay between genetic and caregiving contributions to disorganized attachment. In order to increase the incidence and range of key measures of demographic risk, maternal disrupted communication, and infant disorganized attachment behavior, data were aggregated across the low-income, high-risk US sample ($N=42$), from which the AMBIANCE coding system was developed, and the middle income, low-risk Hungarian sample ($N=96$), which was the source of the previous reports of an association between DRD4 and infant disorganization. For purposes of the current study, DRD4 genotypes were collected from the US sample, and videotaped parent–infant interaction data previously collected from the Hungarian sample were newly coded using the AMBIANCE system.

We had three aims: (1) to evaluate whether the previously observed relations between infant disorganization and maternal disrupted communication, and infant disorganization and the DRD4 allele, would replicate in the larger and socioeconomically more diverse sample created by combining low- and high-risk samples; (2) to test potential passive and evocative gene–environmental correlations by evaluating whether maternal and infant DRD4 genotypes influence the level of maternal disrupted affective communication; and (3) to test for the presence of a gene–environment interaction effect by evaluating whether the quality of maternal–infant communication moderates the relation between the *DRD4* gene polymorphism and infant disorganization.

METHODS

Participants

One hundred thirty-eight mothers and infants participated in the study, 96 drawn from a study of a low-risk community sample of 103 middle-class Hungarian families with healthy, full-term, first-born infants (45 female, 58 males, BW >2500 g) and 42 from a study of 56 low-income families in the US, recruited to over-represent infants at caregiving risk. The Hungarian sample was recruited during the third trimester of pregnancy from antenatal classes to participate in the longitudinal Budapest Infant–Parent Study (BIPS) to investigate infants' social and emotional development. The sample was ethnically homogeneous, Caucasian, of Hungarian origin. Signed informed consent was obtained from the parents for participating in the whole study, and again for 96 children (41 females, 55 males) separately for the genetic investigation. Families of the remaining 7 children (4 girls, 3 boys) either moved away by the time of the DNA sampling or refused the collection of cheek cells. The research protocol was approved by the Research Ethics Committee of the Institute of Psychology of the Hungarian Academy of Sciences. At the time of the infant's birth, all mothers were married or living with a partner. Mothers' mean age was 27.3 years ($SD = 3.5$, range = 19–38), 53% of the mothers and 59% of the fathers had a college or university degree, and a further 43% and 31%, respectively, had completed secondary school education. US families were recruited during the first 18 months of the infant's life, as part of a study of the impact of social risk factors on infant development. All families met US federal poverty guidelines at study entry and 52% of families had been referred for clinical home-visiting services due to service providers' concerns about the quality of care for the infant. Signed informed consent was obtained from all families for participation in the study. As part of a multiphase longitudinal study, genetic samples were collected when the infants were young adults, aged 18–22 years. Of 56 infants relocated in young adulthood, 10 refused participation, and 4 specifically refused cheek cell collection, yielding genetic data for 42 families. The research protocol was approved over all phases of the study by the Institutional Review Board of the Cambridge Health Alliance. At the time of entry into the infant study, 19% were single parents; mothers' mean age was 26.2 years ($SD = 5.6$, range = 16–41); ethnic composition: 35 Caucasian (83%), 7 Black (17%), Asian (0%); 64% of mothers had completed high school and 36% had some post-secondary education.

Measures

Cumulative demographic risk. The cumulative risk index included: (1) low level of maternal education (less than completed high school); (2) early motherhood (mother younger than 20 at first birth); (3) unplanned pregnancy with the target child; (4) birth of an additional infant before age 18 months of the target child; (5) three or more children under age 6 years in the family; (6) single parenthood; (7) two or more moves within the last year; and (8) low level of parental occupation. Hungarian and US occupation data were coded on a matching 6-level scale ranging from unemployed to major professional (1 = *unemployed*, 2 = *unskilled worker*, 3 = *skilled worker*, 4 = *technician/semiprofessional*, 5 = *lesser professionals*, 6 = *higher executive/major professional*) and a score ≤3 was used as a cut-off for dichotomizing socioeconomic risk as reflected by parental occupation. If both parents were employed, the higher level occupation was chosen for risk coding. Hungarian and US samples differed significantly in overall maternal education, occupational level, and single parenthood. Each of the above 8 demographic variables was coded according to absence (0) or presence (1) of risk, and then summed to create the cumulative index.

Attachment classification. In both studies, the infant's attachment to the mother was assessed in the standard Strange Situation assessment (Ainsworth, Blehar, Waters, & Wall, 1978). Assessments were conducted at 12 months of age in the Hungarian study and at 18 months in the US study. Infant attachment classifications have been found to be stable over this age period (Waters, 1978). Classification was carried out blind to infant genotype (Lakatos et al., 2000; Lyons-Ruth, Connell, Grunebaum, & Botein, 1990). Videotapes were evaluated for attachment security as described by Ainsworth et al. (1978). In the Hungarian study, infant disorganized behavior was rated on a 9-point ordinal scale (D-scale) and infants with a D-score ≥ 5 were classified as disorganized (Main & Solomon, 1990). The original 5-point version of the D-scale, later replaced by a 9-point version, was used in the US study, where the cut-off value for D classification was ≥ 4. The later 9-point version of the scale uses the same criteria for classification as the earlier 5-point version, but adds a broader range of description for variations within the D classification (values 5–9). To use D-scale values in regression analyses, the Hungarian D-scale scores were adjusted to the US scale by matching the cut-off values (5 was recoded as 4) and compressing values 3–4 and 5–9 by using half-points.

Maternal behavior. Disrupted maternal affective communication was assessed in the Strange Situation using the AMBIANCE coding scheme (Lyons-Ruth et al., 1999). Five types of maternal atypical behavior were recorded: (1) Disorientation (e.g., appears confused, hesitant, or frightened with infant; incongruous affect); (2) Negative-intrusive behavior (e.g., mocks or teases infant); (3) Role confusion (e.g., draws attention to self when infant is in need); (4) Withdrawal (e.g., fails to initiate interaction, does not greet infant after separation); and (5) Affective communication errors (including contradictory affective communications, e.g., talks in inviting voice but physically blocks infant's access, and failures to respond to clear infant cues). Based on both the frequency and quality of the behavior observed, a 7-point rating was assigned by the coders on the Level of Disrupted Communication Scale. "*Disrupted*" classification was defined by a score of 5–7 on the qualitative level scale. Reliabilities and procedures for coding of the US tapes have been published previously (Lyons-Ruth et al., 1999). A recent meta-analysis of the relation of the AMBIANCE scale to infant attachment disorganization has shown an effect size of .35 ($N = 384$; Madigan et al., 2006). In addition, meta-analytic test–retest data for the AMBIANCE coding of disrupted maternal communication yielded a stability coefficient of .56 ($N = 203$), for periods ranging from 8 months to 5 years. For the current study, two Hungarian researchers were trained to reliability. Following the detailed coding protocol, 101 Hungarian mothers were rated on the 7-point Level of Disrupted Communication Scale and classified as Disrupted or Not-disrupted. Reliability coefficients: Disrupted Scale score, $r_i = .74$; classification kappa = .60; classification agreement 80%. Coders were naïve to coding procedures for attachment security or disorganization, and they were blind to both genetic data and infant attachment classifications. Regarding potential common method variance from coding both infant disorganization and maternal behavior from the same assessment, meta-analytic data confirm that AMBIANCE codes have similar magnitudes of association with infant disorganization when coded in a separate laboratory observation (Madigan et al., 2006).

DRD4 genotype. For both the US and Hungarian samples, DNA was isolated from buccal epithelial cells as described elsewhere (Boor et al., 2002), except that Purgene DNA Purification kits (Gentra – Minneapolis, MN, USA) were used for DNA isolation in the US sample. For the US sample, DNA samples were collected during the 19-year follow-up study and Schleicher & Schuell (Keene, NH, USA) IsoCode ID kits were used for 6 samples that were collected by mail. Purified DNA samples from the US were transported and analyzed for the DRD4 exon III repeat polymorphism in Hungary. Genotyping for the DRD4 exon III repeat polymorphism was carried out as described previously (Barta et al., 2001; Ronai et al., 2000) and blind to behavioral data for both studies.

Control for genetic stratification. In the ethnically homogeneous Hungarian sample, DRD4 genotypes in both the infant and the parental groups were in Hardy–Weinberg equilibrium, reflecting a random combination and stable frequencies of the DRD4 repeat alleles in this population (Gervai et al., 2005; Lakatos et al., 2000). For the ethnically heterogeneous US sample, 40 random marker polymorphisms distributed evenly along the human genome (1–3 markers per chromosome) were analyzed. Population stratification was investigated using the L-POP population substructure detection software (Purcell & Sham, 2004). This program employs the latent class analysis model, which is based on the assumption that unlinked markers are in Hardy–Weinberg equilibrium and in linkage equilibrium in a homogenous population (Pritchard, Stephens, Rosenberg, & Donnelly, 2000). Based on this approach, no significant population stratification was observed in the US study sample.

Analytic plan. First, the low-social-risk Hungarian data were tested for replication of the association between maternal disrupted communication and infant disorganized attachment to the mother. Further analyses were carried out on the combined Hungarian–US data set. Combining the high-risk US longitudinal and the low-risk Hungarian samples with complete data sets ($N=42$ and 96, respectively) increased power and also resolved the problem of restriction of range present in both samples by increasing the variability of demographic risk, maternal atypical behavior and infant attachment behavior. Control variables were first assessed for their relation to maternal disrupted communication and infant disorganization, and, if significantly associated, were included as covariates in further multivariate modeling. Continuous scales for infant disorganization and maternal disrupted communication were used in regression models to preserve maximum power. Parallel secondary analyses assessed whether the pattern of results also held when attachment disorganization was analyzed as a categorical variable, using the standard well-validated scale point for classification as disorganized.

Imputing missing data. All 138 infants had genotype data, but there were missing data on other variables of interest in 8 cases. Attachment data were available for all but 3 US infants. AMBIANCE data were missing for 4 mothers in the US sample. Two Hungarian tapes had sound problems, so these could not be coded for maternal atypical behavior. To impute complete data for the samples, we used the Multiple Imputation (MI) procedure of the SAS/STAT, designed to follow the multiple imputation guidelines by Rubin (1987) and Schafer (1997). The procedure creates multiply imputed data sets for incomplete multivariate data. It uses methods that incorporate appropriate variability across the *m* imputations. In the present study, five data sets were generated. Regression analyses were then computed for each data set with unstandardized coefficients (B) and their standard errors pooled and analyzed through the SAS/STAT's MIanalyze procedure.

RESULTS

Relations between maternal disrupted communication and disorganized attachment in the low-social-risk Hungarian sample

The first set of analyses revealed that the rated level of disrupted maternal communication in the Hungarian sample was significantly associated with the level of infant disorganization, Spearman's rho $=.21$, $p<.05$, replicating previous results in the US sample (Lyons-Ruth et al., 1999). As shown in Table 1, the relation found for the continuous ratings was also significant using dichotomized classification cut-off points both for disrupted communication and for infant disorganization.

Study sample comparisons

Frequencies of gender categories were nearly identical, with 57.3% and 57.1% males in the Hungarian and US samples, respectively. Table 2 summarizes data for key measures for the two samples. There was no significant difference in the frequency of the DRD4 7-repeat genotype by sample, $\chi^2(1,\ N=138)=1.97$, $p=.23$. As expected, the samples differed by demographic risk status, $t(136)=-9.78$, $p<.001$, by extent of infant disorganization $t(136)=-4.41$, $p<.001$, and by extent of maternal disrupted communication, $t(136)=-3.19$, $p<.002$.

Univariate analyses for control variables

Gender. Gender was first assessed for its relation to infant disorganization, DRD4 7-repeat

TABLE 1
Infant attachment classification and maternal disrupted communication in the Hungarian sample

	Infant attachment classification		
	Organized (N =80)	*Disorganized (N =21)*	*Strength of association*
Level of disrupted communication	3.38±1.50	4.14±1.80	η =.20; *p* <.05
Maternal disrupted classification			
Not disrupted	75% (60)	52% (11)	Φ =.20; *p* <.05
Disrupted	25% (20)	48% (10)	

allele, and maternal disrupted communication. Consistent with meta-analytic findings (van IJzendoorn et al., 1999), there was no significant effect of gender on infant disorganization, $t(136) = 0.18$, $p > .85$. Also consistent with meta-analytic findings (Madigan et al., 2006), child gender was unrelated to level of maternal disrupted communication, $t(136) = 1.27$, $p > .20$. There was also no relation between presence of the 7-repeat DRD4 allele and infant gender $\chi^2(1, N = 138) = 0.13$, $p > .70$. Relations with gender were also nonsignificant *within* the national groups, so gender was not considered further in the analyses.

Cumulative demographic risk. As expected from prior literature, there were significant effects of cumulative demographic risk both on levels of infant disorganization, Spearman rho = .37, $p < .001$, and on maternal disrupted communication, rho = .25, $p < .005$. There was no association between cumulative demographic risk and presence of the 7-repeat allele in the infant, $t(136) = 1.25$, $p > .20$. Since variability of the cumulative risk index was considerably restricted within both national groups (see Table 2), correlations within the samples were not significant (infant disorganization, rho = .07 and −.08, $p > .40$; maternal disrupted communication, rho = .02 and −.01, $p > .80$, for the Hungarian and the US samples, respectively). Cumulative demographic risk, therefore, was controlled for in later analyses of genetic and caregiving effects.

TABLE 2
Frequencies and means (standard deviations) of key variables in the Hungarian and US samples

Variable	*Hungarian (N =96)*	*US (N =42)*
Frequency of DRD4 7-repeat genotype	38.5%	26.2%
Cumulative demographic risk	0.27 (0.53)	2.60 (1.50)
Level of infant disorganization	2.13 (1.46)	3.43 (1.25)
Level of maternal disrupted communication	3.51 (1.56)	4.45 (1.65)

Assessment of gene–environment correlation

There was no significant relation between the presence of the DRD4 7-repeat allele in the *infant* and the level of maternal disrupted communication, $t(136) = -.26$, $p > .70$. Therefore, infant genotype did not appear to elicit disturbed forms of maternal interaction. Relation between the presence of the DRD4 7-repeat allele in the *mother* and the level of maternal disrupted communication was not significant either, $t(132) = 1.27$, $p > .20$, so maternal DRD4 genotype could not account for the correlation between maternal disrupted communication and infant disorganization.

Assessment of genetic effects and gene–environment interaction on infant disorganization

Level of infant disorganization was regressed on presence of the DRD4 7-repeat allele and level of maternal disrupted communication, with level of cumulative demographic risk entered first into the equation as a control variable. Main effects, as well as the interaction term between infant DRD4 genotype and maternal disrupted communication, were tested.

With cumulative demographic risk controlled, the main effect of maternal disrupted communication was significant, $t(134) = 3.18$, $p < .002$, B = .37. The main effect of infant genotype was not significant, $t(134) = 1.02$, $p > .30$. The interaction term was entered on the third step and was significant, $t(133) = -2.18$, $p = .03$, B = .35.

The genotype by level of maternal disrupted communication interaction term was further explored as recommended (Aiken & West, 1991, p. 130) by computing regression analyses for the relation between maternal disruption and infant disorganization separately for infants who did not carry the DRD4 7-repeat allele ($N = 90$) and for those who did ($N = 48$). Among infants who did *not* carry the allele, with demographic risk controlled, there was a significant relation between level of disorganization and maternal disrupted communication, $t(87) = 4.35$, $p < .0001$, $B = .37$. Among infants who *did* carry the 7-repeat allele, with demographic risk controlled, there was no relation between maternal disruption and infant disorganization, $t(45) = 0.13$, $p > .80$. Results (not shown) were similar without controlling for demographic risk.

We also computed regression analyses for the relation between DRD4 7-repeat allele and infant disorganization separately within maternal groups categorized as non-disrupted and disrupted at the standard classification point of 5 or above on the 7-point scale (Lyons-Ruth et al., 1999). For infants whose mothers were non-disrupted ($N = 88$), with demographic risk controlled, there was a significant relation between the level of disorganization and the DRD4 genotype, $t(85) = 2.31$, $p < .025$, $B = .69$. Among infants whose mothers displayed high levels of disrupted behavior ($N = 50$), with demographic risk controlled, the relation between the level of disorganization and DRD4 genotype was not significant, $t(45) = -1.20$, $p > .20$.

Because cumulative demographic risk was also a correlate of infant disorganization, a demographic risk by genotype interaction term was entered into the primary regression analysis to test whether a similar gene–environment interaction effect on dysregulation would be associated with cumulative demographic risk. This term did not reach significance, $t(132) = -1.61$, $p = .11$, indicating some specificity to the role of early care in the interaction with infant genotype.

Since the categorical classification of infant disorganization is widely used and variability at the lower ranges of the disorganization scale may have less prognostic significance than variation across the classification point, the analysis was repeated regressing the dichotomous disorganized classification on the same set of variables as above. After controlling for demographic risk as before, the genotype by maternal disruption interaction term was again significant, $t(133) = 2.10$, $p < .04$, $B = .60$. This interaction is displayed graphically in Figure 1. A final follow-up analysis confirmed that the relation between disorganized attachment and maternal disrupted behavior was significant only in the 7-absent groups both for the Hungarian and US samples, $t(56) = 2.03$, $p < .05$, $B = .19$ and $t(28) = 2.84$, $p < .01$, $B = .39$, respectively.

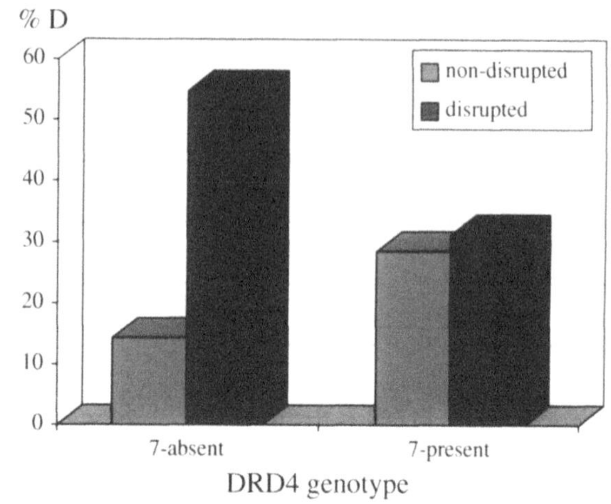

Figure 1. Maternal disrupted communication and disorganized infant attachment by DRD4 genotype.

DISCUSSION

In relation to the first aim of the current work, maternal disrupted communication was related to infant disorganization, both in the newly coded Hungarian data set and in the socioeconomically diverse combined Hungarian–US sample. However, with a more diverse sample, the main effect of the infant DRD4 7-repeat genotype, previously found to relate to infant disorganized attachment in the Hungarian sample alone, was not significant. Infant attachment disorganization was related to cumulative demographic risk, as well as to maternal disrupted communication, but, after controlling for demographic risk, the link between maternal disrupted communication and infant disorganization still remained significant.

In relation to the second aim, neither *maternal* nor *infant* DRD4 7-repeat genotype accounted for maternal disrupted affective communication with the infant.

Most importantly, in relation to the third aim, the strong relation between maternal disrupted communication and infant disorganization was moderated by the infant's DRD4 7-repeat genotype. The nature of this moderation was that attachment disorganization was only strongly related to the quality of mother–infant communication in the larger group of infants with the 7-absent genotype. Only in this group did we find the predicted relationship between maternal caregiving behavior and infant disorganization. Among infants who carried the 7-repeat allele, infant disorganization was unrelated to the level of maternal disrupted communication, suggesting that infants with the 7-repeat allele were less sensitive to regulation by the caregiving relationship.

It is important to note that the data presented here moderate the previous findings relating both caregiving and *DRD4* gene polymorphism to attachment disorganization. The current findings do not argue for main effects on disorganized attachment. On the contrary, genetic variation appears to modulate infant sensitivity to care. Infant attachment behavior among those with the more frequent genotype (7-absent) was strongly related to quality of care. Sensitivity to quality of care promoted secure attachment behavior under conditions of sensitive care and vulnerability to disorganization only when parent–infant communication was disrupted. Conversely, the 7-repeat allele appeared to mute the infant's responsiveness to care, providing some protection in the context of very disrupted interactions but providing less regulation in the context of higher quality care.

These results are consistent with, but also extend, previous findings from both samples. First, the relation between attachment disorganization and maternal disrupted communication previously described for the US sample as well as other samples (Lyons-Ruth et al., 1999; Madigan et al., 2006) was replicated in the Hungarian sample. Second, the addition of the maternal caregiving data to the Hungarian sample newly revealed the significant moderation by maternal behavior of our previously reported genetic effect on attachment (Gervai et al., 2005; Lakatos et al., 2000, 2002). Third, the relation between maternal behavior and infant disorganization was conditional on the 7-absent genotype not only in the combined sample but also in both samples analyzed separately.

What these findings imply about attachment is that secure attachment is aided not only by the parent's sensitivity to the infant's cues, but also by the infant's sensitivity to the parent's sensitivity. For parental sensitivity to provide effective regulation of infant stressful arousal, the infant must register the sensitive and contingent nature of parental responses to the infant's cues. If the infant is inattentive to these contingencies, the regulatory effect will be lessened. We would hypothesize that infants who carry the DRD4 7-repeat allele are relatively *less* attentive or *less* sensitive to the more subtle social rewards involved in affective communications, while infants with the more common short forms of the DRD4 allele are relatively *more* sensitive. These results are consistent with a number of other findings describing heterogeneity of behavioral responses to early rearing conditions due to specific genotypes. In the Dunedin study, Caspi and colleagues (2002, 2003) found that the relation of childhood maltreatment with later psychological maladjustment was moderated by genetic factors. The functional polymorphism of the regulatory region of the monoamine oxidase A (*MAOA*) gene has been shown to moderate the relation between early maltreatment and later antisocial behavior (Caspi et al., 2002), and a regulatory polymorphism of the serotonin transporter (*5-HTT*) gene has been shown to moderate the effect of early maltreatment on adult depression (Caspi et al., 2003). Both findings have since been replicated (Foley et al., 2004; Kaufman et al., 2004; Kim-Cohen et al., 2006). The homologous *MAOA* and *5-HTT* gene polymorphisms in rhesus monkeys also moderated the effects of adverse rearing conditions on infants' aggressive behavior (Newman et al., 2005), on CNS function (Bennett et al., 2002), on early behavioral development (Champoux et al., 2002), and on hormonal responses to separation stress (Barr et al., 2004). Although links of temperament and personality with *DRD4* gene polymorphisms have been reported previously (Ebstein, 2006), there is very sparse data demonstrating *DRD4* gene–environment interaction effects. In a small low-social-risk twin sample (Bakermans-Kranenburg & van IJzendoorn, 2006), low maternal sensitivity in infancy predicted higher levels of mother-reported externalizing behavior problems at 2–3 years of age, but only if infants carried the 7-repeat allele of the *DRD4* gene. Since this study used an earlier measure of maternal sensitivity that has not shown a robust relation to disorganization of attachment in infancy (van IJzendoorn et al., 1999) and a very different outcome measure, results can not be meaningfully compared to results of the current study.

Animal studies have shown that mesocorticolimbic dopamine is involved in both reward-related learning and in mammalian maternal caregiving (Insel, 2003), but the role of dopamine in the infant's complementary attachment behavior is much less clear. The mesolimbic dopamine system has been implicated in the processing of natural and artificial rewards by mediating the hedonic aspects of rewarding stimuli and acting as a learning signal for behavioral reinforcement (Schultz, 1998). Dopamine may be integral to the formation of mother–infant attachment through specific pathways that enhance the hedonic value and salience of stimuli associated with the mother, perhaps especially emotional communicative signals. We may hypothesize that the development of the mother–infant bond may be supported and guided by the motivational/rewarding aspects of maternal social behavior and infant emotional learning. The role of the D4 dopamine receptor in emotional learning (both in encoding and expression) in the medial prefrontal cortex is supported by the results of a recent animal study (Laviolette, Lipski, & Grace, 2005). It has also been shown that administration of a D2 receptor antagonist in prairie voles blocks male–female bonding by removing the hedonic value of stimuli associated with the sexual partner (Young & Wang, 2004), congruent with the hypothesis that dopamine plays a role in attachment processes. In this neural model of pair bond formation, dopamine reward is activated in social recognition circuits involving the prefrontal cortex, the nucleus accumbens and the medial amygdala, so that social cues associated with a specific partner acquire enhanced hedonic value. It may be hypothesized that functional variations in the *DRD4* gene expressed preferentially in brain regions of the reward circuit (prefrontal/cingulate cortex, nucleus accumbens and ventral tegmental area) modulate sensitivity to maternal stimuli resulting in differential sensitivity to caregiving behavior. In view of the present results on moderation of infant sensitivity to maternal care by the *DRD4* gene variation, further work is needed to explore interrelations among infant attachment and genetic variations affecting the neurotransmitter systems involved in attachment processes and distress regulation.

REFERENCES

Aiken, S. L., & West, S. G. (1991). *Multiple regression: Testing and interpreting interactions.* Thousand Oaks, CA: Sage.

Ainsworth, M. D., Blehar, M. C., Waters, E., & Wall, S. (1978). *Patterns of attachment: A psychological study of the Strange Situation.* Hillsdale, NJ: Lawrence Erlbaum Associates, Inc.

Asghari, V., Sanyal, S., Buchwaldt, S., Paterson, A., Jovanovic, V., & Van Tol, H. H. (1995). Modulation of intracellular cyclic AMP levels by different human dopamine D4 receptor variants. *Journal of Neurochemistry, 65*, 1157–1165.

Bakermans-Kranenburg, M. J., & van IJzendoorn, M. H. (2004). No association of dopamine D4 receptor (DRD4) and −521 C/T promoter polymorphisms with infant attachment disorganization. *Attachment and Human Development, 6*, 211–218.

Bakermans-Kranenburg, M. J., & van IJzendoorn, M. H. (2006). Gene–environment interaction of the dopamine D4 receptor (DRD4) and observed maternal insensitivity predicting externalizing behavior in preschoolers. *Developmental Psychobiology, 48*, 406–409.

Barr, C. S., Newman, T. K., Becker, M. L., Parker, C. C., Champoux, M., Lesch, K. P., et al. (2003). The utility of the non-human primate; model for studying gene by environment interactions in behavioral research. *Genes, Brain and Behavior, 2*, 336–340.

Barr, C. S., Newman, T. K., Shannon, C., Parker, C., Dvoskin, R. L., Becker, M. L., et al. (2004). Rearing condition and rh5-HTTLPR interact to influence limbic-hypothalamic-pituitary-adrenal axis response to stress in infant macaques. *Biological Psychiatry, 55*, 733–738.

Barta, C., Ronai, Z., Nemoda, Z., Szekely, A., Kovacs, E., Sasvari-Szekely, M., et al. (2001). Analysis of dopamine D4 receptor gene polymorphism using microchip electrophoresis. *Journal of Chromatography A, 924*, 285–290.

Belsky, J., Jaffee, S. R., Sligo, J., Woodward, L., & Silva, P. A. (2005). Intergenerational transmission of warm-sensitive-stimulating parenting: A prospective study of mothers and fathers of 3-year-olds. *Child Development, 76*, 384–396.

Bennett, A. J., Lesch, K. P., Heils, A., Long, J. C., Lorenz, J. G., Shoaf, S. E., et al. (2002). Early experience and serotonin transporter gene variation interact to influence primate CNS function. *Molecular Psychiatry, 7*, 118–122.

Bokhorst, C. L., Bakermans-Kranenburg, M. J., Fearon, R. M. P., van IJzendoorn, M. H., Fonagy, P., & Schuengel, C. (2003). The importance of shared environment in mother–infant attachment security: A behavioral genetic study. *Child Development, 74*, 1769–1782.

Boor, K., Ronai, Z., Nemoda, Z., Gaszner, P., Sasvari-Szekely, M., Guttman, A., et al. (2002). Noninvasive genotyping of dopamine receptor D4 (DRD4) using nanograms of DNA from substance-dependent patients. *Current Medical Chemistry, 9*, 793–797.

Carlson, E. A. (1998). A prospective longitudinal study of attachment disorganization/disorientation. *Child Development, 69*, 1107–1128.

Caspi, A., McClay, J., Moffitt, T. E., Mill, J., Martin, J., Craig, I., et al. (2002). Role of genotype in the cycle of violence in maltreated children. *Science, 297*, 851–854.

Caspi, A., Moffitt, T. E., Morgan, J., Rutter, M., Taylor, A., Arseneault, L., et al. (2004). Maternal expressed emotion predicts children's antisocial behavior problems: Using monozygotic-twin differences to identify environmental effects on behavioral development. *Developmental Psychology, 40*, 149-161.

Caspi, A., Sugden, K., Moffitt, T. E., Taylor, A., Craig, I. W., Harrington, H., et al. (2003). Influence of life stress on depression: Moderation by a polymorphism in the 5-HTT gene. *Science, 301*, 386-389.

Champoux, M., Bennett, A., Shannon, C., Higley, J. D., Lesch, K. P., & Suomi, S. J. (2002). Serotonin transporter gene polymorphism, differential early rearing, and behavior in rhesus monkey neonates. *Molecular Psychiatry, 7*, 1058-1063.

Ebstein, R. P. (2006). The molecular genetic architecture of human personality: Beyond self-report questionnaires. *Molecular Psychiatry, 11*, 427-445.

Fan, J., Fossella, J., Sommer, T., Wu, J., & Posner, M. I. (2003). Mapping the genetic variation of executive attention onto brain activity. *Proceedings of the National Academy of Sciences USA, 100*, 7406-7411.

Faraone, S. V., Perlis, R. H., Doyle, A. E., Smoller, J. W., Goralnick, J. J., Holmgren, M. A., et al. (2005). Molecular genetics of attention-deficit/hyperactivity disorder. *Biological Psychiatry, 57*, 1313-1323.

Finkel, D., & Matheny, A. P., Jr. (2000). Genetic and environmental influences on a measure of infant attachment security. *Twin Research, 3*, 242-250.

Fish, E. W., Shahrokh, D., Bagot, R., Caldji, C., Bredy, T., Szyf, M., et al. (2004). Epigenetic programming of stress responses through variations in maternal care. *Annals of the New York Academy of Sciences, 1036*, 167-180.

Fleming, A. S., Kraemer, G. W., Gonzalez, A., Lovic, V., Rees, S., & Melo, A. (2002). Mothering begets mothering: The transmission of behavior and its neurobiology across generations. *Pharmacology Biochemistry and Behavior, 73*, 61-75.

Foley, D. L., Eaves, L. J., Wormly, B., Silberg, J. L., Maes, H. H., Kuhn, J., et al. (2004). Childhood adversity, monoamine oxidase A genotype, and risk for conduct disorder. *Archives of General Psychiatry, 61*, 1-7.

Fossella, J., Sommer, T., Fan, J., Wu, Y., Swanson, J. M., Pfaff, D. W., et al. (2002). Assessing the molecular genetics of attention networks. *BMC Neuroscience, 3*, 14.

Francis, D. D., Champagne, F. A., Liu, D., & Meaney, M. J. (1999). Maternal care, gene expression, and the development of individual differences in stress reactivity. *Annals of the New York Academy of Sciences, 896*, 66-84.

Gervai, J., Nemoda, Z., Lakatos, K., Ronai, Z., Toth, I., Ney, K., et al. (2005). Transmission Disequilibrium Tests confirm the link between *DRD4* gene polymorphism and infant attachment. *American Journal of Medical Genetics, Part B (Neuropsychiatric Genetics), 132B*, 126-130.

Gunnar, M. (2005). Attachment and stress in early development. In C. S. Carter, et al. (Eds.), *Attachment and bonding: A new synthesis* (pp. 245-255). Cambridge, MA: MIT Press.

Insel, T. R. (2003). Is social attachment an addictive disorder? *Physiology and Behavior, 79*, 351-357.

Kaufman, J., Yang, B.-Z., Douglas-Palumberi, H., Houshyar, S., Lipschitz, D., Krystal, J., et al. (2004). Social supports and serotonin transporter gene moderate depression in maltreated children. *Proceedings of the National Academy of Sciences of the USA, 101*, 17316-17321.

Kazmi, M. A., Snyder, L. A., Cypess, A. M., Graber, S. G., & Sakmar, T. P. (2000). Selective reconstitution of human D4 dopamine receptor variants with Gi alpha subtypes. *Biochemistry, 39*, 3734-3744.

Kim-Cohen, J., Caspi, A., Taylor, A., Williams, B., Newcombe, R., Craig, I. W., et al. (2006). MAOA, maltreatment, and gene-environment interaction predicting children's mental health: New evidence and a meta-analysis. *Molecular Psychiatry, 11*, 903-913.

Lakatos, K., Nemoda, Z., Toth, I., Ronai, Z., Ney, K., Sasvari-Szekely, M., et al. (2002). Further evidence for the role of the dopamine D4 receptor gene (DRD4) in attachment disorganization: Interaction of the III exon 48 bp repeat and the −521 C/T promoter polymorphisms. *Molecular Psychiatry, 7*, 27-31.

Lakatos, K., Toth, I., Nemoda, Z., Ney, K., Sasvari-Szekely, M., & Gervai, J. (2000). Dopamine D4 receptor (DRD4) gene polymorphism is associated with attachment disorganization. *Molecular Psychiatry, 5*, 633-637.

Laviolette, S. R., Lipski, W. J., & Grace, A. A. (2005). A subpopulation of neurons in the medial prefrontal cortex encodes emotional learning with burst and frequency codes through a dopamine D4 receptor-dependent basolateral amygdala input. *Journal of Neuroscience, 25*, 6066-6075.

Liu, D., Diorio, J., Tannenbaum, B., Caldji, C., Francis, D., Freedman, A., et al. (1997). Maternal care, hippocampal glucocorticoid receptors, and hypothalamic-pituitary-adrenal responses to stress. *Science, 277*, 1659-1662.

Lyons-Ruth, K., Bronfman, E., & Parsons, E. (1999). Maternal frightened, frightening, or atypical behavior and disorganized infant attachment patterns. In J. I. Vondra & D. Barnett (Eds.), Atypical attachment in infancy and early childhood among children at developmental risk. *Monographs of the Society for Research in Child Development, 64*, 67-96.

Lyons-Ruth, K., Connell, D., Grunebaum, H., & Botein, S. (1990). Infants at social risk: Maternal depression and family support services as mediators of infant development and security of attachment. *Child Development, 61*, 85-98.

Lyons-Ruth, K., Easterbrooks, M. A., & Cibelli, C. D. (1997). Infant attachment strategies, infant mental lag, and maternal depressive symptoms: Predictors of internalizing and externalizing problems at age 7. *Developmental Psychology, 33*, 681-692.

Madigan, S., Bakermans-Kranenburg, M., van IJzendoorn, M., Moran, G., Pederson, D., & Benoit, D. (2006). Unresolved states of mind, anomalous parental behavior, and disorganized attachment: A

review and meta-analysis of a transmission gap. *Attachment and Human Development*, *8*, 89–111.
Maestripieri, D. (2005). Early experience affects the intergenerational transmission of infant abuse in rhesus monkeys. *Proceedings of the National Academy of Sciences of the USA*, *102*, 9726–9729.
Main, M., & Hesse, E. (1990). Parent's unresolved traumatic experiences are related to infant disorganized/disoriented attachment status: Is frightened and/or frightening parental behavior the linking mechanism? In M. Greenberg, D. Cicchetti, & E. M. Cummings (Eds.), *Attachment in the preschool years: Theory, research, and intervention* (pp. 161–182). Chicago: University of Chicago Press.
Main, M., & Solomon, J. (1990). Procedures for identifying infants as disorganized/disoriented during the Ainsworth Strange Situation. In M. T. Greenberg, D. Cicchetti, & E. M. Cummings (Eds.), *Attachment in the preschool years: Theory, research, and intervention* (pp. 121–160). Chicago: University of Chicago Press.
Meaney, M. J. (2001). Maternal care, gene expression, and the transmission of individual differences in stress reactivity across generations. *Annual Reviews of Neuroscience*, *24*, 1161–1192.
Moss, E., Bureau, J.-F., Cyr, C., Mongeau, C., & St-Laurent, D. (2004a). Correlates of attachment at age 3: Construct validity of the preschool attachment classification system. *Developmental Psychology*, *40*, 323–334.
Moss, E., Cyr, C., & Dubois-Comtois, K. (2004b). Attachment at early school age and developmental risk: Examining family contexts and behavior problems of controlling-caregiving, controlling-punitive, and behaviorally disorganized children. *Developmental Psychology*, *40*, 519–532.
Muller, J. M., Brunelli, S. A., Moore, H., Myers, M. M., & Shair, H. N. (2005). Maternally modulated infant separation responses are regulated by D2-family dopamine receptors. *Behavioral Neuroscience*, *119*, 1384–1388.
Nachmias, M., Gunnar, M., Mangelsdorf, S., Parritz, R. H., & Buss, K. (1996). Behavioral inhibition and stress reactivity: The moderating role of attachment security. *Child Development*, *67*, 508–522.
Newman, T. K., Syagailo, Y. V., Barr, C. S., Wendland, J. R., Champoux, M., Graessle, M., et al. (2005). Monoamine oxidase A gene promoter variation and rearing experience influences aggressive behavior in rhesus monkeys. *Biological Psychiatry*, *57*, 167–172.
NICHD Early Child Care Research Network. (2001). Child care and children's peer interaction at 24 and 36 months: The NICHD study of early child care. *Child Development*, *72*, 1478–1500.
NICHD Early Child Care Research Network. (2005). Duration and developmental timing of poverty and children's cognitive and social development from birth through third grade. *Child Development*, *76*, 795–810.
O'Connor, T. G., & Croft, C. M. (2001). A twin study of attachment in preschool children. *Child Development*, *72*, 1501–1511.
Ogawa, J. R., Sroufe, L. A., Weinfield, N. S., Carlson, E. A., & Egeland, B. (1997). Development and the fragmented self: Longitudinal study of dissociative symptomatology in a nonclinical sample. *Development and Psychopathology*, *9*, 855–879.
Pritchard, J. K., Stephens, M., Rosenberg, N. A., & Donnelly, P. (2000). Association mapping in structured populations. *American Journal of Human Genetics*, *67*, 170–181.
Purcell, S., & Sham, P. (2004). Properties of structured association approaches to detecting population stratification. *Human Heredity*, *58*, 93–107.
Ronai, Z., Guttman, A., Nemoda, Z., Staub, M., Kalasz, H., & Sasvari-Szekely, M. (2000). Rapid and sensitive genotyping of dopamine D4 receptor tandem repeats by automated ultra-thin-layer gel electrophoresis. *Electrophoresis*, *21*, 2058–2061.
Rubin, D. B. (1987). *Multiple imputation for nonresponse in surveys.* New York: Wiley.
Rutter, M. (2006). *Genes and behavior: Nature-nurture interplay explained.* Oxford, UK: Blackwell.
Rutter, M., Moffitt, T. E., & Caspi, A. (2006). Gene–environment interplay and psychopathology: Multiple varieties but real effects. *Journal of Child Psychology and Psychiatry*, *47*, 226–261.
Schafer, J. L. (1997). *Analysis of incomplete multivariate data.* London: Chapman & Hall.
Schmidt, L. A., Fox, N. A., Perez-Edgar, K., Hu, S., & Hamer, D. H. (2001). Association of DRD4 with attention problems in normal childhood development. *Psychiatric Genetics*, *11*, 25–29.
Schoots, O., & Van Tol, H. H. (2003). The human dopamine D4 receptor repeat sequences modulate expression. *Pharmacogenomics Journal*, *3*, 343–348.
Schultz, W. (1998). Predictive reward signal of dopamine neurons. *Journal of Neurophysiology*, *80*, 1–27.
Spangler, G., Schieche, U., Ilg, U., Maier, U., & Ackermann, C. (1994). Maternal sensitivity as an external organizer for biobehavioral regulation in infancy. *Developmental Psychobiology*, *27*, 425–437.
Thompson, R. A. (1999). Early attachment and later development. In J. Cassidy & P. R. Shaver (Eds.), *Handbook of attachment* (pp. 265–286). New York: Guilford Press.
Van Craenenbroeck, K., Clark, S. D., Cox, M. J., Oak, J. N., Liu, F., & Van Tol, H. H. (2005). Folding efficiency is rate-limiting in dopamine D4 receptor biogenesis. *Journal of Biological Chemistry*, *280*, 19350–19357.
van IJzendoorn, M. H., Goldberg, S., Kroonenberg, P. M., & Frenkel, O. J. (1992). The relative effects of maternal and child problems on the quality of attachment: A meta-analysis of attachment in clinical samples. *Child Development*, *63*, 840–858.
van IJzendoorn, M. H., Schuengel, C., & Bakermans-Kranenburg, M. J. (1999). Disorganized attachment in early childhood: Meta-analysis of precursors, concomitants, and sequelae. *Development and Psychopathology*, *11*, 225–249.
Van Tol, H. H., Wu, C. M., Guan, H. C., Ohara, K., Bunzow, J. R., Civelli, O., et al. (1992). Multiple dopamine D4 receptor variants in the human population. *Nature*, *358*, 149–152.
Waters, E. (1978). The reliability and stability of individual differences in infant–mother attachment. *Child Development*, *49*, 483–494.

Watts, V. J., Vu, M. N., Wiens, B. L., Jovanovic, V., Van Tol, H. H., & Neve, K. A. (1999). Short- and long-term heterologous sensitization of adenylate cyclase by D4 dopamine receptors. *Psychopharmacology, 141*, 83–92.

Weinfield, N. S., Whaley, G. J., & Egeland, B. (2004). Continuity, discontinuity, and coherence in attachment from infancy to late adolescence: Sequelae of organization and disorganization. *Attachment and Human Development, 6*, 73–97.

Young, L. J., & Wang, Z. (2004). The neurobiology of pair bonding. *Nature Neuroscience, 7*, 1048–1054.

Psychology Press
Taylor & Francis Group

Only connect? Communication, identification, and autism

R. Peter Hobson
Tavistock Clinic and University College London, London, UK

Anthony Lee
Tavistock Clinic, London, UK

Jessica A. Hobson
Tavistock Clinic and University College London, London, UK

In this paper, we elaborate a theoretical position and report an empirical study on a specific form of interpersonal engagement: the propensity to *identify with* the subjective orientation of another person. On the basis of a hypothesis that individuals with autism have a relative lack of this form of intersubjective connectedness (Hobson, 1993, 2002), we predicted that children and adolescents with autism would contrast with matched participants without autism ($n = 12$ per group) in specific aspects of communication when someone requested them to "Get Pete to do this" and demonstrated actions in Pete's absence. As predicted, on blind ratings of videotapes of participants' communication, those with autism achieved lower scores on four indices of identification that were selected a priori: emotional engagement, sharing experience in joint attention, communication of style, and shifting in communicative role. The two groups were almost completely separate on a composite measure of identification. We consider the implications of these findings for typical and atypical development.

Advances in social-developmental neuroscience depend not only upon increasingly penetrating techniques to assess neurological functioning, but also upon more refined characterization of the processes of social perception and responsiveness that allow human beings to enter others' experiential worlds. One priority is to specify the mechanisms that effect interpersonal engagement between young children and others and give rise to specifically human forms of sharing experience, symbolic thinking, and understanding of the mind (Hobson, 2002).

In the present paper, we unfold a theoretical perspective and report an empirical study on this topic. From a theoretical point of view, our aim is to explicate the notion of *identifying with* another person. From an empirical standpoint, we report a study in developmental psychopathology designed to illuminate how such a process might structure social responsiveness and adjustment in person-with-person communication. More specifically, we investigated several putatively interrelated facets of communicative impairment among children hypothesized to suffer a relative

Correspondence should be addressed to: R. Peter Hobson, Developmental Psychopathology Research Unit, Tavistock Clinic, 120 Belsize Lane, London NW3 5BA, UK. E-mail: r.hobson@ucl.ac.uk

The Economic and Social Research Council (ESRC) provided financial support for this project.

We gratefully acknowledge the parents, staff, and students of Edith Borthwick School, Essex, UK, and Helen Allison School, Kent, UK, whose participation made this study possible. We thank Verity Gill for her assistance with ratings. This paper was written in part while the first author was a Fellow at the Center for Advanced Study in the Behavioral Sciences, Stanford, CA, USA.

www.psypress.com/socialneuroscience DOI:10.1080/17470910701376852

lack in the propensity to identify with other people—those with early childhood autism.

The empirical study builds upon previous research on social engagement, joint attention, imitation and communication among children with autism, but is unique in two respects. The first respect is that we devised a novel approach to study these phenomena. Our aim was to evaluate how participants accomplish the task of acting as a go-between within a communicative setting. We set up conditions in which participants received a bodily expressed message from one person, and were given the task of transmitting this message to someone else. In order to do so, they needed to perceive (and arguably, assimilate) the movements and/or actions demonstrated by one person, and then communicate to another how he or she should do what they themselves had just been shown. Our aim in devising this procedure was to invoke a series of steps involving interpersonal sensitivity and communicative role taking, each of which depends on identifying with the psychological stance of someone else. Such identification may augment (a) an individual's ability to register and take on board what someone else is communicating, (b) translate the message into expressive form, and then (c) adopt the role of teacher in passing on the communication to someone else who may need to be identified with as recipient of the message.

It is relevant to observe that from the perspective of common sense, the procedure just described appears simple and seamless: all one has to do is to copy what one has been shown, and pass it on. Yet we predicted that in virtue of the need to identify with each of the adult testers—not only to pick up and transmit the bodily expressed gestures and actions, but also to adopt appropriate communicative roles and adjust to the stance of the communicative partners—this would present considerable difficulty for children with autism. The reason, according to our hypothesis, is that the children have a limited propensity to identify with the attitudes of others.

The second special feature of the study is in the way we conceptualize, and therefore rate and analyze, the process of identifying with others. We shall begin by offering a preliminary theoretical discussion of this process.

On identifying with others

The principal source of the idea that human beings identify with other people was Freud, and the notion of identification has been elaborated in detail by subsequent generations of psychoanalysts. There are several levels on which identification is said to work, but for the present purposes, we restrict discussion to a relatively "basic" level that Freud seemed to have in mind when he reflected upon the basis for empathy. Freud wrote: "Identification is known to psychoanalysis as the earliest expression of an emotional tie with another person" (Freud, 1921/1955, p. 105). We consider the process of identifying with others as perceiving, registering and assimilating another person's attitude or psychological orientation from that person's bodily anchored stance, such that this becomes a potential attitude or orientation when translated into a perspective from one's own vantage-point. Shortly we shall provide examples of how aspects of this process have been subject to controlled investigation, but for now we highlight how the structure of identification might be critical for empathy, joint attention, and communication.

The crux is that when one person (say, P1) identifies with the attitudes of someone else (P2), the structure of P1's experience takes a particular form: the attitudes of the someone else are experienced by P1 *as* those of P2. There is a paradox here, in that this means that P1's experience comes to encompass something of P2's experience in such a way that P1 is emotionally affected, without this meaning that P1's experience becomes P2's experience. After all, the setting for P2's feelings is experienced in ways that are specific to P2, just as the setting for P1's experience is also special in arising through P1 perceiving and responding to P2's experience. In most, if not all, instances of such emotional responsiveness there would be several other differences in the qualities of experience entertained by P1 and P2, for example those that pertain to personal meanings, or the ways P1 and P2 deal with feelings. Yet it is critical both for the phenomenology of relating to others and for the developmental implications of such relations, that there is differentiation *within* P1's experience with respect to attitudes derived from the perception of P2's bodily expressed mental state.

This implicit structure to certain forms of social engagement and experience may be considered through the paradigmatic case of sharing experiences (Hobson, 1991). In the case of sharing experiences through joint attention, for example, the philosopher Campbell (2002) has described how, in sharing, it is constitutive of your experience that the other is, with you, attending to something. In other cases, for example when one is frightened of another person's anger, one may experience (in the one experience) conflicting attitudes of fear-towards-an-angry-other-person. The partitioning implicit in such seemingly unified states becomes evident in other settings, for example in the play that one observes in young children who re-enact troubling encounters from the perspectives of *both* protagonists in the exchanges. In each case, experiences in relation to other persons have a special quality in which one has affective responses that are co-ordinated with but not subsumed by the other person's bodily expressed feelings. This is what interpersonal engagement means.

Before we consider the implications of such structuring of social experience for the phenomenology and patterning of psychological functioning, we need to make two important theoretical points. First, the account we have just given does not necessarily entail that one person conceptualizes someone else as another person. This is pivotal for how identification operates in the earliest phases of life, because infants do not have concepts of people. According to our hypothesis, the experience of "otherness" as one component of infants' own interpersonal experience is a vital part of what enables growth towards understanding persons-with-minds. Once developed, conceptual understanding never completely replaces the more intuitive and unconceptualized modes of identification on which it is founded.

Second, the propensity to identify with others is a source of cognitive differentiation. This is true not only for the conceptual differentiation of self and other, but also for several (perhaps not all) forms of perspective taking and symbolic thinking (Hobson, 2002; Werner & Kaplan, 1963; Wolf & Gardner, 1981). The reason is that through identifying with the attitudes of others towards a shared world, infants come to grasp the distinction between person-anchored attitudes and the objects and events that are at the focus of those attitudes; that is, more or less, they acquire the ability to differentiate between thoughts and things. They also learn the nature of predication in which a topic is distinguished from whatever is said about it.

In each of these respects, the process of identification is not merely cognitive in nature. It is also a motivational and paradigmatically emotional process. We are *drawn to* and *moved by* our affective engagements with other people. In virtue of the structure of identification, we are also moved towards assuming the psychological stances and perspectives of others.

Having explicated our view of how the propensity to identify with others is critical for both the intensity and the quality of typically developing children's emotional engagement with others, we need to elaborate further on how this structures their sensitivity towards and adjustment to other individuals' psychological orientation. If, as we have argued elsewhere (e.g., Hobson, 2002), both sharing experiences and co-ordinating one's own attitudes with the attitudes of others are necessary underpinnings for learning *through* other people's ways of relating to the world (Tomasello, 1999), as well as for sensitivity to people's communicative intentions, then identification is likely to play a critical role in shaping almost all forms of human communication. This theoretical perspective contrasts with more cognitive and/or representation-based analyses of communication. The latter tend to focus upon children's cognitive appreciation of what people intend in communicating. An account in terms of identification does not ignore the significance of such understandings, but it is *also* orientated to a person's interest in, sensitivity towards, engagement with, and adjustment to what others afford by way of psychological connectedness through their bodily expressions.

The case of autism

Children with autism have profound impairments in interpersonal engagement, creative symbolic play, and flexible, context-sensitive communication and thinking. Elsewhere we have presented detailed clinical as well as experimental material to suggest that these cardinal features of the syndrome are rooted in impairments in emotional perception, engagement, and responsiveness to other people, especially as such interpersonal functioning is dependent on the propensity to identify with other people (Hobson, 1993). Here we illustrate some recent studies on autism that point to the role of identification in affective engagement and other aspects of nonverbal communication, including joint attention and imitation, and highlight the potential importance of the very basic form of role taking that identification promotes. Each of these domains comes into focus in the empirical study of communication that we shall be reporting.

Emotion perception and engagement. There is now a body of evidence (see Hobson, 2005, for a summary) that children and adolescents with autism are atypical in the ways they perceive and express emotion, and in their engagement with the attitudes of other persons. This profoundly influences their relatedness with others in relation to a world that can be co-referenced and communicated about.

In one study, for example, Wimpory and colleagues (Wimpory, Hobson, Williams, & Nash, 2000) interviewed parents of very young children who were referred with difficulties in relating to and communicating with others. During the first two years of life, according to the parents, not one of the infants with autism had shown frequent and intense eye contact, engaged in turn taking with adults, or used noises communicatively, whereas half of the control children were reported to show each of these kinds of behavior. There were also fewer infants with autism who greeted or waved to their parents, who raised their arms to be picked up, who directed feelings of anger and distress towards people, who were sociable in play, or who enjoyed and participated in lap games.

A question arises about the source and significance of these limited forms of person-with-person engagement. Very recently, Hobson and colleagues have completed a set of controlled studies on self-awareness and expressions of social emotions such as coyness and guilt among children and adolescents with autism (Hobson, Chidambi, Lee, & Meyer, 2006). Among the findings were that parents of children with autism reported how their offspring were often affected by the moods of others, albeit often in a rather diffuse manner, and they also showed jealousy and a relatively non-social form of pride. Despite this, few showed clear concern for or guilt towards other people. When children were tested in potentially guilt- and coyness-inducing situations, these elicited affective and even self-conscious responses among the children and adolescents with autism, just as they did among matched participants without autism. For the most part, however, the reactions of those with autism lacked certain person-centered qualities of the feelings such as guilty looks towards the owner of an object that fell apart in their hands, or re-engagement looks towards the person in relation to whom they might be feeling coy.

Our suggestion is that this relative dearth of person-centered feelings holds the key to what is atypical in the social behavior and experience of persons with autism. We propose that individuals with the syndrome have a specific limitation in the self–other structure to socioemotional relations, a structure that is usually afforded by identifying with other bodily endowed persons. How does such a hypothesis square with evidence concerning other aspects of social interaction?

Joint attention. The Wimpory et al. (2000) study also provided evidence of what lack of engagement with another person's attitudes might mean for interpersonally co-ordinated relations with reference to objects and events in the environment. For example, not one of the infants with autism but at least half those in the control group were reported to offer or give objects to others in the first two years of life. The same was true of pointing at objects or following others' points. Few children with autism were said to show objects to others, and not one was said to have looked between an object of interest and an adult, for example when the infant wanted something out of reach. Such retrospective reports are complemented by quasi-experimental observational studies, an especially revealing instance being a study of social referencing. Sigman, Kasari, Kwon, and Yirmiya (1992; also Charman et al., 1997) tested 30 young autistic children with a mean age of under 4 years, closely matched non-autistic retarded and typically developing children. These children's responses were videotaped when an adult pretended to hurt herself by hitting her finger with a hammer, simulated fear towards a remote-controlled robot, and pretended to be ill by lying down on a couch for a minute, feigning discomfort. The children with autism not only showed less empathic responsiveness towards another person for his or her own sake, but they were also less hesitant than the control participants in playing with the robot.

The point we would stress here is that, often, young children with autism appear to be relatively unmoved in relation to another person's plight, and also unmoved to adopt a stance towards the world that corresponds with the stance of someone else (see also Mundy, 1995). Critically, the nature of the movement in question is towards assuming attitudes that have their source in another individual.

Further aspects of non-verbal communication. Consider the lack of greeting and person-focused anger and distress reported by parents in the Wimpory et al. (2000) study, and the following observations of older children and adolescents who were videotaped as they greeted an unfamiliar person and later took their leave (Lee & Hobson, 1998). Compared with matched mentally retarded control children, only half as many of the children and adolescents with autism gave spontaneous expressions of greeting and few smiled. In the farewell episode, half the individuals without autism but only three of those with autism made eye contact and said a goodbye. A telling observation was that all five participants with autism who gave a wave of farewell when the investigator called "Goodbye" as the child left through the door, did so with partial and often ill-directed flaps of the hand that were not 3co-ordinated with eye contact.

In our view, what one sees here is not simply that the participants with autism manifested a relative lack of affective contact with the person of the investigator, a fact that was reflected in specific "subjective" ratings of interpersonal engagement. In addition to this, the morphology of the waves suggested that the participants with autism had failed to identify with the communicative acts of other people waving. It appeared that they had been unable to assimilate into their own gestural repertoire what they had witnessed in people waving goodbye to others (or, indeed, to the children themselves), so that their other-person-directed gestures of farewell were hardly waves at all. We have recently reported how in conversation, autistic children and adolescents showed very few episodes of nods and shakes of their heads when listening to another person talking, in contrast to control participants who (in our view) were nodding in accordance with perspectives aligned with (García-Pérez, Lee, & Hobson, in press-b).

Imitation. The evidence we have presented thus far is suggestive, but only suggestive, that children with autism might have a limited propensity to identify with other people. Although we take the view that the most important source of evidence for this idea is that it serves to integrate and explain a broad set of phenomena in autism—a point that is central to the study we shall be describing in the second part of this paper—the single most persuasive body of evidence comes from recent research in the domain of imitation.

In the first of a series of investigations by our research group, Hobson and Lee (1999) tested matched groups of children with and without autism for their ability to imitate a person demonstrating four novel goal-directed actions on objects in two contrasting "styles." The results were that the children with autism were able to copy the demonstrator's actions, for example in pressing down a toy policeman's head to make him move, but they contrasted with control participants insofar as very few adopted the demonstrator's style of acting upon the objects (here, pressing with two fingers or with a wrist, or in other cases adopting a gentle or harsh style). In our view, this revealed a contrast between children's ability to observe and copy intended actions per se, relatively intact in autism, and the propensity to identify with and thereby imitate a *person's* expressive mode of relating to objects and events in the world.

This conclusion was buttressed by a further result from the study. One condition involved the demonstrator holding a pipe-rack against his own shoulder in order to strum it with a stick, either harshly or gently. Here a substantial majority of the control participants identified with the demonstrator and positioned the pipe-rack against their own shoulders before strumming it. By contrast, most of the children with autism positioned it at a distance in front of them, on the table. The children with autism neither assumed the manner nor the self-orientation with which the other person executed actions, even though they copied the actions per se.

To recap, our claim is that through identifying with the attitudes of another person, one participates in the other person's state while at the same time maintaining awareness of otherness in the person with whom one is affectively involved. According to our hypothesis, this process is operative not only in certain forms of imitation, as just illustrated, but also in a special quality of intersubjective engagement. In a recent study, Hobson and Hobson (in press) have presented evidence for just such a link between imitation and sharing of experience: those relatively few children with autism who imitated someone else's self/other-orientation in a set of actions designed to test this aspect of imitation *also* tended to be those individuals who manifested "sharing looks" to the demonstrator, that is, looks that appeared to reflect affective engagement with the person. This was also the case among participants without autism.

Role shifting. Finally, it is directly relevant to the present study that children and adolescents with autism have a lesser propensity to shift person-anchored roles in the course of communication. Hobson and Meyer (2005) gave a "sticker test" in which children needed to communicate to another person where on her body she should place her sticker badge. Over half the children with autism, but not a single child in the control group, *never* pointed to themselves. Instead the children with autism tended to point to the investigator. In a second condition, when the children's communication followed the investigator pointing to her own body to refer to the participant's body, only 3 of the 17 non-autistic children did *not* make a self-referential gesture to the investigator, yet 11 of the 17 children with autism still did not do so.

Therefore in making reference to their own body when trying to convey something to their communicative partner about her body, the children without autism anticipated that the other person would identify with their act of identifying with her body. The children with autism rarely communicated in this way, nor did they identify with the other person using such gestures towards themselves.

We believe that these simple observations reflect something that is profound and far reaching in its developmental implications, not least for the development of thinking and language. If we turn to role shifting in language, it is widely acknowledged that children with autism have special difficulties with pragmatics, and often fail to appreciate how language is modified according to social and non-social context in order to achieve mutual understanding between people (e.g., Tager-Flusberg, 2000). Those who are more verbally able often fail to adjust their language to fit with the social and communicative needs of the moment, so that their speech is based on their own immediate perspective, rather than being attuned to that of the listener. To take one instance that reveals a relative lack of self–other structure to the communicative experience of children with autism: not only according to clinical descriptions (Bosch, 1971; Kanner, 1943), but also according to systematic reports by teachers (Lee, Hobson, & Chiat, 1994) and parents (Hobson et al., 2006), they may be contrasted with matched children without autism in being especially prone to confusions in the use of the personal pronouns "I" and "you." As we have argued elsewhere (e.g., Hobson, 1993), to learn the use of "I," "mine," or "yours," young children need to identify with the attitudes expressed by a person who is observed to employ these terms. When the children want to communicate such attitudes from their own vantage point, they use the terms to express themselves just as they had witnessed others use the term to express *them*selves. In effect, we would argue, this is a linguistic counterpart of what we captured in the "sticker test."

Neuroscientific perspectives

We shall not attempt to detail how the perspective we have been developing dovetails with recent evidence concerning the neurofunctional underpinnings of interpersonal responsiveness (e.g., Decety & Chaminade, 2003; Gallese, 2001), because we anticipate the issue will be covered elsewhere in this volume. However, we note that there is emerging evidence, both from fMRI findings (e.g., Dapretto et al., 2006) and EEG patterns of mu frequency suppression (Oberman et al., 2005), that such functioning appears to be atypical among children with autism (also Gallese, 2006; Williams, Whiten, Suddendorf, & Perrett, 2001). Here there seems to be impressive convergence between typical and atypical neurological organization subserving communicative functions, and what may be discerned in the structures of typical and atypical interpersonal engagement. However, it remains to be seen how far neurofunctional findings have captured the essence of intersubjective co-ordination—and here we stress the pivotal importance of identification with person-centered *attitudes*—and the degree to which they are revealing the underpinnings of, or perhaps neurodevelopmental sequelae to, such interpersonal connectedness and differentiation.

Hypothesis and predictions

As already outlined, it is our hypothesis that there is a social-affective *and* social-motivational *and* social-cognitive process that may be specifically impaired in autism: the process of identifying with other people. Given that this process is considered to structure a range of socioemotional and social-communicative interactions, we consider that one source of evidence for its impairment in autism is concurrent deficits in specific aspects of emotional engagement, joint attention, imitation, and communication. It is just this coherent expression of seemingly diverse deficits that we explore in the present investigation.

We predicted that in the setting of a "Get Pete to do this" task, participants with autism would contrast with matched participants without autism in showing lesser degrees of (a) emotional engagement with the investigators, (b) forms of joint attention that implicate sharing of experiences, (c) communication of styles of action, and (d) role shifting from that of the learner to that of the teacher. We also anticipated that when these measures were combined in a composite index of identifying with someone else, there would be a substantial group difference. Although we also made lower-level, more behavioral ratings of participants' social interactions to fill out a picture of group similarities and differences, as well as to anticipate a potential criticism that some of our measures appeared to involve overly subjective ratings, these were not the target of our predictions.

METHOD

Participants

Participants were 12 adolescent males who had been diagnosed with autism and 12 adolescents (9 boys and 3 girls) with learning difficulties or developmental delay. We verified fulfillment of *diagnostic* criteria for autism on the basis of (a) observations of participants in their classroom settings, where scores for participants on the Childhood Autism Rating Scales (CARS, on which a score of 30 or above is generally taken to indicate autism; Schopler, Reichler, & Renner, 1988) were $M = 34.63$, $SD = 2.67$, range = 30–38.5, and (b) interviews with teachers using a checklist based on DSM-IV criteria (American Psychiatric Association, 1994) to ensure a comprehensive review of the clinical features. All participants with autism displayed characteristic patterns of impairment in social interaction and communication, along with repetitive or stereotyped interests and activities.

Participants were matched according to chronological age and verbal mental age (Table 1). Estimates of verbal ability were based on performance on the British Picture Vocabulary Scales (BPVS; Dunn, Dunn, & Whetton, 1982), the British version of the Peabody Picture Vocabulary Scale. The BPVS is a standardized, widely used measure of receptive vocabulary that also appears to assess a relative area of weakness for persons with autism (Jarrold, Boucher, & Russell, 1997; Lockyer & Rutter, 1970). Verbal intelligence as calculated using BPVS verbal age-equivalents relative to chronological age was very similar in the participants of each group (for participants without autism $M = 36.06$, $SD = 6.90$, range = 26–48, and for those with autism $M = 38.85$, $SD = 8.83$, range = 21–51).

Procedure

We established a setting in which we expected individuals without autism to identify with a person who demonstrated something to them, and then to identify with and therefore adjust their communication to the recipient of their own demonstration. Two investigators conducted the videotaped sessions in quiet rooms in the participants' schools. While the first investigator (E1, the second author) demonstrated each action to the participant, the second investigator (E2, the first author) was sent outside the testing room and as a result, was not present during the demonstration. Prior to each demonstration, participants were instructed to "Watch" and "Get Pete to do this" (where Pete was E2, not present during the demonstration).

Therefore once the investigators had introduced the task as one in which participants would be asked to get Pete (E2) to do things, E1 told E2

TABLE 1
Participant characteristics

	Chronological age			*Verbal mental age (BPVS)*		
	M	*SD*	*Range*	*M*	*SD*	*Range*
With autism ($n = 12$)	15;1	2;8	11;0 to 19;0	5;9	1;3	4;0 to 7;10
Without autism ($n = 12$)	15;0	2;1	11;11 to 17;11	5;5	1;1	3;6 to 7;2

Note: Ages presented in "years;months".

to leave the room. Then he presented a participant with the first of a series of six actions. Following this demonstration, E2 was invited to re-enter the room, where he stood looking at the participant expectantly. With E1 standing in the background (and, if necessary, encouraging the participant to get Pete to do what had been shown), the participant conveyed the message to E2, who to begin with, made a standard and mildly humorous error in executing the action. He looked up to the participant and when necessary, asked if this was correct. Except in the case of the first two actions (raising hands above one's head and stacking blocks), E2 then made an improved but stylistically inaccurate attempt. After a final pause to allow the participant to make further comment, E2 carried out the action accurately, and looked to the participant with satisfaction. Throughout, E2 conveyed his attempts to understand what the participant meant to convey, with a serious but also light-hearted attitude.

This procedure was repeated for each of six actions in turn. Overall, three actions involved goal-directed use of objects (e.g., using a mechanical arm to place a cloth frog into a waste bin), two were non-goal-directed involving the body (e.g., raising hands above head), and three included a form of expressive style (e.g., placing hands on hips in proud, assertive stance). Details of the demonstrations and the clumsy first and second attempts of E2, in order of administration, were as follows:

1. *Raise hands above head*: The first investigator held up both arms, bent at the elbows with palms facing the participant, above his own head. The second investigator's single incorrect action was to raise one knee and bend one arm with palm facing away and down.
2. *Stack blocks*: The investigator stacked a set of six blocks to create a vertical tower. The second investigator's single incorrect action was to line the blocks up in a horizontal row on the chair.
3. *Place hands on hips in proud stance*: The first investigator assumed a proud stance by assertively moving his legs astride while gripping each hip with his hands. The second investigator's first incorrect action was to lift one leg behind his body while gently touching his waist with one hand. His second improved, but stylistically inaccurate, attempt was to move his legs astride while gripping each hip with his hands, but in a gentle manner.
4. *Mechanical arm and cloth animals*: The first investigator used a mechanical arm with hand-held grip to lift two cloth animals from the floor and place them, one at a time, into an empty waste bin. The second investigator's first incorrect action was to attempt to use the handle of the mechanical arm to scoop up the cloth animals. His second, improved, attempt was to lift the animal with the claw but then to have difficulty squeezing the handle to release the animal from the claw into the waste bin.
5. *Jump aggressively on to cloth frog*: The first investigator jumped aggressively with both feet onto a cloth frog in the center of the room. The second investigator's incorrect action was to jump over the frog. His improved, but stylistically inaccurate, attempt was to step lightly on to the frog with one foot.
6. *Polish shoe vigorously*: The first investigator used a cloth to polish his shoe vigorously. The second investigator's first incorrect action was to polish the child's shoe, instead of his own shoe. His improved, but stylistically inaccurate, attempt was to polish his own shoe daintily.

Ratings

Two sets of ratings were completed by two independent judges, one of whom was blind to participant diagnoses and the hypotheses and predictions of the study. The blind judge's ratings were employed for the results and statistical analyses. The principal set of ratings were completed when judges took an overview of the entire set of six actions for each participant. Prior to this, the judges had completed a set of more specific ratings for each of the demonstrated actions, as described below. These ratings were conducted in order to ground the judges' global ratings as well as to capture particular aspects of behavior, but were not the target of our predictions.

Principal ratings (with interrater reliabilities). The following items were all rated on scales with scores ranging from 0 to 2 and, as we have stressed, concerned judgments of participants over the whole sequence of six trials:

1. The participant appeared *emotionally engaged* with E1 and/or E2 during the interaction, for example, through showing humor and pride (weighted $\kappa = .72$, substantial).
2. The participant engaged in joint attention of a kind that involved *sharing of experience* in relation to E2's actions with at least one of the testers. Judges were informed that they might rate such sharing according to participants' verbal and nonverbal communicative expressions (weighted $\kappa = .54$, moderate).
3. The participant integrated *stylistic and expressive* aspects of E1's behavior in his or her own expression during communication of the actions (weighted $\kappa = .56$, moderate).
4. The participant showed an ability to *shift roles* in assuming the communicative role and stance previously occupied by E1 as instructor, and fulfilling this role *both* in (a) demonstrating the action and (b) showing awareness of and helpfulness towards the responses of E2 in the role of the participant as learner (weighted $\kappa = .72$, substantial).

Ratings of each demonstration in turn. Participants' behavior in relation to the demonstrated actions were rated as follows. Two of the more specific ratings (conducted for each demonstration in turn) served as background variables. These provided an indication of participants' attentiveness to the demonstrator's actions, and the propensity to imitate those actions in communicating to E2. Therefore in a setting where participants might have tried to communicate solely through the spoken word, or have failed to communicate at all, these background measures established a framework for subsequent analyses. Here, and for subsequent ratings, we employ the intra-class correlation as a measure of interrater reliability, given that the measures concern mean scores across several ratings, or percentage values where appropriate.

Background variables:

1. Attentiveness to the demonstration by E1. This was rated on a scale of 0 (*poor*) to 2 (*good*) for each action (ICC = .69 for mean scores across the six actions).
2. Use of copied actions to communicate message to E2. This was rated as either present or not for each of the six actions (ICC = .93 on frequency of use of imitation across the six actions).

More specific ratings:

1. Emotional reactions to the errors and correct actions of E2. These were rated as either present or absent. For each of the six trials, a single rating was made for responses to E2's errors (in four trials, this covered two such errors), and another was made for responses to E2's ultimate success. The percentage of ratings in which participants showed an emotional reaction was computed (ICC = .84).
2. Eye contact (looking to E2). This was rated as being present or absent: (i) immediately following the demonstration by the participant; (ii) following the incorrect action(s) made by E2; and (iii) following E2's final success. Thus, there were three opportunities in each of the six trials to receive credit for looking. We computed the percentage of these ratings on which eye contact was made (ICC = .68).
3. Accuracy of imitation. This was rated only if copying was used to convey the action, on a scale of 0 (*poor*) to 2 (*good*). The ICC was .72 for the mean accuracy score across the relevant (up to 6) actions.
4. Quality of communication following the incorrect and correct actions of E2. This was rated on a scale of 0 (*poorly*) to 2 (*well*) following E2's responses on each trial, and again a single rating was made across both of E2's incorrect actions when he made more than one, and a single rating for whether a participant conveyed that E2 had finally got it right. Here we calculated the mean score for the two ratings each across the six trials (ICC = .90).

RESULTS

Although we had made directional predictions for the principal measures—those concerned with emotional engagement, sharing experiences in joint attention, communication of style, and communicative role shifting—we analyzed the data using two-tailed non-parametric tests of significance. According to our hypothesis, each index of identification was not independent of the others, and so it is not appropriate to adjust p values as if there were multiple independent comparisons. However, if one were to adopt a conservative approach and consider each of the principal measures independently, one might choose to accept $p < .01$ as significant. It should also be recalled that despite having made directional predictions, we applied two-tailed tests of significance.

Background variables

We begin by considering the two specific measures that serve to frame the remainder of the results. On ratings of attentiveness towards E1's demonstrations of the actions, every participant had a mean attention score between 1.50 and 2 (possible range = 0 to 2: for the group with autism, $M = 1.85$, $SD = 0.19$, without autism $M = 1.99$, $SD = 0.05$). The groups were also similar in the likelihood of using imitation as a way of communicating to E2. Nine participants with autism and 11 without autism used imitation of E1 in their instructions to E2 on at least five of the six actions, and although three participants with autism and one without autism used imitation on fewer than five of the six actions, all participants did so at least once. These results are helpful in indicating that, albeit sometimes with prompting, participants of both groups were able to make *some* imitative communication to E2.

Principal results

Emotional engagement. There was a highly significant group difference on interpersonal emotional engagement ($U = 7.50$, $z = 3.95$, $p < .001$). Seven of the twelve participants with autism but not a single individual in the control group received a score of 0. This score reflected emotional disconnection as evidenced by boredom, distress, emotional indifference, or unusual emotional responses such as laughter out of context. There were five participants with autism and three without autism who received scores of 1, indicating some engagement with the examiners through occasional displays of positive affect. Not a single participant with autism, but nine of those without autism, received a score of 2 for having fun being together with E1 and E2 as manifested in joint humor, enthusiasm, and/or pride during the interaction.

Experience sharing in joint attention. Here again there was a highly significant group difference ($U = 15.00$, $z = 3.54$, $p < .001$). Seven participants with autism received scores of 0 for not sharing the experience with the examiners, whereas not a single participant in the comparison group received this score. Five participants with autism and six of those in the comparison group were given scores of 1 for looking at E1 or E2 occasionally following imitative acts, but not really sharing the experience. No participant with autism but six in the comparison group received a score of 2 for spontaneously looking to and engaging with E2 and/or E1 about E2's performance.

The communication of style. Our next rating concerned the participants' propensity to adopt and integrate the expressive style shown by E1 in their subsequent communication to E2. Again, there was a marked group difference ($U = 21.0$, $z = 3.25$, $p < .01$). Three participants with autism and none without autism failed to communicate style at all, seven of those with autism and two in the comparison group were rated as showing an approximation of style, and only two participants with autism compared with ten of those without autism were rated as having communicated style clearly.

Communicative role shifting. There was a significant group difference in role shifting ($U = 17.00$, $z = 3.46$, $p < .01$), that is, the propensity to adopt the role of teacher in relation to E2 as a learner. Seven children with autism but no participant in the comparison group received a score of 0, and only two of those with autism and ten in the comparison group "clearly" showed this change in communicative stance.

Identification composite. Within each group considered separately, and with a single exception, incorrelations among individuals' scores for emotional engagement, experience sharing, communication of style, and communicative role shifting were positive (range of Spearman's rho .25 to .7). The one exception out of the 12 analyses occurred between the communication of style and role shifting among participants without autism (where the correlation was $-.20$), but this result was unsurprising given that on each of these measures, 10 out of the 12 participants achieved maximum scores.

To arrive at a composite measure of identification, we summed individual participants' scores for emotional engagement, experience sharing, the communication of style, and role shifting, yielding a range of possible scores from 0 to 8.

Figure 1 illustrates how there was an almost complete separation between the groups on this measure, with the majority of the participants with autism scoring between 0 and 2 and the majority of those without autism scoring between 6 and 8.

Scores on the identification composite were neither significantly associated with chronological nor verbal mental age for either group of participants (for group with autism, the correlations were −.44 with chronological age and .29 for verbal mental age, and for the group without autism, −.39 for chronological age and −.20 for verbal mental age). For example, on the identification composite the six participants with autism in the lower range of verbal mental age (48–62 months) scored a mean of 1.8 and those in the upper range (69–94 months) a mean of 2.8, and these scores may be contrasted with the corresponding values of 6.9 and 7.0, respectively, among the lower- and upper-ability control participants.

More specific ratings

We conducted these ratings to flesh out details of participants' responses, rather than to be the target of our principal predictions (although we did anticipate that in each case there would be group differences). Therefore, here we provide statistics for illustrative purposes only.

Emotional responses to the investigator's behavior. Here the data were in the form of percentages, and concerned the proportion of times that participants showed affective responses to the actions of E2. On specific ratings of responses to the failures and successes of E2, participants with autism were less likely to show the characteristic emotional reactions such as humor and pride that were shown by the control participants: for participants with autism $M = 42\%$, $SD = 35\%$, range = 0 to 92%, and for those without autism $M = 88\%$, $SD = 14\%$, range = 55 to 100%, $t(22) = -4.17$, $p < .001$.

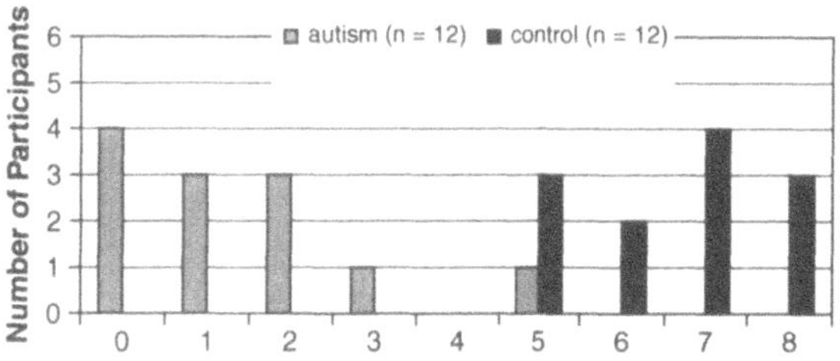

Figure 1. Number of participants with each score on the *Identification Composite* (range 0 to 8). This score is the sum of each individual's scores for emotional engagement, experience sharing, communication of style, and communicative role taking.

Eye contact. On specific ratings of tendency to make eye contact with E2 immediately following their own demonstrations, as well as following the successes and failures of E2, participants with autism did so approximately 50% of the time ($M = 51\%$, $SD = 36\%$, range = 0 to 100%), compared with over 80% for the comparison group ($M = 86\%$, $SD = 20\%$, range = 50 to 100%), $t(22) = -2.88$, $p < .01$.

Imitation. For accuracy of imitation across the six items (possible range 0 to 2), participants with autism were rated as being less accurate than the control group (for group with autism, $M = 1.6$, $SD = 0.32$, for group without autism, $M = 1.96$, $SD = 0.08$, $U = 18.00$, $p < .01$). No participant had a mean score below 1, however, suggesting that they all imitated the actions of E1 with some degree of accuracy.

Communication. On specific ratings of the quality of communication following the incorrect and then the final correct actions of E2, on a range of mean scores 0–2 across the six trials, participants with autism were less likely to communicate effectively, $M = 0.59$, $SD = 0.64$, compared with comparison group, $M = 1.42$, $SD = 0.38$, $t(22) = -3.86$, $p < .001$. Eight of the twelve participants with autism had mean scores of less than 1 out of 2, whereas this was the case for only one participant in the comparison group.

DISCUSSION

There are two parts to the brief discussion that follows. First, we consider the results of the present study and note potential implications for our understanding of autism. Second, we return to consider the nature and significance of the propensity to identify with the attitudes of others, both for interpersonal engagement and for the development of communication and interpersonal understanding. We also suggest that the process of identification may be distinctive to human beings, and critical for distinguishing the specialness of human social-cognitive development.

In the present study, all participants were attentive to E1's demonstration, and were able to understand enough of the "Get Pete" task to communicate with E2 (Pete) in *some* manner. In most cases participants copied E1's actions in order to convey what E2 should do. Despite this, there were significant group differences in each of the respects that were anticipated on the basis of our hypothesis that individuals with autism are limited in the propensity to identify with other people. Participants with autism were less likely (a) to engage emotionally with the experimenters, (b) to share with either of the testers their experience of E2's responsiveness, (c) to communicate to E2 the expressive styles of action demonstrated by E1, and (d) to assume the role of teacher and adjust their own demonstrations to the needs of E2 as learner.

When we computed the composite index of identification, the group difference was striking: apart from a single individual with autism who achieved a score equal to the lowest-scoring three participants without autism, there was a complete separation of the two groups. It appears that our approach captured something of the essence of the communicative impairments of children with autism.

The more specific ratings filled in details of this picture, most notably in relation to the prevalence of participants' emotional responses to and eye contact with the investigator as he struggled (playfully) to learn what to do, their potential abilities in imitating what was demonstrated, and the effectiveness of their communication. Although the participants with autism could copy actions, their engagement was limited in depth and their communication relatively constrained.

Given the way in which the two groups of participants were matched, it is clear that the group differences that emerged were out of keeping with at least one index of linguistic cognitive ability. Thus the group contrasts in affective and communicative styles were not merely a reflection of "general" mental retardation. Rather, the participants with autism showed highly distinctive patterns of personal relatedness and communication. Even within the groups considered separately, there was not a close tie between cognitive ability and the aspects of communication we were studying, in that scores were neither correlated with chronological nor verbal mental age. Therefore it would appear that the propensity to engage in these forms of affective and communicative interaction is not constrained by cognitive ability, at least not among participants of the age and ability that we were testing.

If one steps back to consider what these results might signify, it is relevant to note that the methodological approach of the study was in some ways unconventional. Our aim was to investigate how the process of identifying with others might configure communication, and to gather evidence either for or against the hypothesis that the propensity to identify with others is relatively weak among individuals with autism. Instead of following the more familiar procedure of seeking correlations among scores on a battery of separate and abstracted tests of emotional responsiveness, joint attention, imitation, role taking and so on, we took a complementary tack and attempted to study a single set of communicative transactions so that we could analyze these through a variety of indices *of the very same process*, namely that of identifying with the attitudes of another person. Such an approach amounted to a dissection of the process of communication itself, one that preserved rather than dismembered the forms of intersubjective engagement and adjustment that such communication typically entails.

Correspondingly, we interpret the results as suggesting that persons with autism have a weaker propensity to identify with the subjective experiences of others and, beyond this, as reflecting how such a process is profoundly important for individuals' abilities to engage in communicative exchanges characterized by mutual adjustment and emotional sharing. Having said this, one might question just how far it is specifically in virtue of their limited propensity to identify with others, that persons with autism have communicative impairments—and whether those aspects of communication that do *not* implicate identification are relatively spared.

This is a question that can be addressed adequately only through more focused studies involving direct comparisons among particular aspects of communication in appropriately matched groups of participants. In the introduction, we referred to a number of such studies yielding evidence that, when compared with matched children without autism, those with autism have been (a) less inclined to imitate a person's style and self-orientation than to copy goal-directed actions (Hobson & Lee, 1999), (b) less likely to communicate with reference to self/other-orientation and more likely to employ a "geometric" framework (Hobson & Meyer, 2005; also Meyer & Hobson, 2004), (c) less likely to demonstrate any "sharing" looks when imitating (Hobson & Hobson, 2007), (d) less likely to nod their heads in accord with another person during conversation, and less likely to be rated as highly engaged, despite being similar on counts of some components of nonverbal communication (García-Pérez, Lee, & Hobson, in press-b), and (e) less likely to sustain role taking in recounting different versions of a story, even when they were relatively proficient in producing narratives (García-Pérez, Hobson, & Lee, in press-a). In each case, the specificity of the group differences could be interpreted as manifestations of these children's limited ability to engage with, assimilate, assume and maintain the bodily anchored stance of other persons through identification, whereas other aspects of imitation, communication, and language are relatively spared.

At certain points, we have referred to identification as a "mechanism," and at other times as a process. Perhaps we should qualify, as well as clarify, what we mean. The principal qualification is that there is development in how one person may identify with another, or indeed with such social characteristics such as a race or religion. For the purposes of the present study, our focus was upon a level of engagement with, and assimilation of, another person's attitudes. The clarification concerns the role of "identifying with" at different explanatory levels. We take the view that at a psychological level of explanation, there is a configuration of human relatedness that does indeed unfold as a process with the properties we have been stressing. At the same time, the process of "identifying with" can be viewed as a mechanism, insofar as it explains why social engagement takes the form, and has the developmental implications, that it does. For example, it characterizes why human forms of sharing are special and why role reversals in imitation and communication occur as they do; and beyond the scope of the present paper, identifying-with is needed if early social engagement is to yield context-sensitive creative symbolizing and in person-centered feelings such as concern and guilt (Hobson, 1993; Hobson et al., 2006).

These considerations link with another, namely the question of whether identification is an all-or-none process. This is relevant for understanding how humans differ from nonhuman primates, insofar as it has been claimed that the process of identification is unique to humans (Hobson, 2002). It is also important for our account of autism. Our view is that for particular individuals, identifying-with might occur weakly or not at all, in that the organization of and motivation for self–other engagement and experience might be partial or tenuous. We suppose that a weak rather than absent propensity to identify-with is typical among children with autism. If so, it may be possible to foster affected children's interest in, and ability to engage with, other people as selves. Yet we also take the view that there is a threshold effect, such that if an individual has less than a certain degree of social experience structured by identifying with the attitudes of others in relation to a shared world (a form of experience also limited among congenitally blind children), that individual is at risk for developing the syndrome of autism. In the case of nonhuman primates, on the other hand, there appears to be an absence or near absence of identification, and this explains why individuals of these species do not manifest "sharing" forms of pointing, role-reversing communication, symbolic functioning, and much else besides.

In the latter regard, the present study illustrates how certain forms of *interpersonal* flexibility and co-ordination integral to communication bear an intimate relation to certain forms of *intra-individual* flexibility and co-ordination of thinking and attitude. On the one hand, it is possible to take the view that here we see how the difficulties encountered by children with autism in achieving creative, context-sensitive thinking and in adopting a variety of perspectives are manifest in their limited abilities to meet the demands of communicating effectively to someone else. On the other hand, one may surmise that the children's limited ability to identify with the stances and psychological orientations of others might not only correspond with, but also *underlie*, their relatively one-track and inflexible modes of thought and attitude. For almost two decades now, we have been marshalling arguments in favor of this latter, developmentally grounded perspective on autism (e.g., Hobson, 1989, 1990, 1993). If a range of intra-psychological functions such as understanding other people (so-called Theory of Mind), creative symbolic functioning, and aspects of cognitive flexibility and executive functioning develop in part through the interiorization of interpersonal processes in the manner suggested by Vygotsky (1962), then we may find the source of many of the features of the syndrome of autism in the children's lesser propensity to identify with the attitudes of other people (Hobson, 2002). An implication is that facilitating the development of communicative skills among individuals with autism may involve more than teaching specific forms of behavior, and instead require a focus on affective relatedness in order to foster more person-centered engagement (e.g., Greenspan & Wieder, 2006; Gutstein, 2000).

The writings of several other authors appear to bear upon aspects of the present account, which was founded upon the psychoanalytic concept of identification (Freud, 1921/1955). For example, in Trevarthen's (1998) reprise of his long-held views that young infants have self–other differentiation and engage in true communication with others, he cites the ideas of Bråten, who suggests that from very early in life there is "a 'virtual other' complement to the bodily self that offers a participant (companion) space in which others may be included in felt immediacy" (Bråten, 1998, p. 106). Indeed, Bråten (1998) drew upon evidence that was only provisional at the time of writing to suggest that children with autism might have difficulty with the kind of "altercentric participation" that we consider is essential to the process of identifying with others. Meltzoff (2007a, 2007b) continues to develop his "like me" framework for social cognition, reasoning that young infants imitate others through a process of active intermodal mapping in which "representations of the observations of others are 'tagged' so as to differentiate them from the representation of one's own motor acts" (Meltzoff, 2007a, pp. 37–38). This is a critical matter, and pivotal to our own view that there is differentiation as well as connectedness within the process of identifying with others. Where the present orientation differs from that of Meltzoff, is that (with qualifications) he still writes in terms of the child coming to understand others through applying a "like me" *analogy*, whereas the process of identifying-with is posited as a grounding for, not the outcome of, such reasoning (and is different, too, from Meltzoff's 2007b, p. 129, alternative formulation that "the infant recognizes '*that* looks like *this* feels'").

Finally, we return to implications of the present study for neuroscientific approaches to social functioning, and more specifically to understanding how people are able to enter others' mental worlds and to develop concepts of self and others as persons-with-minds. We have already noted how recent neurofunctional evidence is consonant with elements of the account of social engagement we have been elaborating. We hope that the present study adds grist to the social neuroscience research mill, not least in highlighting how this needs to encompass a relational as well as a cognitive dimension.

More specifically, we may be led to envisage more precisely what is entailed in processes of interpersonal simulation or mirroring, especially with regard to the human propensity to be moved to adopt other-person-centered feelings. There appears to be a relational self–other structure to the registration, representation, and evocation of person-anchored psychological stances, where the person is not necessarily a particular named individual but, rather, an individuated center of subjective orientation. Here we anticipate that although researchers who study nonhuman primates will continue to find neurofunctional correlates of simulation of conspecifics' actions and feelings, there will also prove to be neuroscientific correlates of identification that are unique to human beings. The reason we say this is that according to our view (Hobson, 2002; and see Tomasello, Carpenter, Call, Behne, & Moll, 2005, for a related perspective), the structure of identification not only distinguishes the social relations of human beings from those of other primates, but also holds the key to understanding the developmental underpinnings of much that is special in the thinking and language of *Homo sapiens*.

A further implication of this perspective is that we should be acutely aware of the need to interpret neurofunctional evidence, whether from typical or atypical populations, in developmental perspective. After all, both the brain and the mind become organized over time, and what appear to be localized functions in later childhood and beyond may have become so on the basis of an individual's social and nonsocial experiences earlier in life. We believe some Theory of Mind theorizing has been led astray by its failure to appreciate how concepts of mind, *and* a person's propensity and ability to apply such concepts appropriately (not least in the course of communication), might be grounded in modes of interpersonal relatedness that are configured by preconceptual relational processes that operate from very early in life, that are paradigmatically emotional in nature, and that both link and differentiate self and other as embodied persons (e.g., Hobson, 1990; Hobson et al., 2006).

REFERENCES

American Psychiatric Association. (1994). *Diagnostic and statistical manual of mental disorders* (4th rev. ed.). Washington, DC: APA.

Bosch, G. (1970). *Infantile autism* (D. Jordan & I. Jordan, Trans.). New York: Springer-Verlag.

Bråten, S. (1998). Infant learning by altercentric participation: The reverse of egocentric observation in autism. In S. Bråten (Ed.), *Intersubjective communication and emotion in early ontogeny* (pp. 105–124). Cambridge, UK: Cambridge University Press.

Campbell, J. (2002). *Reference and consciousness.* Oxford, UK: Oxford University Press.

Charman, T., Swettenham, J., Baron-Cohen, S., Cox, A., Baird, G., & Drew, A. (1997). Infants with autism: An investigation of empathy, pretend play, joint attention, and imitation. *Developmental Psychology, 33*, 781–789.

Dapretto, M., Davies, M. S., Pfeifer, J. H., Scott, A. A., Sigman, M., Bookheimer, S. Y., et al. (2006). Understanding emotions in others: Mirror neuron dysfunction in children with autism spectrum disorders. *Nature Neuroscience, 9*, 28–30.

Decety, J., & Chaminade, T. (2003). When self represents the other: A new cognitive neuroscience view on psychological identification. *Consciousness and Cognition, 12*, 577–596.

Dunn, L. M., Dunn, L., & Whetton, C. (1982). *British Picture Vocabulary Scale.* Windsor, UK: NFER-Nelson.

Freud, S. (1955). Identification. In J. Strachey (Ed.), *The standard edition of the complete psychological works of Sigmund Freud* (Vol. xviii, pp. 105–110). London: Hogarth. (Originally published 1921)

Gallese, V. (2001). The 'shared manifold' hypothesis: From mirror neurons to empathy. *Journal of Consciousness Studies, 8*, 33–50.

Gallese, V. (2006). Intentional attunement: A neurophysiological perspective on social cognition and its disruption in autism. *Cognitive Brain Research, 1079*, 15–24.

García-Pérez, R. M., Hobson, R. P., & Lee, A. (in press-a). Narrative role-taking in autism. *Journal of Autism and Developmental Disorders.*

García-Pérez, R. M., Lee, A., & Hobson, R. P. (in press-b). On intersubjective engagement: A controlled study of nonverbal communication in autism. *Journal of Autism and Developmental Disorders.*

Greenspan, S. I., & Wieder, S. (2006). *Engaging autism: Helping children relate, communicate and think with the DIR floortime approach.* Cambridge, MA: Da Capo.

Gutstein, S. (2000). *Autism Aspergers: Solving the relationship puzzle.* Arlington, TX: Future Horizons.

Hobson, R. P. (1989). On sharing experiences. *Development and Psychopathology, 1*, 197–203.

Hobson, R. P. (1990). On the origins of self and the case of autism. *Development and Psychopathology, 2*, 163–181.

Hobson, R. P. (1991). Against the theory of 'Theory of Mind'. *British Journal of Developmental Psychology, 9*, 33–51.

Hobson, R. P. (1993). *Autism and the development of mind.* Hove, UK: Lawrence Erlbaum Associates Ltd.

Hobson, R. P. (2002). *The cradle of thought.* London: Pan Macmillan.

Hobson, R. P. (2005). Autism and emotion. In F. R. Volkmar, R. Paul, A. Klin, & D. Cohen (Eds.), *Handbook of autism and developmental disorders* (Vol. 1, pp. 406–422). Hoboken, NJ: Wiley.

Hobson, R. P., Chidambi, G., Lee, A., & Meyer, J. A. (2006). Foundations for self-awareness: An exploration through autism. *Monographs of the Society for Research in Child Development, 71*(Serial No. 284), 1–166.

Hobson, J. A., & Hobson, R. P. (in press). Identification: The missing link between imitation and joint attention? *Development and Psychopathology.*

Hobson, R. P., & Lee, A. (1999). Imitation and identification in autism. *Journal of Child Psychology and Psychiatry, 40*, 649–659.

Hobson, R. P., & Meyer, J. A. (2005). Foundations for self and other: A study in autism. *Developmental Science, 8*, 481–491.

Jarrold, C., Boucher, J., & Russell, J. (1997). Language profiles in children with autism: Theoretical and methodological implications. *Autism: The International Journal of Research and Practice, 1*, 57–76.

Kanner, L. (1943). Autistic disturbances of affective contact. *Nervous Child, 2*, 217–250.

Lee, A., & Hobson, R. P. (1998). On developing self-concepts: A controlled study of children and adolescents with autism. *Journal of Child Psychology and Psychiatry, 39*, 1131–1141.

Lee, A., Hobson, R. P., & Chiat, S. (1994). I, you, me and autism: An experimental study. *Journal of Autism and Developmental Disorders, 24*, 155–176.

Lockyer, L., & Rutter, M. (1970). A five to fifteen year follow-up study of infantile psychosis: IV Patterns of cognitive ability. *British Journal of Social and Clinical Psychology, 9*, 152–163.

Meltzoff, A. (2007a). The "like me" framework for recognizing and becoming an intentional agent. *Acta Psychologica, 124*, 26–43.

Meltzoff, A. (2007b). "Like me": A foundation for social cognition. *Developmental Science, 10*, 126–134.

Meyer, J. A., & Hobson, R. P. (2004). Orientation in relation to self and other: The case of autism. *Interaction Studies, 5*, 221–244.

Mundy, P. (1995). Joint attention and social-emotional approach behavior in children with autism. *Development and Psychopathology, 7*, 63–82.

Oberman, L. M., Hubbard, E. M., McCleery, J. P., Altschuler, E. L., Ramachandran, V., & Pineda, J. A. (2005). EEG evidence for mirror neuron dysfunction in autism spectrum disorders. *Cognitive Brain Research, 24*, 190–198.

Schopler, E., Reichler, R. J., & Renner, B. R. (1988). *The Childhood Autism Rating Scale (CARS)*. Los Angeles, CA: Western Psychological.

Sigman, M. D., Kasari, C., Kwon, J. H., & Yirmiya, N. (1992). Responses to the negative emotions of others by autistic, mentally retarded, and normal children. *Child Development, 63*, 796–807.

Tager-Flusberg, H. (2000). Language and understanding minds: Connections in autism. In S. Baron-Cohen, H. Tager-Flusberg, & D. J. Cohen (Eds.), *Understanding other minds: Perspectives from developmental cognitive neuroscience* (2nd ed., pp. 124–149). Oxford, UK: Oxford University Press.

Tomasello, M. (1999). *The cultural origins of human cognition*. Amherst, MA: Harvard University Press.

Tomasello, M., Carpenter, M., Call, J., Behne, T., & Moll, H. (2005). Understanding and sharing intentions: The origins of cultural cognition. *Behavioral and Brain Sciences, 28*, 675–691.

Trevarthen, C. (1998). The concept and foundations of infant intersubjectivity. In S. Bråten (Ed.), *Intersubjective communication and emotion in early ontogeny* (pp. 15–46). Cambridge, UK: Cambridge University Press.

Vygotsky, L. S. (1962). *Thought and language* (E. Hanfmann & G. Vaker, Trans.). Cambridge, MA: MIT Press.

Werner, H., & Kaplan, B. (1963). *Symbol formation*. Hillsdale, NJ: Lawrence Erlbaum Associates, Inc.

Williams, J. H. G., Whiten, A., Suddendorf, T., & Perrett, D. I. (2001). Imitation, mirror neurons and autism. *Neuroscience & Biobehavioral Reviews, 25*, 287–295.

Wimpory, D. C., Hobson, R. P., Williams, J. M. G., & Nash, S. (2000). Are infants with autism socially engaged? A controlled study of recent retrospective parental reports. *Journal of Autism and Developmental Disorders, 30*, 525–536.

Wolf, D., & Gardner, H. (1981). On the structure of early symbolism. In R. L. Schiefelbusch & D. D. Bricker (Eds.), *Early language: Acquisition and Intervention* (pp. 287–327). Baltimore: University Park Press.

SOCIAL NEUROSCIENCE, 2007, 2 (3–4), 336–352
Psychology Press
Taylor & Francis Group

The self as a moral agent: Linking the neural bases of social agency and moral sensitivity

Jorge Moll
National Institutes of Health, Bethesda, Maryland, USA, and LABS-D'Or Hospital Network, Rio de Janeiro, Brazil

Ricardo de Oliveira-Souza
LABS-D'Or Hospital Network, and Gaffrée e Guinle University Hospital, Rio de Janeiro, Brazil

Griselda J. Garrido
Instituto Israelita de Ensino e Pesquisa Albert Einstein, São Paulo, Brazil

Ivanei E. Bramati
LABS-D'Or Hospital Network, Rio de Janeiro, Brazil

Egas M. A. Caparelli-Daquer
LABS-D'Or Hospital Network, and Gaffrée e Guinle University Hospital, Rio de Janeiro, Brazil

Mirella L. M. F. Paiva, Roland Zahn, and Jordan Grafman
National Institutes of Health, Bethesda, Maryland, USA

The human brain is inherently able to understand the world in moral ways, endowing most of us with an intuitive sense of fairness, concern for others, and observance of cultural norms. We have argued that this *moral sensitivity* ability depends on a sophisticated integration of cognitive, emotional, and motivational mechanisms, which are modulated by individual experience in different cultural *milieus*. Different lines of investigation on agency and morality have pointed to overlapping neural systems. Therefore, understanding the relationships between morality and agency may provide key insights into the mechanisms underlying human behavior in several clinical and societal settings. We used functional MRI to investigate the contribution of agency and of specific moral emotions to brain activation using action scripts. Results showed that *emotionally neutral agency* recruited neural networks previously associated with agency, intentionality and moral cognition, encompassing ventral and subgenual sectors of the medial prefrontal cortex (PFC), insula, anterior temporal cortex and superior temporal sulcus (STS). Compared to emotionally neutral agency, different categories of moral emotions led to distinct activation patterns: (1) prosocial emotions (guilt, embarrassment, compassion) activated the anterior medial PFC and STS, with (2) empathic emotions (guilt and compassion) additionally recruiting the mesolimbic pathway; (3) other-critical emotions (disgust and indignation) were associated with activation of the amygdala—parahippocampal and fusiform areas. These findings indicate that agency related to norm-abiding social behaviors of emotionally neutral scripts share neural substrates both with the "default

Correspondence should be addressed to: Jorge Moll, Cognitive Neuroscience Section, NINDS, National Institutes of Health, Bethesda, Maryland, USA. E-mail: mollj@ninds.nih.gov

This research was supported by the LABS–D'Or Hospital Network and PRONEX–CNPq, and by FAPESP (project # 03/11794–6).

www.psypress.com/socialneuroscience DOI:10.1080/17470910701392024

mode" of brain function and with the moral sensitivity network. Additional activation in specific components of this network is elicited by different classes of moral emotions, in agreement with recent integrative models of moral cognition and emotion.

INTRODUCTION

Moral experience is inextricably tied to our "sense of agency," or the subjective experience of being the cause of our own actions or attributing agency to the actions of others. The sense of agency and related phenomena, such as intentionality and perspective taking, are also closely allied to the subjective experience of personal responsibility: without agency we would not feel, or could not be held, responsible for our actions. Our natural propensity to attribute cause to our and to other people's behavior allows us to make moment-to-moment predictions on the intentions of others and to update them continuously as the interpersonal scenarios change.

The sense of *agency* constitutes a major dimension of moral experience, particularly in what concerns the emotions that accompany it, or "the moral emotions." Whereas basic emotions, which we share with most mammalian species, spring from ideas and perceptions endowed with personal relevance, moral emotions are linked to the interest either of society as a whole, or at least of persons other than the judge or agent (Haidt, 2003). Moral emotions include guilt, anger, indignation, shame, regret, gratitude, pride, embarrassment, envy, pity, contempt, awe, and jealousy. Thus, guilt is experienced when we recognize ourselves as the cause of another person's misfortune. In contrast, if we witness someone being accidentally hurt, compassion is experienced instead. We feel indignation when we get hurt, especially if that was done intentionally. But indignation can also be experienced by witnessing someone being hurt intentionally by a third-party agent. In a more prosaic context, a *faux pas* occurs when one *unintentionally* violates an accepted social rule ("Sorry, I really didn't mean it . . . "). In this case, embarrassment is created by the dissonance between the *intended* (a witty remark intended to provoke amusement at a dinner party) and the actual outcome (ridiculing someone). From the aspects mentioned above, it is already apparent that agency cannot be reduced to a single dimension; agency can be constructed at different cognitive and, as recent neuroscientific studies have started to show, at different neural levels (Decety & Sommerville, 2003). Relevant to the present study, the sense of agency is automatically evoked when we watch the protagonist of a movie in action, or when reading a novel, as if we were actually performing such actions. We are also able to immediately feel immersed in the contexts of an action scenario, by way of simulation mechanisms (Gallese, Keysers, & Rizzolatti, 2004; Iacoboni, 2005). Here we will refer to this phenomenon as "*social agency*." Beyond this more "basic" agency mechanism, we are also able to quickly apprehend causal relations among agents. For example, whether you spill coffee on another person because of your clumsiness or because someone bumped into you, the more elementary sense of agency in a social situation is present; but the "causal agent" leading to coffee spillage will differ between the two situations, as well as the associated emotional experiences. Whereas both types of agency attributions are relevant, in the present study we shall focus on the more elementary "social agency" construct; the relationships between causality, agency and moral emotions will need to be further addressed in future studies.

Moral emotions play a central role in the guidance of social behavior in everyday life (Moll, de Oliveira-Souza, & Eslinger, 2003; Schulkin, 2004). They are culturally ubiquitous (Fessler, 1999) notwithstanding the ample diversity of local norms and values within societies. Moral emotions can be both discriminated from, or clustered according to, more basic components including attachment, agency, intentionality and prediction of outcomes (Moll, de Oliveira-Souza, Zahn, & Grafman, in press). Emotions such as disgust, sadness and fear have generally been enlisted as basic emotions, but they can acquire "moral" overtones and thus qualify as moral depending on societal and cultural contexts (Moll, Zahn, de Oliveira-Souza, Krueger, & Grafman, 2005b). Although a final taxonomy has not as yet been agreed upon, moral emotions can be tentatively clustered in superordinate and subordinate categories. A *prosocial* cluster including guilt, embarrassment, compassion, and gratitude, promotes cooperation, helping, reparative actions and social conformity (Moll et al., in press). A subclass of prosocial emotions, the *empathic* emotions, does so by invoking a feeling of social attachment—guilt, gratitude and compassion (Batson et al., 1991; Eisenberg, 2000). On the other hand, *other-critical* emotions[1] (disgust, contempt, indignation) are experienced when others violate norms or one's "rights," promoting aggression, punishment, group dissolution and social reorganization (Allport 1954; Moll et al., 2005a).

A distributed brain network is involved in moral experience regardless of stimuli (pictures, written statements, narratives) and tasks (passive viewing of morally-relevant stimuli, active moral judgments) employed in functional imaging studies (Moll et al., 2005b). The key regions for moral sensitivity include the prefrontal cortex (orbitofrontal cortex, OFC, and anterior sectors of the prefrontal cortex, PFC), the temporal poles, superior temporal sulcus (STS) region and the temporo-parietal junction, and limbic-subcortical structures (ventral striatum, hypothalamus, basal forebrain). It has been proposed that moral experiences comprise "Gestalt" states (Moll et al., 2005b) arising from the temporal synchronization of neural activity within this network (Singer, 2001). The engagement of these areas by moral appraisals independently of task contingencies originated the idea of *moral sensitivity* as a mechanism by which moral significance is automatically attributed to ordinary events (Moll et al., 2002b). Remarkably, functional imaging studies on agency and intentionality have shown that several of the aforementioned regions are similarly recruited (Blakemore & Frith, 2004; den Ouden, Frith, Frith, & Blakemore, 2005; Frith & Frith, 2005).

Although lawyers and philosophers have long been aware of the mutual dependence of morality and agency, investigations on the neurobiology of moral emotions and agency have so far evolved largely in parallel (Chorvat & McCabe, 2004; Goodenough & Prehn, 2004; Zeki & Goodenough, 2004). Most reports have either focused on agency without controlling for the experience of specific emotions (Blakemore, Winston, & Frith, 2004; Decety & Sommerville, 2003) or, conversely, on moral emotions without assessing the differential contribution of agency to brain activation patterns (Heekeren et al., 2005; Moll, de Oliveira-Souza, Bramati, & Grafman, 2002a; Moll et al., 2002b; Moll, Eslinger, & de Oliveira-Souza, 2001; Takahashi et al., 2004). While it is a truism that neuroimaging studies focusing on moral emotions have inspired important insights into the neural bases of morals, further advance will certainly depend on unraveling the contribution of agency, among other mechanisms, to brain activation in moral cognition paradigms. Specifically, the sense of social agency, or of feeling to be an agent in a real or imagined social situation, might potently engage social cognition networks that overlap with those described for the experience of moral emotions during the presentation of morally relevant scenarios.

In the present study we attempted to circumvent some of these limitations by employing written statements describing action scenarios ("scripts") that independently addressed (1) the contribution of agency to brain activation independently of emotional processing, and (2) the effects of distinct classes of moral emotions—namely, prosocial and other-critical emotions while controlling for agency. We predicted that, in comparison to scripts devoid of agency, emotionally neutral agency scripts would engage fronto-temporal areas involved with agency and intentionality, such as the anterior and medial sectors of the PFC and the STS/temporo-parietal junction (Blakemore et al., 2004; Decety & Sommerville, 2003; Frith & Frith, 2003). We also expected that emotional social scripts would additionally recruit brain systems linked to specific cognitive and motivational components (Moll et al., 2005b, in press): (1) the anterior PFC, given its purported role in prospective assessments, especially for prosocial emotions; (2) anterior temporal cortex and STS region, which code for abstract social semantic knowledge and perceptual features; (3) limbic–paralimbic structures linked to (a) social attachment and affiliative behaviors for empathic emotions, including the basal forebrain, ventral striatum, medial OFC and subgenual area, or to (b) aversive emotions, aggression and response reversals for other-critical emotions, including the amygdala, dorsal ACC and lateral OFC (Blair, Morris, Frith, Perrett, & Dolan, 1999; Botvinick, Nystrom, Fissell, Carter, & Cohen, 1999; Kringelbach & Rolls, 2004; O'Doherty, Kringelbach, Rolls, Hornak, & Andrews, 2001; Rolls, Hornak, Wade, & McGrath, 1994). Given our focus on social neuroscience, only the *social* aspects of agency will be dealt with in this investigation, leaving aside the more basic forms of sensorimotor agency (Ruby &

[1] These emotional clusters can further be decomposed into a more fine-grained taxonomy (Moll et al., in press).

Decety, 2001). Therefore, here we will refer to "agency" according to its social dimension, defined as the capacity to perceive and influence one's own behavior and to construct appropriate courses of actions (Bandura, 2006).

METHODS

Subjects. Participants in the fMRI study (total = 12; 6 women; age = 28.5 ± 9.6 years; education 14.4 ± 2.2 years; mean ± *SD*) were right-handed healthy volunteers free of neurological or psychiatric conditions, and were not using centrally active medications. All were native Portuguese speakers born in Rio de Janeiro. Ten other volunteers were scanned, but their functional data were not analyzed due to signal loss in the ventromedial PFC/basal forebrain and/or temporal regions ($n = 7$) or due to suspected clinical depression ($n = 3$). They were recruited from local universities and professional settings, and informed consent was obtained according to procedures approved by the LABS-D'Or ethics committee. Participants were not paid for their participation, but received a CD containing their structural MRI data as an incentive. We conscientiously adopted a low tolerance to signal loss on echoplanar (functional) datasets, therefore discarding a significant fraction of participants because we wanted to probe the role of critical areas of the ventromedial PFC, temporal poles and limbic structures. This evaluation was performed on the individual "implicit masks," created on the first EPI volume of each individual, after registration and normalization to the SPM template.

Stimuli and task. Written stimuli describing action scenarios ("scripts") were employed to elicit the experience of agency and moral emotions. These scripts controlled for the following variables: (1) occurrence of intentional or unintentional interpersonal harm; (2) target of harmful consequences; (3) feeling of attachment to the victim; and (4) agency. The specific contribution of agency to brain activation was explored by the use of two emotionally neutral control conditions: one involving agency and the other being devoid of agency and social context. Scripts were developed based on two preliminary behavioral studies involving the scoring of a larger sample of stimuli according to the elicitation of basic and moral emotions (Moll & Oliveira-Souza, unpublished data, available upon request from authors). Scripts that reliably (~80% agreement across 48 participants of the behavioral pre-study) evoked specific emotions (guilt, embarrassment, compassion, indignation and disgust), were selected for the fMRI study. In addition, emotionally neutral social scripts were selected by showing that participants consistently scored them as devoid of emotional content (≈80%). Scripts belonging to different conditions were structurally similar grammatically and each comprised two short sentences, the first describing a situation and the second an outcome (e.g., a guilt script: "Your mom called you and said she didn't feel well. You ignored her, and the next day she died"). Indignation scripts were further sorted into two categories according to whether they implied victimization of self ("indignation-self") or others ("indignation-other"). Emotional scripts were in the first person, with participants acting as agents in social situations, thus invoking the sense of agency. This aspect was strengthened by asking subjects to imagine themselves as the main characters of each script. It is important to emphasize, however, that only this more basic level of agency—taking part in a social scenario—was addressed. The role of "causal agency" and intentionality, or being the agent of specific intentional or non-intentional actions, which themselves lead to specific consequences, was not explored here. In addition to the emotional conditions, two emotionally neutral conditions were included: one comprising agency scripts, in which one was depicted as agent in a social situation (e.g., a neutral agency script: "Your mom asked you for an address, but you didn't have it. You left for work and she used the phonebook"), and the other including scripts describing non-social actions in the third person (e.g., a no agency script: "There was an empty shelf in a furniture store. The next day, it was full of new items which were put on display"). See Table 1 for samples of scripts.

Prior to scanning, participants were requested to read and understand samples of scripts, which were presented through a goggle LCD screen inside the scanner. They were asked to press a button as soon as they finished reading each script, which could be comfortably done within the maximum allowed time of 6 seconds. A black screen with a central fixation star followed for an additional 6 seconds after each script. During this time, participants were instructed to keep thinking about the presented script. The simple task

TABLE 1
Examples of stimuli for each experimental condition (translated from Portuguese)

Condition	*Sample of scripts*
Guilt	
	You tricked your child into climbing a tree. He slipped, fell down and broke both of his legs.
	You took your nephew to a park and he got lost in the crowd. Despite a long search, you couldn't find him.
	Your dad called and said he needed to see you. You didn't call back, and he had a stroke the next day.
	You had a few drinks at a bar and decided to drive home. You lost control of the car and hit a mother and her baby.
	Some money was missing and you decided to fire an old employee. Later you found out he was innocent.
Embarrassment	
	Your boss asked for a favor, but you told him you were too busy. He then heard you telling jokes over the phone.
	You made a prejudiced remark about homosexuals at a private party. You forgot the host was gay.
	You gave a gift that you had received from someone else to a friend. Your friend found the card addressed to you.
	You made some nasty remarks about an ex-classmate to a friend. He then told you he had married her.
	You were picking your nose at a stop sign. You noticed a friend looking at you from the next car.
Compassion	
	You visited your neighbor on a weekend day. He told you how his son became paralyzed during a football game.
	You entered your son's room in the leukemia clinic. He was completely bald and in pain.
	You received a goodbye letter from your sister. She wanted to be alone in the final stage of her terminal disease.
	You talked to an old lady on a bus. She told you how she lost her husband and children in an accident.
	You went with your sister to the doctor. She was informed that her son would go blind in a few years.
Disgust	
	Your friend returned from the woods. He said he saw a bunch of dead frogs covered with flies.
	You were petting your neighbor's dog. You touched something wet and noticed an infected sore on his ear.
	In a hospital, you saw a nurse carrying a tray. You peeked into it and saw it was filled with vomit.
	A child came walking in your direction. He suddenly threw-up the remains of the food he had eaten for lunch.
	Your neighbor came to help you wash your car. When you looked at the engine you saw pieces of a dead rat.
Indignation–other	
	At work, a colleague said that her boss had abused her. She was afraid of reporting it to the police.
	On a bus, you saw a man rob an old woman. Before leaving, he slashed her face with a knife.
	From the window you saw a young man driving a motorcycle. You saw him hit a child and then drive away.
	You saw your neighbor leave his toddler alone at home. You noticed he returned drunk two days later.
	Driving downtown, you saw a man hit his wife. She cried for help before collapsing unconscious.
Indignation–self	
	You went to a restaurant with a colleague. Looking at the kitchen, you saw rats crawling all over the place.
	At lunch you went out to have a sandwich. After the first bite, you noticed that it was rotten.
	You were late to work and the traffic was jammed. You saw an illegally parked car blocking the street.
	Walking to your car, you dropped your keys. You helplessly saw them falling into a deep hole.
	You sat at a restaurant table. In the neighboring table, you saw a man was spitting on the floor.
Neutral agency	
	You heard your neighbors chatting in the apartment next door. They were talking about business and related topics.
	You made a couple of remarks during a meeting. Your co-workers also gave their opinion about the issue.
	Your mom asked you for an address, but you didn't have it. You left for work and she used the phonebook.
	You went to the post office to mail a package. The clerk told you it would arrive within three days.
	A traffic officer gave you the right of way and you stepped on the gas. He then signaled for the other cars to cross.
No agency	
	There was an empty shelf in a furniture store. Next day, it was full of new items which were put on display.
	A new building was built in a recently grounded area. All windows faced an internal patio.
	Certain European movies lack a soundtrack. But the Italian ones often feature elaborate musical scores.
	The weather channel advisory announced a severe storm. Next day there were only few cars on the road.
	A tree grew so large that its roots broke the sidewalk. The maintenance department was able to fix the problem.

requirement of a button press after reading each script minimized any possible interference with executive-attentional processes and made it easier for participants to imagine themselves in the described scenarios. Response times (RTs) were recorded, and essentially reflected the reading times for individual scripts. The study design allowed for both (1) exploring the effects of agency, by comparing the neutral agency condition to the no agency condition, and (2) probing the brain correlates of the additional contribution of specific moral emotions to agency, by comparing individual emotional conditions to the neutral agency condition.

After scanning, participants were given a booklet containing the full list of scripts, which appeared one after the other in a different order from the one used in the scanner. At the top of each sheet, two lists of emotions, "basic" (list A) and "moral" (list B) were shown. Using the emotional tags contained in these lists, participants scored each script according to the presence and intensity of (a) a "basic" emotion (chosen from the list A: happiness, fear, disgust, anger, surprise, sadness, and neutral), and (b) of a moral emotion (chosen from the list B: guilt, contempt, embarrassment, compassion, pride, indignation, gratitude and neutral). Participants were told not to assign intensity when selecting the "neutral" option. This procedure allowed us to assess which emotions were predominantly associated with each experimental script. Note that disgust here appears in the basic emotion list because we adhered to the conventional classification of basic and moral emotions. Participants scoring >10 in the Beck Depression Inventory were excluded from the study.

Functional MRI data acquisition. Functional MRI data were collected using a 1.5 Tesla clinical scanner (Vision, Siemens, Erlangen, Germany) and a receiver head coil in Barra D'Or Hospital. Head motion was restricted using foam pads. Functional imaging was performed with T2* blood-oxygenation-level-dependent contrast (BOLD) echoplanar imaging (TR/TE = 3000/66 ms; 25 oblique slices prescribed with 30 degrees from the anterior–posterior commissure line; slice thickness/gap = 2.6/1 mm; Matrix = 64 × 64, FOV = 240 mm, flip angle = 90 degrees), covering all the brain except the top edge of the medial superior parieto-occipital cortex. The use of oblique and thin slices greatly reduced signal loss related to intravoxel dephasing and susceptibility artifacts in the OFC and anterior temporal cortex, brain stem and diencephalic structures (Bellgowan, Bandettini, van Gelderen, Martin, & Bodurka, 2006; Deichmann, Gottfried, Hutton, & Turner, 2003). Structural images were collected with a MPRAGE sequence (TR/TE = 14/4 ms). A total of 485 functional volumes were acquired (the first 5 volumes were discarded to allow for signal stabilization).

Data analysis. Behavioral data analysis was carried out using SPSS (SPSS Inc., Chicago, USA; http://www.spss.com). Alpha was set to $p < .05$ for all behavioral analyses. The fMRI data analysis was performed with Statistical Parametric Mapping software (http://www.fil.ion.ucl.ac.uk/spm/software/spm2). The fMRI study design consisted of an event-related paradigm, with each event lasting 12 seconds. Preprocessing steps included correction for slice-timing and head movement, registration of EPI datasets to the first volume, non-linear spatial normalization to the SPM canonical templates, and spatial smoothing (FWHM = 8 mm). A general linear model (GLM) was applied by performing multiple linear regressions of the BOLD response time courses in each voxel. Regressors included in the design matrix represented the following conditions: neutral agency, neutral no-agency, guilt, embarrassment, compassion, disgust, indignation–self and indignation–other. We included the hemodynamic response function (HRF) and its temporal and dispersion derivatives. The regressors were convolved with the HRF and its derivatives and these partial effects were summed up. The random-effects analyses of the group data enabled generalization of the statistical inferences to the population level. Statistical effects for conditions of interest were obtained by performing one-sample *t*-tests on the first-level contrast images. Common effects across conditions were assessed by performing conjunctions of random effects. *A priori* regions of interest were defined on the basis of anatomical landmarks established in previous studies (Moll et al., 2005b). For these regions of interest, a statistical threshold of $p = .005$ (uncorrected) and a minimum spatial extent of five contiguous voxels was set. Non-*a priori* regions of interest were evaluated at a threshold of $p = .05$, corrected for multiple comparison across whole-brain gray matter. Confirmatory small volume correction (SVC) analyses were performed on activations falling within regions of interest by searching for the local maxima within each region and using local maxima as the center coordinates for spherical regions of interest with predetermined radiuses (Worsley et al., 1996). The regions of interest and sphere radiuses were as follows: medial OFC (7 mm), subgenual area (8 mm), lateral OFC (8 mm), anterior medial PFC (10 mm), frontopolar cortex (12 mm), anterior cingulate cortex (12 mm), ventral striatum-hypothalamic region (12 mm), midbrain ventral tegmental area (8 mm), anterior insula (10 mm),

STS region (14 mm), temporal poles (15 mm) and amygdala (7 mm). Montreal Neurological Institute (MNI) coordinates provided by SPM2 were converted to Talairach coordinates (using Matthew Brett's formula; http://www.mrc-cbu.cam.ac.uk/Imaging/Common/mnispace.shtml) for entry into the Talairach Daemon Client software (Research Imaging Center, San Antonio; http://ric.uthscsa.edu/).

RESULTS

Behavioral results

RTs were not registered for 5 participants due to technical reasons. Though RTs varied across individuals ($p < .02$) and categories of emotions ($p < .03$), a repeated measures ANOVA did not reveal significant interactions between individual reaction times and emotional category, $F(42, 420) = 1.30$, $p = .11$. With the exception of indignation–self, which was correctly selected on only 60% of trials, assignment of stimuli to intended categories ranged from 74 to 90%, and did not differ from each other after excluding the indignation–self category, $F(6, 66) = 1.29$, $p = .28$. Percentages of correct classification for each condition are shown in Table 2. Activation results for the indignation–self condition, thus, must be interpreted with caution. There was an overall difference in mean intensity of emotional experience when all emotional conditions were entered in a repeated-measures ANOVA, $F(5, 55) = 8.72$, $p < .001$, which was explained by a slightly higher intensity of guilt and indignation–other scripts. Emotional intensity did not differ among the other emotional conditions, $F(3, 33) = 1.51$, $p = .23$). Mean intensity of emotional experience and standard deviation for each emotional category (calculated from scorings on scripts assigned to the correct emotional category) were as follows (intensity: 0–4 on a Likert scale): guilt 3.55 ± 0.46, embarrassment 3.08 ± 0.49, compassion 3.20 ± 0.58, disgust 2.88 ± 0.63, indignation–other 3.57 ± 0.49, and indignation–self 2.94 ± 0.58.

Brain activation results

Effects of agency. We first explored the activation patterns of agency without emotional experience. For this purpose, the neutral agency condition was directly compared to the no agency condition. Stimuli used in both these conditions were structurally equivalent, but differed in terms of the presence of agency. The neutral agency condition evoked activation in the mid-STS and middle temporal gyrus, lateral OFC, medial OFC–subgenual cortex, dorsal ACC, and anterior insula (Table 3, Figure 1).

Effects of prosocial and other-critical emotions. To investigate the contribution of specific emotions, each emotional condition was directly contrasted to the neutral agency condition (Tables 4–9). Prosocial moral emotions showed important commonalities in their neural organization. More specifically, guilt, embarrassment and compassion were all associated with activation of the anterior medial PFC and STS/temporo-parietal junction (Figure 2a). The "empathic" emotion subset, represented by guilt and compassion, was specifically associated with activation of the mesolimbic reward pathway, including the ventral tegmental area of the midbrain and of ventral striatum (particularly the ventral pallidum). These effects survived conjunction analysis of random-effects for guilt and compassion conditions compared to neutral agency ($p < .01$). Distinct activation patterns emerged

TABLE 2
Number of scripts assigned to the expected or "correct" category ("hit rate")

Condition	*Hit rate ± SD (max = 15)*	*% correct*	*Expected category*
Guilt	12.16 ± *2.08*	81	Guilt
Embarrassment	12.33 ± *2.23*	82	Embarrassment
Compassion	13.50 ± *2.31*	90	Compassion
Disgust	12.08 ± *1.98*	81	Disgust
Indignation–other	11.17 ± *2.59*	74	Indignation
Indignation–self	8.95† ± *1.99*	60	Indignation
Neutral agency	12.08 ± *2.54*	81	Neutral
No agency	13.08 ± *2.19*	87	Neutral

Note: †Hit rate for "indignation–self" stimuli was significantly lower. Stimuli belonging to other categories did not differ in how often they were assigned to the correct emotional category.

TABLE 3
Activation results for the "neutral agency" condition compared to "no agency" condition

Brain region (BA)	Talairach coordinates x	y	z	t-value	k	P-value	SVC P-value (FDR)
Subgenual cortex (25)	6	11	−11	5.5	7	<.001	.012
Middle frontal gyrus (10)	22	55	21	4.16	65	.001	.042
Superior temporal sulcus/middle temporal gyrus (21)	53	−12	−16	4.97	101	<.001	.05
Insula	36	18	−8	4.07	50	.001	.037
Lateral orbitofrontal cortex (47)	−28	19	−13	3.88	35	.001	.031
Anterior cingulate cortex (23/24)	−4	9	30	3.79	12	.002	.195
Medial orbitofrontal cortex (10/11)	12	42	−11	3.41	6	.003	.11

Notes: BA = Brodmann areas; *k* = number of activated voxels in a given cluster; SVC = small volume correction; FDR = false discovery rate.

TABLE 4
Activation results for the "guilt" condition compared to "neutral agency"

Brain region (BA)	Talairach coordinates x	y	z	t-value	k	P-value	SVC P-value (FDR)
Insula	−46	6	13	5.68	59	<.001	.01
Medial frontal gyrus (10)	2	63	12	5.45	73	<.001	.01
Lateral orbitofrontal cortex (46/47)	44	52	−9	4.52	16	<.001	.019
Midbrain ventral tegmental area/medial thalamus	2	−12	−6	3.71	28	.002	.045
Superior temporal sulcus (21)	63	−46	7	4.53	63	<.001	.056
Anterior-superior temporal gyrus (38)	−51	7	−12	4.12	15	.001	.253

Notes: BA = Brodmann area; *k* = number of activated voxels in a given cluster; SVC = small volume correction; FDR = false discovery rate.

for other-critical emotions (disgust, indignation–other and indignation–self). These emotions activated mainly the ACC, lateral OFC, and the parahippocampal/amygdala region (Figure 2b). It is noteworthy that activation of anterior sectors of the PFC was only observed for the indignation–other condition, but was absent for the disgust and indignation–self conditions.

Direct comparison of prosocial and other-critical emotions. A clearer picture of the effects of prosocial and other-critical emotions was

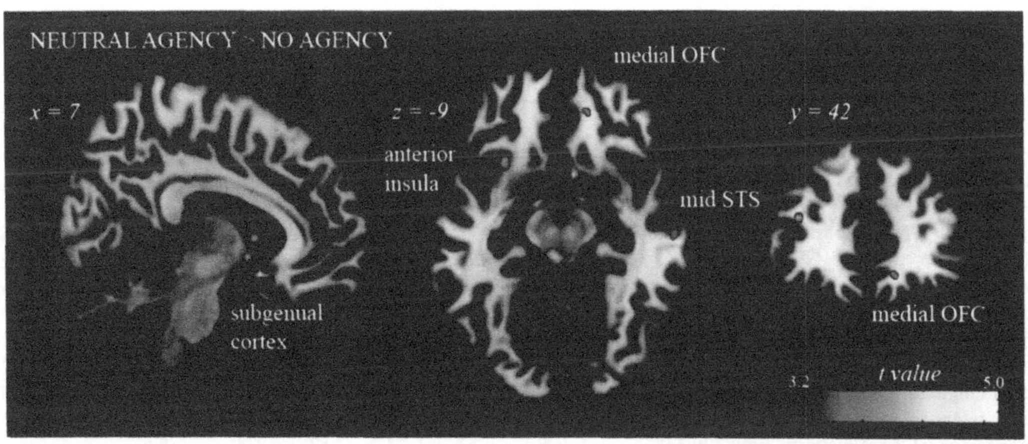

Figure 1. Functional MRI results for the comparison neutral agency *versus* no agency conditions. Activated regions included the ventral subgenual cortex and medial orbitofrontal cortex (OFC), the anterior insula and mid portion of the superior temporal sulcus (STS).

TABLE 5
Activation results for the "embarrassment" condition compared to "neutral agency"

Brain region (BA)	*Talairach coordinates* x	y	z	*t-value*	*k*	*P-value*	*SVC P-value (FDR)*
Inferior parietal lobule/temporo-parietal junction (39/40)	−55	−36	22	7.76	147	<.001	.003
Superior temporal sulcus/temporo-parietal junction (21)	−57	−46	14	7.11	124	.001	.011
Medial frontal gyrus (10)	4	61	12	4.32	24	.001	.066
Anterior cingulate cortex (24/32)	−8	13	33	3.68	7	.002	.27
Insula/inferior frontal gyrus (44)	−46	3	13	3.59	16	.002	.099

Notes: BA = Brodmann areas; *k* = number of activated voxels in a given cluster; SVC = small volume correction; FDR = false discovery rate.

TABLE 6
Activation results for the "compassion" condition compared to "neutral agency"

Brain region (BA)	*Talairach coordinates* x	y	z	*t-value*	*k*	*P-value*	*SVC P-value (FDR)*
Middle temporal gyrus/temporo-parietal junction (37/39)	−42	−60	11	5.33	33	<.001	.03
Medial frontal gyrus (10)	6	65	10	5.25	117	<.001	.005
Parahippocampal gyrus/amygdala (28)	22	−18	−16	5.23	63	<.001	.009
Anterior cingulate cortex (23/24)	−12	−8	39	4.49	26	<.001	.032
Ventral pallidum	12	2	−3	4.3	17	.001	.05
Superior temporal sulcus	−57	−44	16	5.46	47	<.001	.053
Middle temporal gyrus/temporo-parietal junction (22)	51	−50	3	4.55	42	<.001	.112
Superior frontal gyrus (10)	−26	48	25	4.51	42	<.001	.063
Midbrain ventral tegmental area	4	−12	−9	4.27	24	.001	.102
Middle temporal gyrus (21)	−63	−14	−14	4.00	14	.001	.26
Cerebellum, tuber	42	−73	−25	5.39	219	.05*	—

Notes: BA = Brodmann areas; *k* = number of activated voxels in a given cluster; SVC = small volume correction; FDR = false discovery rate.*Whole brain correction for multiple corrections, $p < .05$.

TABLE 7
Activation results for the "disgust" condition compared to "neutral agency"

Brain region (BA)	*Talairach coordinates* x	y	z	*t-value*	*k*	*P-value*	*SVC P-value (FDR)*
Supramarginal gyrus/temporo-parietal junction (40)	−48	−28	25	5.28	171	>.001	.018
Insula	−42	−34	20	5.28	84	>.001	.013
Amygdala	24	−6	−11	3.99	26	.001	.029
Middle temporal gyrus/fusiform gyrus (37)	−44	−58	0	5.87	605	>.001	.05*
Lateral orbitofrontal cortex (47)	38	33	−3	3.94	11	.001	.061
Anterior cingulate cortex (24)	−10	10	37	4.07	38	.001	.062
Anterior cingulate cortex (24)	−2	30	15	3.4	12	.003	.124
Superior temporal gyrus (38)	42	1	−14	3.87	10	.001	.435
Inferior temporal gyrus/fusiform gyrus (20/37)	50	−57	−4	4.83	334	.05*	—

Notes: BA = Brodmann areas; *k* = number of activated voxels in a given cluster; SVC = small volume correction; FDR = false discovery rate. *Whole brain correction for multiple corrections, $p < .05$.

TABLE 8
Activation results for the "indignation–self" condition compared to "neutral agency"

	Talairach coordinates						
Brain region (BA)	*x*	*y*	*z*	*t-value*	*k*	*P-value*	*SVC P-value (FDR)*
Lateral orbitofrontal cortex (47)	44	43	1	5.67	106	<.001	.024
Anterior cingulate cortex (23/24)	6	7	25	3.94	15	.001	.044
Parahippocampal gyrus/amygdala (36)	28	−19	−24	3.7	7	.002	.031
Subgenual cortex (25)	2	13	−6	4.21	9	.001	.126

Notes: BA = Brodmann areas; k = number of activated voxels in a given cluster; SVC = small volume correction; FDR = false discovery rate.

provided by directly contrasting these conditions (guilt+embarrassment+compassion vs. disgust+ indignation–self + indignation–other). This supplementary analysis excluded the contribution of the neutral agency condition. The anterior medial PFC and temporal poles were more consistently activated by prosocial emotions, whereas the dorsal ACC, lateral OFC and the ventral temporo-occipital cortex responded more consistently to other-critical emotions (Figure 3a and b). Comparison of empathic emotions (guilt + compassion) to other-critical emotions showed strong effects on the midbrain and of ventral striatum (Figure 3c).

DISCUSSION

The main results of this study can be summarized as follows. First, exposure to scripts involving agency, as compared to scripts devoid of agency, elicited activation in anterior PFC, lateral OFC, medial OFC and subgenual cortex, and mid-STS. These regions have been implicated in moral sensitivity in functional imaging and lesion studies (Moll et al., 2005b), as well as in various aspects of social cognition, including empathy and inference of feeling states of others (Decety & Jackson, 2004; Ruby & Decety, 2004), and spontaneous self-referential thinking (Gusnard, Akbudak, Shulman, & Raichle, 2001). Second, scripts evocative of moral emotions, compared to neutral agency, elicited additional cortico-limbic activation. As predicted (Moll et al., 2005b, in press), cognitive-motivational components shared by different groups of emotions were reflected in specific activation patterns: prosocial emotions (guilt, embarrassment and compassion) activated anterior and medial PFC regions, with empathic emotions (guilt and compassion) additionally activating the mesolimbic reward pathway (ventral tegmental area of the midbrain and ventral striatum), which are important for social attachment. Other-critical emotions (disgust, indignation–other and indignation–self), in contrast, evoked more activity in brain regions associated with aversive responses, such as the parahippocampal area and amygdala, dorsal ACC and lateral OFC. These results are discussed in more detail below.

Effects of social agency. Scripts involving agency evoked activity in the medial OFC-subgenual cortex, lateral OFC, mid-STS, dorsal ACC and insula. Previous studies have consis-

TABLE 9
Activation results for the "indignation–other" condition compared to "neutral agency"

	Talairach coordinates						
Brain region (BA)	*x*	*y*	*z*	*t-value*	*k*	*P-value*	*SVC P-value (FDR)*
Anterior cingulate cortex (24/32)	4	30	20	5.56	190	.05*	.008
Medial frontal gyrus (10)	0	61	12	4.89	56	>.001	.023
Parahippocampal gyrus/amygdala (28/35)	22	−26	−10	3.76	27	.002	.031
Lateral orbitofrontal cortex (47)	54	34	0	3.88	14	.001	.054
Inferior temporal gyrus/fusiform gyrus (20/37)	51	−64	−5	5.28	239	.05*	—

Notes: BA = Brodmann areas; k = number of activated voxels in a given cluster; SVC = small volume correction; FDR = false discovery rate. *Whole brain correction for multiple corrections, $p < .05$.

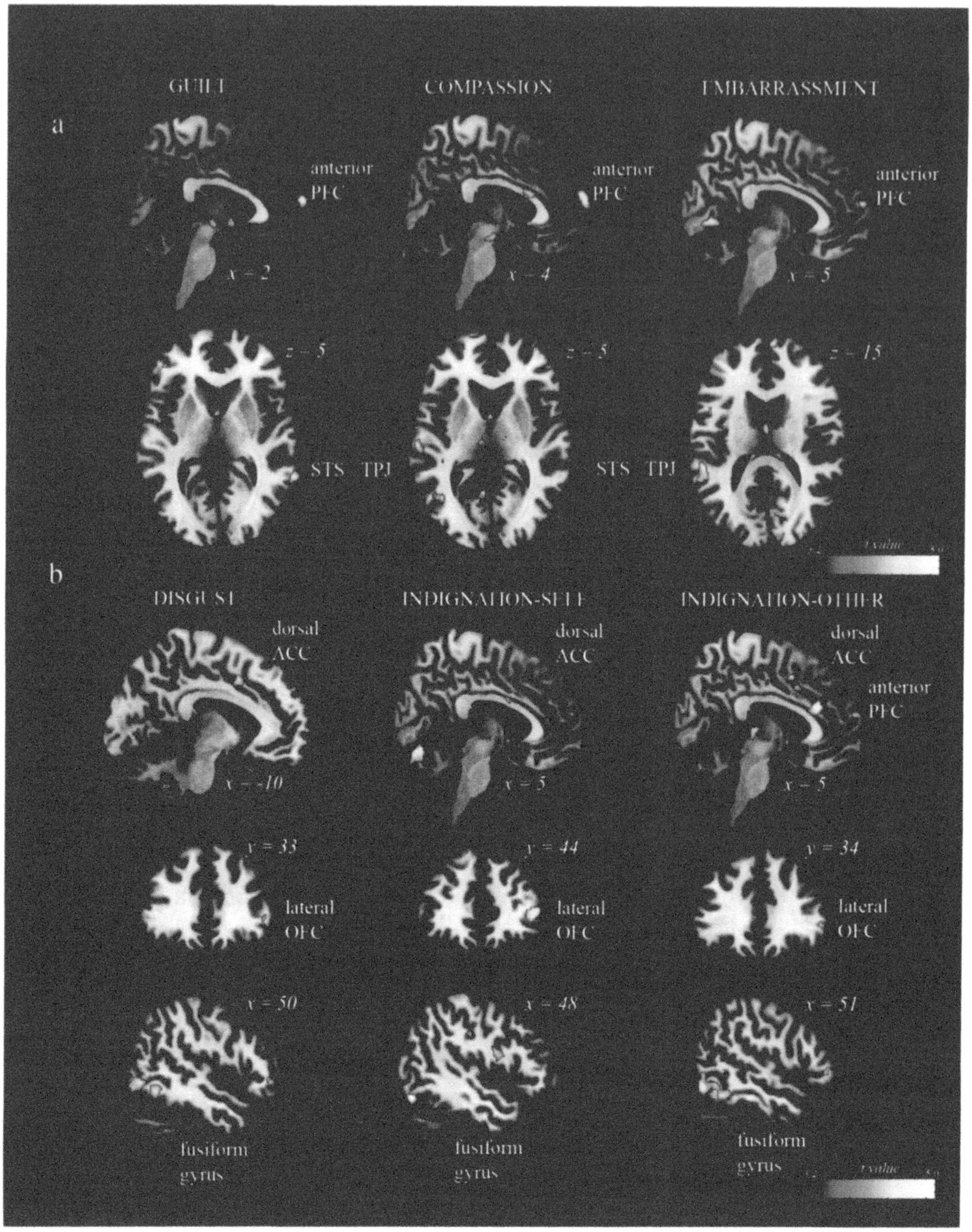

Figure 2. Functional MRI results for specific emotional conditions. (a) Activation of the anterior prefrontal cortex (PFC) and superior temporal sulcus/temporo-parietal junction (STS/TPJ) was observed when comparing guilt, compassion, or embarrassment to neutral agency condition. (b) Activation of the dorsal anterior cingulate cortex (ACC), lateral orbitofrontal cortex (OFC) and fusiform gyrus was observed when disgust, indignation–self or indignation–other were compared to neutral agency condition.

tently demonstrated that the ventral and medial sectors of the PFC (including the medial OFC-subgenual area) show high "resting" baseline metabolic activity (i.e., when subjects are lying in the scanner in the dark or simply looking at a fixation cross). This activity is markedly decreased during the active performance of a wide variety of tasks, including visual and auditory attention, language processing, memory and motor control (Gusnard & Raichle, 2001). This "default mode" state, which takes place when subjects are idle, thinking freely, and undisturbed by demanding cognitive tasks, has been related to "self-referential" mental states (Gusnard et al., 2001). Interestingly, active tasks requiring "self-related" thinking (Happé, 2003; Northoff &

Bermpohl, 2004), and attention to self-relevant emotional stimuli are associated with either increases or less pronounced decreases in activation in medial PFC sectors (Gusnard et al., 2001). It should be emphasized that our neutral agency condition, though requiring simple attentional and executive resources, was an *active* task summoning the sense of agency in social contexts. Therefore, the observed activations cannot be attributed to uncontrolled, spontaneous thoughts, or to "resting" activity. Instead, we interpret this effect as evidence for the recruitment of neural representations linked to emotionally neutral but motivationally relevant action sequences and behavioral contingencies that take place during social navigation (Moll et al., 2005b; Raichle & Gusnard, 2005; Schultz, 2000; Tremblay & Schultz, 1999). Our results additionally show that medial OFC activity is also evoked by prosocial emotions, pointing to a similarity between these emotions and emotionally neutral agency.

Prosocial and other-critical emotions. We found support for dissociable brain patterns for prosocial and other-critical emotions. The observation that the anterior PFC and medial OFC were more active for prosocial emotions (guilt, embarrassment, compassion) is compatible with lesion data pointing to the critical role of these regions in social behavior, both in adulthood and throughout development (Eslinger, Flaherty-Craig, & Benton, 2004). The experience of moral emotions in active social situations is impaired in patients with OFC lesions (Beer, Heerey, Keltner, Scabini, & Knight, 2003). Previous fMRI studies consistently found increased anterior PFC activation during autobiographical recollection of guilt-evoking events (Shin et al., 2000), social scripts involving guilt and embarrassment (Takahashi et al., 2004), "intentional" and "non-intentional" violations of social norms (Berthoz, Armony, Blair, & Dolan, 2002) and moral violations with or without bodily harm (Heekeren et al., 2005). How this is specifically related with a more general capacity to make inferences on self and other's mental states (Ruby & Decety, 2004) is still a matter of future inquiry. We have argued that the *contents* of mental inferences—instead of the *process* through which these inferences are made—might be a more decisive factor in defining which fronto-limbic regions will be recruited (Moll et al., 2005b). This hypothesis is in line with fMRI studies showing that making mechanical, intentional and other types of inferences of causality differ in their neural substrates (den Ouden et al., 2005; Martin & Weisberg, 2003). Yet, these studies are not conclusive on the issue of the role of contents and processes in brain activation, and further investigation is certainly needed.

The stronger anterior PFC response to prosocial emotions may result from higher assessment of social outcomes and more complex assessment of entangled possible future actions, such as repairing damaged social links, helping, correcting mistakes (Christoff & Gabrieli, 2000; Moll et al., 2001, 2005b; Ramnani & Owen, 2004; Wood & Grafman, 2003). Other-critical emotions (disgust, indignation–other, indignation–self), in contrast, led to more consistent activation of the dorsal ACC, lateral OFC and parahippocampal gyrus/amygdala. These regions have been implicated in painful responses in self and others (Eisenberger, Lieberman, & Williams, 2003; Singer, 2004), changes in reinforcement contingencies (Kringelbach, 2005; O'Doherty et al., 2001) and other aversive responses associated with anger and disgust (Blair et al., 1999; Gosselin et al., 2006; Moll et al., 2005a). Such emotions provide the grounds for the separation of "pure from impure" and "good from bad" (Haidt, 2001; Rozin, Lowery, Imada, & Haidt, 1999). Remarkably, except for the indignation–other condition, no significant activity of the anterior PFC or medial OFC was observed for other-critical emotions. This is in agreement with the functional shutdown of these regions in aggressive situations (Pietrini, Guazzelli, Basso, Jaffe, & Grafman, 2000). The finding of anterior medial PFC activation in the indignation–other condition is especially interesting, and may reflect the prosocial function of indignation, when it is felt on behalf of others. Finally, non-predicted activation of extrastriate visual areas (fusiform and other ventral temporal areas) was observed preferentially for other-critical emotions. This could have potentially resulted from a higher imageability of these scenarios, or from other uncontrolled factors across conditions.

Guilt and compassion. Empathic emotions were related to activation of the ventral tegmental area and ventral striatum, which are major components of the mesolimbic reward system. Integration among the various components of this system is largely carried out by the medial forebrain bundle, which provides a com-

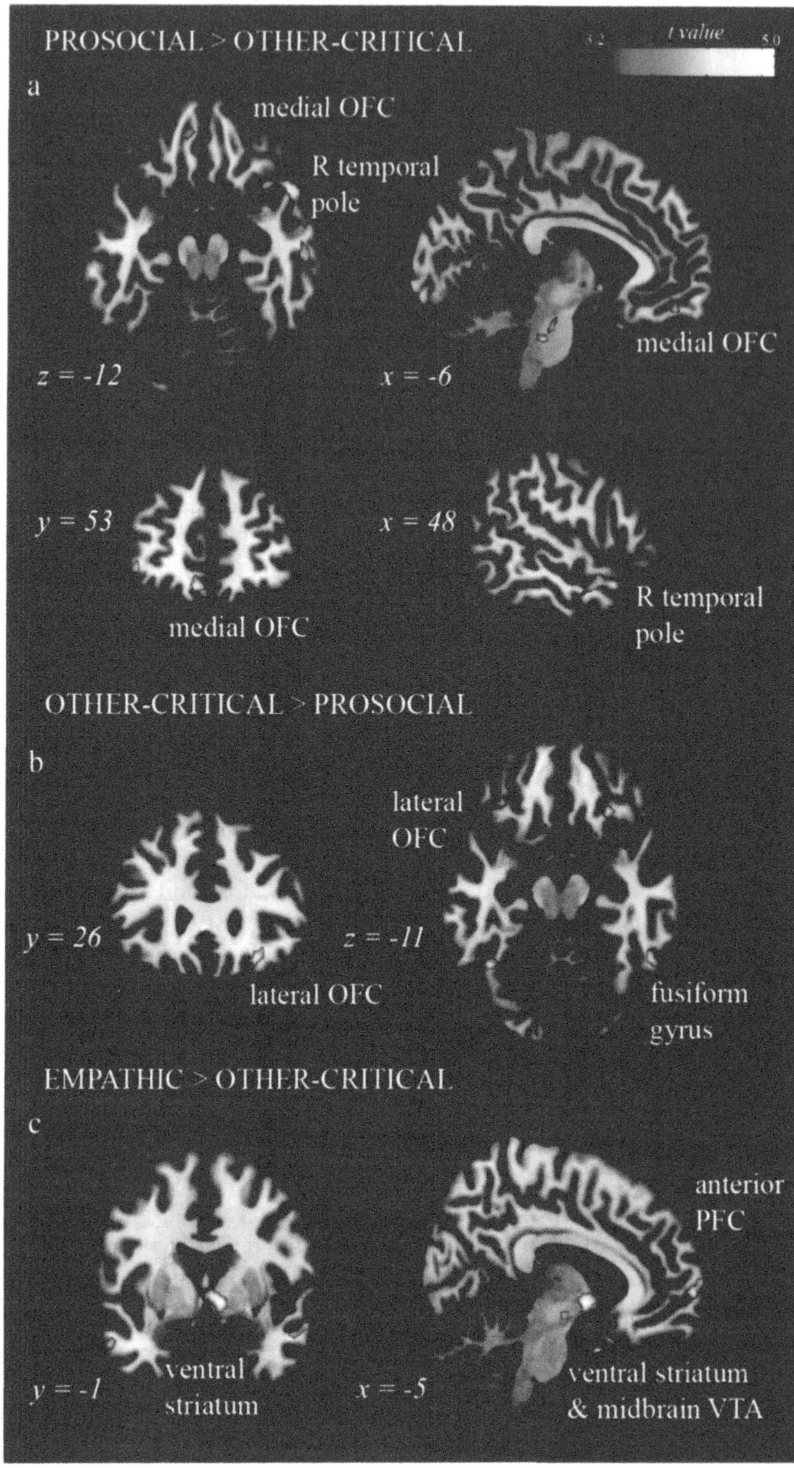

Figure 3. Functional MRI results for direct comparisons among emotional clusters (prosocial, other-critical and empathic emotions). (a) Brain regions more active in response to prosocial emotions (guilt, compassion, and embarrassment) than to other-critical emotions (disgust and indignation). Activated regions included the medial orbitofrontal cortex (OBC) and temporal pole. (b) The opposite comparison, other-critical versus prosocial emotions, led to activation of the lateral OFC and fusiform gyrus. (c) Comparison of 'empathic' emotions (guilt and compassion) to other-critical emotions showed increased activation in the ventral striatum, ventral tegmental area of the midbrain (VTA) and in the anterior prefrontal cortex (PFC).

mon mechanism of reward reinforcement for a large array of behaviors (Nauta 1960; Schultz, Dayan, & Montague, 1997). It might seem contradictory that moral emotions associated with the negative experience of sadness, like guilt and compassion, should ever evoke reward responses. However, different rewards may come in different flavors, and the brain motivational system is capable of finely orchestrating their roles. For instance, feeding, drinking, sexual gratification, and—key for the observed effects of guilt and compassion—affiliative or social bonding (Young & Wang, 2004), can be selectively elicited. Based on this notion, we have anticipated that activation of primitive circuits related to attachment and affiliative behaviors would be centrally involved in the experience of these emotions (Moll et al., in press). Guilt and compassion, according to this view, qualify as empathic emotions as they involve: (1) a more elementary level of embodied empathy (sharing the pain or sadness experienced by another person); (2) feeling socially close, or emotionally attached to that person; and (3) being able to differentiate one's own feeling from those of the suffering person (Batson et al., 1991; Decety & Jackson, 2004; Eisenberg, 2000; Harris, 2003; Moll et al., in press; Tangney, 2002). The ventral and medial PFC sectors are intimately involved in emotional experience and viscero-endocrine regulations, with the subgenual cortex being especially related to social bonding (Freedman, Insel, & Smith, 2000; Insel & Young, 2001) and social reward (Aron et al., 2005; Bartels & Zeki, 2004), as well as to major depression (Drevets, Ongur, & Price, 1998; Liotti et al., 2000; Mayberg et al., 1999; Vogt, 2005) and during altruistic decisions, such as charitable donations (Moll et al., 2006). The lack of activation of the subgenual cortex by guilt and compassion, predicted on the basis of the relationship of these regions to social attachment, can be explained by the fact that this region was already robustly activated by neutral agency, as discussed below.

Role of the OFC in moral sensitivity and agency

The OFC is organized into distinct networks (Kringelbach, 2005; Öngür, Ferry, & Price, 2003); while the medial OFC provides a visceromotor link, the lateral OFC is involved in multimodal sensory processing. The activation of these OFC sectors in the neutral agency condition suggests that their role in social cognition is not necessarily "emotional." By associating knowledge of sequences of events with current motivational contingencies, the OFC (together with more anterior sectors of the medial PFC) sets us free from making conscious moral judgments at every fold of our social space. Moral sensitivity is thus automatically engaged in social contexts even in the absence of overt emotional provocation: the automatic recruitment of these action–outcome and motivationally salient representations during social intercourse enables us to perceive and respond flexibly and in a contextually sensitive fashion to subtle cues indicating violations of social expectations (Moll et al., 2002b; Schulkin, 2000). Violations or enforcement of these expectations, even when subtle or only mentally entertained, lead to the experience of specific moral emotions.

Limitations of the present investigation and future directions

The present study suffered from two main limitations, which are addressed below. The first one concerns the design of the stimuli: although experimental conditions were better controlled as compared to most previous studies of moral emotion and judgment, there might be additional confounding factors to be accounted for in future studies. The second limitation was that the experimental design did not allow for testing the influence of causal agency on specific moral emotions. Although script-like scenarios have been successfully employed in several studies and possess a number of advantages—which include the ability to efficiently convey rather constrained information about complex social situations, which is more difficult using pictures, single words or faces—it is more difficult to control for structural and perceptual properties of the stimuli. Therefore, future studies will need to better control for variables such as imageability, number of people in the scenarios and number of actions performed by the protagonist agent and/or by other characters in social scripts. Another important aspect, which will certainly need to be better explored, is the refinement of the role of agency. For instance, in addition to the simple presence or absence of agency, social scenarios could potentially be designed to account for causal agency, intentionality, and types of social outcomes. However, because causal agency operates as a defining parameter for which moral emotion will be experienced in specific agent–recipient relationships, controlling for this is a non-trivial problem. Disentangling these variables will be an important and challenging task for future studies.

CONCLUSION

Moral sensitivity and social agency share a large amount of a specific, yet distributed neural system. When the dynamic equilibrium of this network is disturbed by violations of moral values and expectations, enhanced activity of regions within this distributed circuitry marks the experience of distinctive moral emotions. The present results support the suggestion that complex moral phenomena are mediated through co-activation of specific cortico-limbic networks (Moll et al., 2005b), which rely on three main components: social event sequences and reinforcement contingencies (provided by the OFC and anterior PFC), social semantic and perceptual representations (provided by anterior and postero-superior temporal regions respectively), and central motive states (mediated by limbic structures). Prosocial emotions, which are more loaded with knowledge of social rules and moral values, activate the anterior PFC and STS regions, with empathic emotions additionally recruiting limbic-related regions associated with affiliative reward and social attachment. The shared effects on the medial OFC–subgenual area for agency in norm-abiding social behaviors and prosocial emotions support the hypothesis that moral sensitivity is at the core of spontaneous, or "default-state," mental activity. Future studies will be needed to address the specific relationships among different levels of agency, emotional valence and their relationships to moral emotions.

REFERENCES

Allport, G. W. (1954). *The nature of prejudice*. Boston: Beacon Press.

Aron, A., Fisher, H., Mashek, D. J., Strong, G., Li, H., & Brown, L. L. (2005). Reward, motivation, and emotion systems associated with early stage, intense, romantic love. *Journal of Neurophysiology, 94*, 327–337.

Bandura, A. (2006). Toward a psychology of human agency. *Perspectives on Psychological Science, 1*, 164–180.

Bartels, A., & Zeki, S. (2004). The neural correlates of maternal and romantic love. *NeuroImage, 21*, 1155–1166.

Batson, C. D., Batson, J. G., Slingsby, J. K., Harrell, K. L., Peekna, H. M., & Todd, R. M. (1991). Empathic joy and the empathy–altruism hypothesis. *Journal of Personality and Social Psychology, 61*, 413–426.

Beer, J. S., Heerey, E. A., Keltner, D., Scabini, D., & Knight, R. T. (2003). The regulatory function of self-conscious emotion: Insights from patients with orbitofrontal damage. *Journal of Personality and Social Psychology, 85*, 594–604.

Bellgowan, P. S., Bandettini, P. A., van Gelderen, P., Martin, A., & Bodurka, J. (2006). Improved BOLD detection in the medial temporal region using parallel imaging and voxel volume reduction. *NeuroImage, 29*, 1244–1251.

Berthoz, S., Armony, J. L., Blair, R. J., & Dolan, R. J. (2002). An fMRI study of intentional and unintentional (embarrassing) violations of social norms. *Brain, 125*, 1696–1708.

Blair, R. J., Morris, J. S., Frith, C. D., Perrett, D. I., & Dolan, R. J. (1999). Dissociable neural responses to facial expressions of sadness and anger. *Brain, 122*(5), 883–893.

Blakemore, S. J., & Frith, U. (2004). How does the brain deal with the social world? *Neuroreport, 15*, 119–128.

Blakemore, S. J., Winston, J., & Frith, U. (2004). Social cognitive neuroscience: Where are we heading? *Trends in Cognitive Sciences, 8*, 216–222.

Botvinick, M., Nystrom, L. E., Fissell, K., Carter, C. S., & Cohen, J. D. (1999). Conflict monitoring versus selection-for-action in anterior cingulate cortex. *Nature, 402*, 179–181.

Chorvat, T., & McCabe, K. (2004). The brain and the law. *Philosophical Transactions of the Royal Society of London. Series B: Biological Sciences, 359*, 1727–1736.

Christoff, K., & Gabrieli, J. D. E. (2000). The frontopolar cortex and human cognition: Evidence for a rostrocaudal hierarchical organization within the human prefrontal cortex. *Psychobiology, 28*, 168–186.

Decety, J., & Jackson, P. L. (2004). The functional architecture of human empathy. *Behavioral and Cognitive Neuroscience Reviews, 3*, 71–100.

Decety, J., & Sommerville, J. A. (2003). Shared representations between self and other: A social cognitive neuroscience view. *Trends in Cognitive Sciences, 7*, 527–533.

Deichmann, R., Gottfried, J. A., Hutton, C., & Turner, R. (2003). Optimized EPI for fMRI studies of the orbitofrontal cortex. *NeuroImage, 19*, 430–441.

den Ouden, H. E., Frith, U., Frith, C., & Blakemore, S. J. (2005). Thinking about intentions. *NeuroImage, 28*, 787–796.

Drevets, W. C., Ongur, D., & Price, J. L. (1998). Neuroimaging abnormalities in the subgenual prefrontal cortex: implications for the pathophysiol-

ogy of familial mood disorders. *Molecular Psychiatry, 3*, 220–226.
Eisenberg, N. (2000). Emotion regulation and moral development. *Annual Review of Psychology, 51*, 665–697.
Eisenberger, N. I., Lieberman, M. D., & Williams, K. D. (2003). Does rejection hurt? An fMRI study of social exclusion. *Science, 302*, 290–292.
Eslinger, P. J., Flaherty-Craig, C. V., & Benton, A. L. (2004). Developmental outcomes after early prefrontal cortex damage. *Brain and Cognition, 55*, 84–103.
Fessler, D. (1999). Toward an understanding of the universality of second order emotions. In A. Hinton (Ed.), *Beyond nature or nurture: Biocultural approaches to the emotions* (pp. 75–116). New York: Cambridge University Press.
Freedman, L. J., Insel, T. R., & Smith, Y. (2000). Subcortical projections of area 25 (subgenual cortex) of the macaque monkey. *Journal of Comparative Neurology, 421*, 172–188.
Frith, C., & Frith, U. (2005). Theory of mind. *Current Biology, 15*, R644–R646.
Frith, U., & Frith, C. D. (2003). Development and neurophysiology of mentalizing. *Philosophical Transactions of the Royal Society of London. Series B: Biological Sciences, 358*, 459–473.
Gallese, V., Keysers, C., & Rizzolatti, G. (2004). A unifying view of the basis of social cognition. *Trends in Cognitive Sciences, 8*, 396–403.
Goodenough, O. R., & Prehn, K. (2004). A neuroscientific approach to normative judgment in law and justice. *Philosophical Transactions of the Royal Society of London. Series B: Biological Sciences, 359*, 1709–1726.
Gosselin, N., Samson, S., Adolphs, R., Noulhiane, M., Roy, M., Hasboun, D., et al. (2006). Emotional responses to unpleasant music correlates with damage to the parahippocampal cortex. *Brain, 129*, 2585–2592.
Gusnard, D. A., Akbudak, E., Shulman, G. L., & Raichle, M. E. (2001). Medial prefrontal cortex and self-referential mental activity: relation to a default mode of brain function. *Proceedings of the National Academy of Sciences of the USA, 98*, 4259–4264.
Gusnard, D. A., & Raichle, M. E. (2001). Searching for a baseline: functional imaging and the resting human brain. *Nature Reviews. Neuroscience, 2*, 685–694.
Haidt, J. (2001). The emotional dog and its rational tail: a social intuitionist approach to moral judgment. *Psychological Review, 108*, 814–834.
Haidt, J. (2003). The moral emotions. In R. J. Davidson, K. R. Scherer, & H. H. Goldsmith (Eds.), *Handbook of affective sciences* (pp. 852–870). Oxford, UK: Oxford University Press.
Happé, F. (2003). Theory of mind and the self. *Annals of the New York Academy of Science, 1001*, 134–144.
Harris, N. (2003). Reassessing the dimensionality of the moral emotions. *British Journal of Psychology, 94*, 457–473.
Heekeren, H. R., Wartenburger, I., Schmidt, H., Prehn, K., Schwintowski, H. P., & Villringer, A. (2005). Influence of bodily harm on neural correlates of semantic and moral decision-making. *NeuroImage, 24*, 887–897.
Iacoboni, M. (2005). Neural mechanisms of imitation. *Current Opinion in Neurobiology, 15*, 632–637.
Insel, T. R., & Young, L. J. (2001). The neurobiology of attachment. *Nature Reviews. Neuroscience, 2*, 129–136.
Kringelbach, M. L. (2005). The human orbitofrontal cortex: linking reward to hedonic experience. *Nat Rev Neurosci, 6*, 691–702.
Kringelbach, M. L., & Rolls, E. T. (2004). The functional neuroanatomy of the human orbitofrontal cortex: evidence from neuroimaging and neuropsychology. *Progress in Neurobiology, 72*, 341–372.
Liotti, M., Mayberg, H. S., Brannan, S. K., McGinnis, S., Jerabek, P., & Fox, P. T. (2000). Differential limbic–cortical correlates of sadness and anxiety in healthy subjects: implications for affective disorders. *Biological Psychiatry, 48*, 30–42.
Martin, A., & Weisberg, J. (2003). Neural foundations for understanding social and mechanical concepts. *Cognitive Neuropsychology, 20*, 575–587.
Mayberg, H. S., Liotti, M., Brannan, S. K., McGinnis, S., Mahurin, R. K., Jerabek, P. A., et al. (1999). Reciprocal limbic–cortical function and negative mood: converging PET findings in depression and normal sadness. *American Journal of Psychiatry, 156*, 675–682.
Moll, J., de Oliveira-Souza, R., Bramati, I. E., & Grafman, J. (2002a). Functional networks in emotional moral and nonmoral social judgments. *NeuroImage, 16*, 696–703.
Moll, J., de Oliveira-Souza, R., & Eslinger, P. J. (2003). Morals and the human brain: a working model. *Neuroreport, 14*, 299–305.
Moll, J., de Oliveira-Souza, R., Eslinger, P. J., Bramati, I. E., Mourao-Miranda, J., Andreivolo, P. A., et al. (2002b). The neural correlates of moral sensitivity: a functional magnetic resonance imaging investigation of basic and moral emotions. *Journal of Neuroscience, 22*, 2730–2736.
Moll, J., de Oliveira-Souza, R., Moll, F. T., Ignacio, F. A., Bramati, I. E., Caparelli-Daquer, et al. (2005a). The moral affiliations of disgust: A functional MRI study. *Cognitive and Behavioral Neurology, 18*, 68–78.
Moll, J., de Oliveira-Souza, R., Zahn, R., & Grafman, J. (in press). The cognitive neuroscience of moral emotions. In W. Sinnott-Armstrong (Ed.), *Moral psychology. Vol. 3: Morals and the brain*. Cambridge, MA: MIT Press.
Moll, J., Eslinger, P. J., & de Oliveira-Souza, R. (2001). Frontopolar and anterior temporal cortex activation in a moral judgment task: preliminary functional MRI results in normal subjects. *Arquivos de Neuropsiquiatria, 59*, 657–664.
Moll, J., Krueger, F., Zahn, R., Pardini, M, de Oliveira-Souza, R., & Grafman, J. (2006). Human fronto-mesolimbic networks guide decisions about charitable donation. *Proceedings of the National Academy of Sciences of the United States of America, 103*, 15623–15628.

Moll, J., Zahn, R., de Oliveira-Souza, R., Krueger, F., & Grafman, J. (2005b). Opinion: the neural basis of human moral cognition. *Nature Reviews. Neuroscience, 6*, 799–809.
Nauta, W. J. (1960). Limbic system and hypothalamus: anatomical aspects. *Physiol Rev Suppl, 4*, 102–104.
Northoff, G., & Bermpohl, F. (2004). Cortical midline structures and the self. *Trends in Cognitive Science, 8*, 102–107.
O'Doherty, J., Kringelbach, M. L., Rolls, E. T., Hornak, J., & Andrews, C. (2001). Abstract reward and punishment representations in the human orbitofrontal cortex. *Nature Neuroscience, 4*, 95–102.
Öngür, D., Ferry, A. T., & Price, J. L. (2003). Architectonic subdivision of the human orbital and medial prefrontal cortex. *The Journal of Comparative Neurology, 460*, 425–449.
Pietrini, P., Guazzelli, M., Basso, G., Jaffe, K., & Grafman, J. (2000). Neural correlates of imaginal aggressive behavior assessed by positron emission tomography in healthy subjects. *American Journal of Psychiatry, 157*, 1772–1781.
Raichle, M. E., & Gusnard, D. A. (2005). Intrinsic brain activity sets the stage for expression of motivated behavior. *Journal of Comparative Neurology, 493*, 167–176.
Ramnani, N., & Owen, A. M. (2004). Anterior prefrontal cortex: insights into function from anatomy and neuroimaging. *Nature Reviews. Neuroscience, 5*, 184–194.
Rolls, E. T., Hornak, J., Wade, D., & McGrath, J. (1994). Emotion-related learning in patients with social and emotional changes associated with frontal lobe damage. *Journal of Neurology, Neurosurgery, and Psychiatry, 57*, 1518–1524.
Rozin, P., Lowery, L., Imada, S., & Haidt, J. (1999). The CAD triad hypothesis: a mapping between three moral emotions (contempt, anger, disgust) and three moral codes (community, autonomy, divinity). *Journal of Personality and Social Psychology, 76*, 574–586.
Ruby, P., & Decety, J. (2001). Effect of subjective perspective taking during simulation of action: a PET investigation of agency. *Nature Neuroscience, 4*, 546–550.
Ruby, P., & Decety, J. (2004). How would you feel versus how do you think she would feel? A neuroimaging study of perspective-taking with social emotions. *Journal of Cognitive Neuroscience, 16*, 988–999.
Schulkin, J. (2000). *Roots of social sensitivity and neural function*. Cambridge, MA: MIT Press.
Schulkin, J. (2004). *Bodily sensibility: Intelligent action*. New York: Oxford University Press.
Schultz, W. (2000). Multiple reward signals in the brain. *Nature Reviews. Neuroscience, 1*, 199–207.
Schultz, W., Dayan, P., & Montague, P. R. (1997). A neural substrate of prediction and reward. *Science, 275*, 1593–1599.
Shin, L. M., Dougherty, D. D., Orr, S. P., Pitman, R. K., Lasko, M., Macklin, M. L., et al. (2000). Activation of anterior paralimbic structures during guilt-related script-driven imagery. *Biological Psychiatry, 48*, 43–50.
Singer, T. (2004). Empathy for pain involves the affective but not sensory components of pain. *Science, 303*, 1157–1162.
Singer, W. (2001). Consciousness and the binding problem. *Annals of the New York Academy of Science, 929*, 123–146.
Takahashi, H., Yahata, N., Koeda, M., Matsuda, T., Asai, K., & Okubo, Y. (2004). Brain activation associated with evaluative processes of guilt and embarrassment: An fMRI study. *NeuroImage, 23*, 967–974.
Tangney, J. P. (2002). Self-conscious emotions: The self as a moral guide. In A. Tesser & D. A. Stapel, & J. V. Woods (Eds.), *Self and motivation: Emerging psychological perspectives* (pp. 97–117). Washington, DC: American Psychological Association.
Tremblay, L., & Schultz, W. (1999). Relative reward preference in primate orbitofrontal cortex. *Nature, 398*, 704–708.
Vogt, B. A. (2005). Pain and emotion interactions in subregions of the cingulate gyrus. *Nature Reviews. Neuroscience, 6*, 533–544.
Wood, J. N., & Grafman, J. (2003). Human prefrontal cortex: processing and representational perspectives. *Nature Reviews. Neuroscience, 4*, 139–147.
Worsley, K. J., Marrett, S., Neelin, P., Vandal, A. C., Friston, K. J., & Evans, A. C. (1996). A unified statistical approach for determining significant signals in images of cerebral activation. *Human Brain Mapping, 4*, 58–73.
Young, L. J., & Wang, Z. (2004). The neurobiology of pair bonding. *Nature Neuroscience, 7*, 1048–1054.
Zeki, S., & Goodenough, O. (2004). Law and the brain: introduction. *Philosophical Transactions of the Royal Society of London. Series B: Biological Sciences, 359*, 1661–1665.

SOCIAL NEUROSCIENCE, 2007, 2 (3–4), 353–365

Psychology Press
Taylor & Francis Group

Simulation trouble

Shaun Gallagher
University of Central Florida, Orlando, FL, USA

I present arguments against both explicit and implicit versions of the simulation theory for intersubjective understanding. Logical, developmental, and phenomenological evidence counts against the concept of explicit simulation if this is to be understood as the pervasive or default way that we understand others. The concept of implicit (subpersonal) simulation, identified with neural resonance systems (mirror systems or shared representations), fails to be the kind of simulation required by simulation theory, because it fails to explain how neuronal processes meet constraints that involve instrumentality and pretense. Implicit simulation theory also fails to explain how I can attribute a mental or emotion state that is different from my own to another person. I also provide a brief indication of an alternative interpretation of neural resonance systems.

When it comes to explaining how we understand other people some of the very best contemporary philosophers, psychologists, and neuroscientists are simulationists. Rather than appealing to a theoretical use of folk psychology, they appeal to their own experience as a measure of others' experience. A short list of simulationists—for example, Vittorio Gallese, Alvin Goldman, Robert Gordon, Jane Heal, Susan Hurley, and Marc Jeannerod—however, already suggests some problems. Not everyone on this list understands simulation in the same way. In effect, there are different simulation theories, and although it is important to distinguish them, in this paper I will not be able to address each and every variation. Rather, I will present arguments against what I take to be the main elements of simulation theory. My primary target in this paper, however, is an implicit version of simulation theory based on the recent neuroscience of resonance systems (mirror neurons, shared representations).

Presenting arguments against simulation theory (ST) does not mean that I favor its main competitor, theory theory (TT). I do not think that the theory of mind (ToM) explains our primary and pervasive way of understanding others any more than the concept of simulation does. The alternative to both of these approaches is not something that I will directly argue for here (see Gallagher, 2001, 2004, 2006, for the alternative). But since part of the argument that I develop against simulation feeds into and depends on that alternative, a quick summary may be helpful.

The alternative approach is based on evidence from developmental psychology, neuroscience, and phenomenology, and is composed of three aspects. (1) Starting in infancy, we understand conspecifics based on a direct perception of their emotion-expressive movements, intentional actions, gestures, facial expressions, etc. (in developmental psychology this is referred to as "primary intersubjectivity"; Trevarthen, 1979). "Direct"

Correspondence should be addressed to: Shaun Gallagher, Philosophy and Cognitive Sciences, University of Central Florida, Orlando, Florida 32816, USA. E-mail: gallaghr@mail.ucf.edu

Part of the research for this paper was done at the Center for Subjectivity Research at the University of Copenhagen where I was Visiting Professor and supported by the university's Research Priority Initiative on Body and Mind.

An earlier version of this paper, entitled "La perception d'autrui en action," was presented as a public lecture in the series *Fondements cognitifs de l'interaction avec autrui*. Collège de France (February 2006). My thanks to Alain Berthoz, Jean-Luc Petit, and Alvin Goldman, and to three anonymous reviewers for help in clarifying some of these issues.

www.psypress.com/socialneuroscience DOI:10.1080/17470910601183549

perception here means perception without some further cognitive or inferential step that goes beyond what is perceived, for example, in an attempt to grasp hidden mental states.[1] Thus, in a non-mentalizing way, I am able to see meaning, intention, and emotion in the actions of others, and in their gestures and facial expressions. This provides some limited understanding of others without recourse to the concept of mental states. (2) Around the age of 1 year we begin to develop capacities for shared attention and, on this basis, begin to understand others in terms of their involvements in pragmatic contexts (secondary intersubjectivity; Trevarthen, 1980; Trevarthan & Hubley, 1978). As developing agents we do not stand to the side as third-person observers, we engage in second-person interactions, and this is something we continue to do throughout our normal social life. Extending this idea, the pragmatic situations, social practices, and socially defined roles through which our interactions with others happen turn out to do much of the work of shaping our understanding of others. We understand what they are doing and what their intentions are because their actions are set within environments that help to define meaningful actions. Specific contexts make sense out of their actions, gestures, and expressions. In such cases, which are most cases, we do not try to get into the other person's mind, we try to get into their world, or more precisely, into a world that we already share with them. (3) These abilities for intersubjective perception and pragmatic interaction are carried forward and enhanced, starting around the age of 3 or 4 years, by a developing narrative competency that allows us to use proto-narratives (Lewin, 2005) or narrative practices to frame our everyday interpretations of others (Gallagher & Hutto, in press; Hutto, 2004). We begin to use narratives to put persons and contexts together in ways that generate the more subtle and sophisticated understandings of others that we have as adults. This kind of narrative practice is not based on ToM or folk psychology as understood in TT or ST, but may in fact be the basis for our use of folk psychological concepts (like beliefs, desires, reasons) in those rare cases when we encounter puzzling behavior and attempt to explain it (Hutto, in press). Most often, however, our narrative competency combined with our perceptual abilities and action contexts provides sufficient scaffolding for our understanding of others.

This alternative view, of which I have just given the most minimal account, suggests that these perceptual, emotional, and pragmatic capacities for interaction, together with narrative competency, are, at the personal or behavioral level, sufficient to explain pervasive everyday social understanding. As I have just suggested, this does not rule out the possibility that in rare cases we do take a theoretical stance, or that we sometimes use simulation routines to solve puzzling cases. But these are the rare cases. Ordinarily, in our everyday encounters in the pragmatic and social contexts that characterize our lives, we gain a perceptual grasp of another's contextualized actions, gestures, and expressions, and we understand their speech acts as meaningful and intentional, without looking beyond such meanings to their mental states. And if, instead of directly interacting with others, we are called upon (or we call upon ourselves) to think more deeply about them, our tendency is not to call up a theory or a simulation, but to call up a narrative framework to facilitate our understanding of them.[2]

ARGUMENTS AGAINST EXPLICIT ST

Simulation theory (ST), as an approach to ToM, has been developed in several versions. For the purposes of this paper I will distinguish between explicit and implicit versions of ST. I acknowledge that there are hybrid versions (e.g., Gallese & Goldman, 1998; Goldman, 2006; Jeannerod & Pacherie, 2004) and "radical" versions (e.g., Gordon, 2004), and although an argument against one version will not necessarily work against all of them, I think that much of what I argue against explicit and implicit versions also applies to the other models.

[1] This concept is developed in phenomenologists like Scheler (1923/1954), who calls it "primary perception" (e.g., p. 10) and Merleau-Ponty (1964).

[2] Narratives that frame our social interaction are different from the kind of theories to which TT appeals. The latter involve a reference to unobservable entities, that is, mental states which are theoretically postulated entities comparable to the black holes of astrophysics. If Bruner (1986) is right, narrative is a particular mode of thinking that relates to the concrete and particular; it takes context to be primary in the determination of meaning. In contrast, the theoretical is concerned with the abstract and general, and, in this sense, abstracts away from the particular context in favor of propositional attitudes. See Gallagher and Zahavi, Chapter 9 (in press) for further discussion.

According to explicit versions of ST, simulation is a conscious or introspective process in which I imagine myself in the other's situation and use the model (the simulation) that is generated to predict the other's mental states. Goldman has been a good representative of this position.[3] He argues that simulation is explicit insofar as it involves a conscious introspective use of the imagination to conceptually manipulate propositional attitudes (beliefs, desires). "When a mindreader tries to predict or retrodict someone else's mental state by simulation, she uses pretense or imagination to put herself in the target's 'shoes' and generate the target state" (Goldman, 2005a). This process involves three steps:

1. First, the attributor creates in herself pretend states intended to match those of the target. In other words, the attributor attempts to put herself in the target's "mental shoes."
2. The second step is to feed these initial pretend states, e.g., beliefs, into some mechanism of the attributor's own psychology ... and allow that mechanism to operate on the pretend states so as to generate one or more new states, e.g., decisions.
3. Third, the attributor assigns the output state to the target ... e.g., we infer or project the decision to the other's mind (Goldman, 2005b, pp. 80–81).

The first step—"the attributor creates in herself pretend states intended to match those of the target"—seems troublesome in that the simulator apparently already has some idea of what's going on with the other person. Where does that knowledge come from and why isn't that already the very thing we are trying to explain? Some theorists who combine TT and ST suggest that folk psychology provides some general rules about how people think and behave in certain situations, and that this is what the simulationist can use to generate the pretend mental states needed for the simulation process (e.g., Currie & Ravenscroft, 2003). Alternatively, however, one might appeal to subpersonal mirror resonance processes (discussed below), although one then faces the problem of how to translate these processes into a conceptual grasp of propositional attitudes. It is clearly not enough to suggest that a belief feels different from a desire because it is generated by different subpersonal processes, which are themselves generated by differential activations induced by our perception of the other (cf. Goldman, 2002, pp. 11–12). One needs to say something about the particular content of the belief used in the simulation, and not just that it is a belief rather than a desire.

There are both logical and phenomenological arguments to be made against this explicit version of ST. Some of these arguments have been around for a long time and have been directed against earlier versions of ST, developed under the heading of "inference from analogy." Gilbert Ryle, for example, argued that the logic of simulation isn't correct because the idea of imputing to a variety of others what is true of my simulated action ignores the diversity of their actions. "[T]he observed appearances and actions of people differ very markedly, so the imputation to them of inner processes closely matching [one's own or] one another would be actually contrary to the evidence" (Ryle, 1949, p. 54). A similar objection was raised by Max Scheler (1923/1954). If I project the results of my own simulation onto the other, I understand only myself in that other's situation, but I don't understand the other. In other words, given the large diversity of motives, beliefs, desires, and behaviors in the world, it is not clear how a simulation process based on my own relatively narrow experience (or relatively unique circumstance) can give me a reliable sense of what is going on in the other person's mind, or in their behavior.

Another objection is based on developmental considerations. Infants are capable of understanding the intentions of others. At 18 months of age they are capable of recognizing and completing another person's failed intention (Meltzoff, 1995). Even earlier, infants seem capable of parsing intentions (see Baird & Baldwin, 2001; Baldwin

[3] I'm using Goldman as a representative of explicit simulation, and this was certainly his view in his earlier papers (e.g., 1989, 1992). In his most recent work, Goldman (2006) defends a hybrid version of ST. He distinguishes between high-level and low-level mind reading. I take this distinction to be compatible with the distinction between explicit and implicit versions, respectively. Low-level simulation is "simple, primitive, automatic, and largely below the level of consciousness" (p. 113). High-level simulation includes one or more of the following features: "(a) it targets mental states of a relatively complex nature, such as propositional attitudes; (b) some components of the mindreading process are subject to voluntary control; and (c) the process has some degree of accessibility to consciousness .... The low-level prototype is the mirroring type of simulation process. The high-level prototype is the kind that uses pretense or E[nactment]-imagination" (p. 147).

& Baird, 2001). Whether some animals are capable of understanding intentions is still a matter of debate. In any case, if infants are capable of understanding the intentions of others, then, as Scheler also suggests, it is nonetheless unlikely that they are capable of the complex cognitive processes needed for explicit simulation.

A third argument that can be made against explicit ST is what I call the *simple* phenomenological argument. According to Goldman and other defenders of the explicit version of ST, simulation is not only explicit but also pervasive. That is, we use it all the time, or at least it is the default way of understanding others. Goldman (2002, pp. 7–8) calls this a moderate claim:

> The strongest form of ST would say that all cases of (third-person) mentalization employ simulation. A moderate version would say, for example, that simulation is the *default* method of mentalization . . . I am attracted to the moderate version . . . . Simulation is the primitive, root form of interpersonal mentalization.

If simulation is both explicit and pervasive, then one should have some awareness of the different steps that one goes through as one consciously simulates the other's mental states. But there is no phenomenological evidence for this; there is no experiential evidence that I use such conscious (imaginative, introspective) simulation routines when I interact with or come to understand another person. That is, when we consult our own common experience of how we understand others, we just don't find such processes, and that is puzzling if they are supposed to be explicit and pervasive. This is not to say that we never use simulations, but that itself is telling. If we are confronted with some strange or unaccountable behavior, I may try to understand the other person by running a simulation routine. But I can easily become aware that I am in fact taking this approach, and it is all the more apparent when I do this, simply because it tends to be the exception, and it tends to stand out in its rarity. This clearly tells against the idea that I employ simulation in the usual everyday circumstance, and it goes against any claim that in fact we use simulation routines all the time, but phenomenology constantly and consistently misses this fact. What phenomenology does show is that most of our encounters are not third-person puzzles that require explanation of the other person, which in turn is generated by first-person simulations. Rather, most of our encounters are second-person interactions in which I easily have a sense of what is going on with the other person based on our common pragmatic or socially contextualized interactions, with no cognitive simulation required.

The explicit simulation theorist might argue that we use simulations so much that they become habitual, and for that reason we should not expect them to stand out in our experience. Running simulation routines is like driving a car. Just as we are not explicitly aware of all of our driving habits, in the same way we are not aware that we are simulating. According to the simple phenomenological objection, however, if such processes stay at the personal level, they would remain accessible to conscious reflection, or at least they would become apparent, as unworkable habits, in problematic situations when our habitual strategies break down. Not only does this not happen, but it turns out that in problematic situations when our habitual strategies break down we often find that we employ the very sort of simulation process described by explicit ST, which again suggests that simulation is the exception rather than the rule.

IMPLICIT SIMULATION

ST can easily counter the simple phenomenological argument by moving to the implicit version of simulation. Implicit ST conceives of simulation as a subpersonal process, and this approach has gained more ground in recent years by appealing to the neuroscientific evidence involving subpersonal activation of mirror neurons, shared representations, or more generally, resonance systems (e.g., Fadiga, Fogassi, Pavesi, & Rizzolatti, 1995; Rizzolatti, Fogassi, & Gallese, 2000; Rizzolatti, Fadiga, Gallese, & Fogassi, 1996). On the one hand, implicit versions of ST do not succumb to the simple phenomenological objection since if simulation is subpersonal, it is not something of which we would be aware. On the other hand, the strictly implicit version of ST is actually an argument against the explicit version of ST. That is, if our understanding of others is in fact mediated by an implicit and automatic simulation process, then we have little need for the more explicit version. Indeed, to the extent that implicit ST explains the phenomenological scarcity of explicit simulation, it would support the simple phenomenological argument against explicit simulation as the default model of social understanding. Along this line Gallese (2005, p. 102), who I will take as the clearest representative of implicit ST, states: "Whenever we face situations in which exposure to others' behavior require a response by us, be it active or simply attentive, we seldom engage ourselves in an explicit, deliberate interpretive act. Our understanding of a situation most of the time is immediate, automatic, and almost reflex like." This, of course, doesn't rule out hybrid models that combine implicit and explicit elements (see Goldman, 2006).

Implicit ST appeals to what is now familiar social neuroscience. In broad terms, one's motor system reverberates or resonates in one's encounters with others. My motor system is activated when I perceive another person performing an intentional action, for example. Mirror neurons in the pre-motor cortex and in Broca's area of the human brain are activated both when the subject engages in specific instrumental actions and when the subject observes someone else engage in those actions (Rizzolatti et al., 1996, 2000). Also, specific overlapping neural areas (shared representations), in parts of the frontal and parietal cortexes, are activated when I engage in intentional actions, when I observe some other person engage in that action, and when I imagine myself or another person engaged in that action (e.g., Grèzes & Decety, 2001). These subpersonal mechanisms are said to constitute a simulation of the other's intentions (Gallese, 2001; Gallese & Goldman, 1998). Thus Gallese contends that "when we observe actions performed by other individuals our motor system 'resonates' along with that of the observed agent ... action understanding heavily relies on a neural mechanism that matches [simulates], in the same neuronal substrate, the observed behaviour with the one [the observer could execute] ... " (2001, pp. 38–39).

The neuroscientific evidence for these processes is very strong, and it is not my intention here to deny the neuroscience. But is it appropriate to characterize these processes as simulations, as Gallese goes on to do? The implicit simulation hypothesis is just this: understanding others is achieved by simulating the other's action "with the help of a motor equivalence between what the others do and what the observer does" (2001, p. 39). This is a subpersonal process generated by "automatic, implicit, and nonreflexive simulation mechanisms ..." (Gallese, 2005, p. 117). He refers to his model as the "shared manifold hypothesis" and distinguishes between three levels (Gallese, 2001, p. 45):

1. The *phenomenological level* is the one responsible for the sense of similarity ... that we experience anytime we confront ourselves with other human beings. It could be defined also as the *empathic* level.
2. The *functional level* can be characterized in terms of simulation routines, *as if* processes enabling models of others to be created.
3. The *subpersonal level* is instantiated as the result of the activity of a series of mirror matching neural circuits.

Recall the Ryle–Scheler objection against explicit ST, that since it is based on a projection from experience that is narrowly my own, no inference about the experience of the other person (which may be greatly diversified from mine) is justified. Does a subpersonal simulation lock us up within our own first-person system? Defenders of the implicit version of ST have a good, albeit partial, response to this. There is growing consensus that mirror neurons (and shared representations) are *neutral*—neither first nor third person. They are activated both for my own action and for observation of the other's action: thus, activation of the system simulates the intentional action but not the agent (deVignemont, 2004; Gallese, 2005; Hurley, 2005; Jeannerod & Pacherie, 2004).

In this case, however, the complete subpersonal simulation process, like its explicit cousin, involves a multi-step process. First, we perceive the other's behavior; this is followed immediately by activation of shared representations—in neutral mode registering what Jeannerod and Pacherie (2004) call "naked intentions", and this is followed by a determination of agency (i.e., a specification of who did the action—me or the other person. This final step is accomplished by what Georgieff and Jeannerod (1998) have called the "Who" system, which distinguishes self-action from other-action and which may involve activation in the inferior parietal lobule (Blakemore & Frith, 2003; Decety, 2005; Jackson & Decety, 2004).

It will pay to stop and consider Jeannerod and Pacherie's claims about naked intentions. They assume that an articulation at the level of the neural activations between those activations responsible for (1) registering in the perceiving system the "naked" intention in an action, and (2) registering the agent for the action, means that there is an articulation at the level of experience between the perception of intention and the experience of agency. "We can be aware of an intention, without by the same token being aware of whose intention it is ... something more than the sole awareness of a naked intention is needed to determine its author" (2004, p. 140). The question is this: if in fact the brain can process information about intentions without assigning the intentions to a particular agent, is it legitimate to say that our experience is similarly articulated? Jeannerod and Pacherie suggest that it is:

> When the naked intention one is aware of yields an overt action, the extra information needed to establish authorship may be found in the outside world. The question "Is this intention mine?" would then be answered by answering the question: "Is this my body performing the corresponding action?" (p. 140).

Phenomenologically (experientially), however, intentions in almost all cases come already fully clothed in agent specification. The "who" question does not come up at the level of experience, because the neural systems have already decided the issue. The wonderful thing about the "Who system" is that it is completely neurological and subpersonal—and the results of its activation are hardly ever experientially manifested as "making a decision about who did the action." Rather, the results of its activation are experientially manifested as "X's action" where X is either you or me. Indeed, our direct perception is highly reliable with regard to discriminating between self and non-self. Pathologies and oddly arranged experiments may reveal "who" problems, but in normal ecological behavior it is generally clear whose intention/action it is. Philosophers like Wittgenstein (1958), Shoemaker (1968), and Evans (1982), provide good reasons why the identification question—"Someone is intending to pick up the apple, is it me?"—just doesn't come up.

On this score, I want to suggest that there is no necessary isomorphism between the phenomenological, the functional, and the neuronal levels. So, if the neuronal processes can be defined as involving a step-wise process, this does not mean that a step-wise process necessarily shows up at the level of experience, and *vice versa*. If we can distinguish a step-wise process in experience, that does not mean that we will find exactly the same step-wise structure in the underpinning functional or neuronal processes. This is tied into the concept of multiple realizability. For the purposes of this paper, I will call any claim to the contrary the fallacy of supposed isomorphism.[4]

PROBLEMS WITH IMPLICIT SIMULATION

This brings us to the first set of questions about implicit ST. Specifically I want to argue that an alternative interpretation of neural resonance is possible. Implicit ST interprets neural resonance as simulation, but it could easily be interpreted as part of the neuronal processes that underlie intersubjective *perception* rather than simulation. That is, the articulated neuronal processes that include activation of mirror neurons or shared representations may underpin a non-articulated direct perception of the other person's intentional actions, rather than a distinct mental process of simulating their intentions. This claim requires that we conceive of perception as a temporal phenomenon, and as an enactive sensory-motor phenomenon. Let us examine this possibility.

First, mirror neurons fire 30–100 ms after appropriate visual stimulation (Gallese, private correspondence). This short amount of time between activation of the visual cortex and activation of the pre-motor cortex, raises the question of where precisely to draw the line between perceptual processes and something that would count as a subpersonal simulation. Even if it is possible to draw a line between activation of the visual cortex and activation of the pre-motor cortex, this does not mean that this line distinguishes between perception and simulation as a step-wise process. The simulation interpretation of these processes seems to read into the neuronal level the step-wise processes of simulation, first identified as explicitly conscious simulation processes.

Let us be careful and clear about this. Gallese and the implicit simulationists are not claiming that the step-wise *neuronal* processes (sensory activation of visual cortex followed by mirror system activation) generate a step-wise *conscious* process of perception plus simulation. In this regard they are not committing the fallacy of

[4] I do not mean to rule out all isomorphisms. This only goes to the claim that there is a necessary isomorphism between neuronal, functional, and phenomenological levels.

supposed isomorphism. Gallese contends that the simulation stays implicit, that the mirror system activation itself can be read, functionally, as part of a step-wise simulation process. Nonetheless, I think two issues can be raised in this regard.

The first issue is this: deciding that mirror neurons function as simulations depends on taking a step-wise model that was developed at the explicit, conscious, or personal level, and looking for that step-wise model at the neuronal level. On various versions of ST, simulation involves a step-wise process that begins with perception and ends with some form of inferential understanding. We first see an action that we need to understand; we then simulate it in our own mind or motor system; we then attribute agency for the action, and infer something about the other's experience. But even if neuronal processes that involve information flow from sensory cortex to pre-motor cortex take some time (as much as 100 ms), it is not clear that we should identify these step-wise processes as perception plus simulation, rather than a temporally extended and enactive perceptual process. That is, at least in terms of temporal parameters, the fact that at the neurological level, S (sensory processing) is followed by M (activation of mirror neurons) does not mean that one should think of this as perception followed by simulation.

If we think of perception as an enactive process (e.g., Hurley, 1998; Noë, 2004)—as involving sensory-motor skills, rather than as just sensory input/processing; as an active, skillful, embodied engagement with the world rather than the passive reception of information from the environment—then it may be more appropriate to think of the resonance processes as part of the structure of the perceptual process when perception is of the action of conspecifics. Fogassi and Gallese, despite their simulationist interpretation, put this point clearly: "perception, far from being just the final outcome of sensory integration, is the result of sensorimotor coupling" (2002, p. 27). Mirror activation, on this interpretation, is not the initiation of simulation, it is part of a direct intersubjective perception of what the other is doing.

At the phenomenological level, when I see the other's action or gesture, I see (I *directly perceive*) the meaning in the action or gesture. I see the joy or I see the anger, or I see the intention in the face or in the posture or in the gesture or action of the other. I see it. I do not have to simulate it. And I immediately see that it is *their* action, gesture, emotion, or intention, and it is extremely rare that I would be in a position to confuse it with my own. Although Jeannerod and Pacherie (2004) defend a version of ST, they nicely express the phenomenological alternative: "Perception and action are closely integrated and when we visually perceive actions, we seem to be immediately sensitive to the distinctive properties of intentional behavior" (p. 139). This may be due to the underlying articulated neuronal processes, but it is not clear at all why we should think of these processes on the simulationist model.

Of course the simulationist can accept the phenomenology ("Yes indeed, that is what *seems* to happen") and still hold to the interpretation that the specific subpersonal processes involve simulation. But what precisely justifies this interpretation? After all, what happens on the neurological level is simply a complex sequence of neuronal activations. If we look at those processes from a functionalist perspective already framed by ST, then we tend to read those processes as involving simulation. If, in contrast, we look at those processes from a phenomenological level that suggests a direct perception of the other's intentions, then we tend to read those processes as perceptual without simulation. Can the simulationist offer any convincing evidence that the activation of resonance processes is in fact a simulation?

This brings us to the second issue, which I think counts as an argument against the implicit version of ST. What theorists of implicit simulation (Gallese, Jeannerod, Pacherie), and even critics of implicit ST, like Saxe (2005), call "simulation", is not simulation in any genuine sense of the word. Consider, first, two definitions of "simulation" offered by the Oxford English Dictionary. (1) Simulation is an imitation, in the sense of something *not real*—counterfeit; to simulate means to feign, to pretend. I will call this the pretense definition. (2) Simulation in the sense of a simulator: a model (a thing) that we can *use* or do things with so we can understand the real thing. I will call this the instrumental definition. Both senses of the term appear to be ubiquitous in the literature of both explicit and implicit ST.[5] Consider the following characterizations (italics are mine).

[5] I think it is clear that, given the history of ST, the definition of simulation relevant to ST is first worked out in accounts of explicit ST, and then is uncritically used in accounts of implicit ST.

Instrumental def. Simulation means "*using* one's own evaluation and reasoning mechanisms as a model for theirs ... " (Dokic & Proust, 2002, p. viii).

Pretense def+instrumental def. Simulation involves "pretend states" where, "by pretend state I mean some sort of surrogate state, which is *deliberately adopted* for the sake of the attributor's task ... In simulating practical reasoning, the attributor *feeds* pretend desires and beliefs into her own practical reasoning system" (Goldman, 2002, p. 7).

The surrogation or pretense, however, is of a precise kind. Bernier (2002) makes this explicit as an essential element found in ST.

Instrumental def+pretense def. "According to ST, a simulator who runs a simulation of a target would *use* the resources of her own decision-making mechanism, in an 'off-line' mode, and then the mechanism would be fed with the mental states she would have *if she was in the target's situation*" (Bernier, 2002, p. 34)

For ST, a simulation is not simply a model that we use to understand the other person—theoretical models would suffice if this were all that was required. Even the fact that the model is constituted in our own mechanisms is not sufficient. Rather, I must use the model "as if" I were in the other person's situation. As Gallese puts it:

Pretense def. "[O]ur motor system becomes active *as if* we were executing that very same action that we are observing" (2001, p. 37).

Gordon (2005, p. 96) locates this pretense right at the neuronal level:

Pretense def. The neurons that respond when I see your intentional action, respond "*as if* I were carrying out the behavior ...".

The concept of simulation, as defined by ST, then, clearly needs to meet these two conditions: it is a process that I control in an instrumental way (in the explicit version it is "deliberately adopted"), and it involves pretense (I put myself "as if" in the other person's shoes).

These characterizations are, as I've suggested, ubiquitous in ST.

Instrumental and pretense defs combined. Mental simulation is a cognitive model, "*ability* or *heuristic* or *methodology*" (Jacob, 2002, who cites Gordon for the latter term)—by which I "*engage* in pretense," *put myself* in someone else's shoes, *compare* my experience to their experience, and *predict* their mental state, emotion, or behavior. We *use* ourselves as a model ... I *create* in myself some pretend beliefs ... and so forth.

This is the way simulation is characterized not only by theorists of explicit simulation, but also by theorists of implicit simulation. The pretense condition mentioned by Gallese is accomplished in a simulation considered to be "an interactive *model* of what cannot be known in itself" (2003). At the subpersonal level, the brain in a stepwise fashion is *modeling* the intentional action of others. Gordon (2004), p. 1) suggests that on the "cognitive-scientific" model, "one's own behavior control system is employed as a *manipulable model* of other such systems. (This is not to say that the 'person' who is simulating is the model, rather, only that *one's brain can be manipulated to model other persons*)."

Thus, according to ST, simulation involves the instrumental use of a first-person model to form third-person "as if" or "pretend" mental states. For subpersonal processes, however, both of these characterizations fail. (1) If simulation is characterized as a process that I (or my brain) instrumentally uses or controls, if this is what simulation is, then it seems clear that what is happening in the implicit processes of motor resonance is not simulation. We, at the personal level, do not *do* anything with the activated brain areas—in fact, we have no instrumental access to neuronal activation, and we can not use it as a model. Nor does it make sense to say that at the subpersonal level the brain itself is *using* a model or methodology, or *initiating* pretend states, or that one set of neurons makes use of another set of neurons as a model in order to generate an understanding of something else. In precisely the intersubjective circumstances that we are considering, these neuronal systems do not take the initiative; they do not activate themselves, but are activated by the other person's action. The perception of the other person's action automatically activates in our brain the same areas that are activated when we engage in similar action. The other person *has an effect on us.* The other *elicits* this activation. This is not a simulation, but a perceptual elicitation. It is not us (or our brain) *doing* it, but the other who does this to us.[6]

Furthermore, (2) in subpersonal processes there is no pretense, and this is the case whether we consider neuronal processes as vehicles or in terms of the content that they might represent. As vehicles, neurons either fire or do not fire. They do not pretend to fire. More to the point, however, what these neurons represent or register cannot be pretense in the way required for ST. As we saw, the mirror system is neutral with respect to the agent; there is no first- or third-person specification involved. In that case, they do not register *my* intentions as pretending to be *your* intentions; there is no "as if"—there is no neuronal subjunctive—because there is no "I" or "you" represented.

In response to this criticism it might be argued that no one is claiming that there is pretense in the firing of neurons, but rather that the neuronal processes underpin a personal-level experience of the other that qualifies as a pretense state.[7] But if, as we saw, implicit simulation theorists like Gallese contend that the simulation stays implicit, and that the mirror system activation itself can be read, functionally, as a simulation process, then on the ST definition of simulation, those subpersonal processes themselves must involve pretense. Alternatively, if the claim is that certain neuronal processes generate pretense states at the personal level, then the simulation becomes explicit and other objections hold.

Goldman and Sripada (2005) acknowledge a discrepancy between the ST definition of simulation and the working of subpersonal mirror processes. "Does [Gallese's] model really fit the pattern of ST? Since the model posits unmediated resonance, it does not fit the usual examples of simulation in which pretend states are created and then operated upon by the attributor's own cognitive equipment (e.g., a decision-making mechanism), yielding an output that gets attributed to the target . . .". To address this discrepancy they propose a minimal definition of simulation:

> However, we do not regard the creation of pretend states, or the deployment of cognitive equipment to process such states, as essential to the generic idea of simulation. The general idea of simulation is that the simulating process should be similar, in relevant respects, to the simulated process. Applied to mind reading, a minimally necessary condition is that the state ascribed to the target is ascribed as a result of the attributor's instantiating, undergoing, or experiencing, that very state. In the case of successful simulation, the experienced state matches that of the target. This minimal condition for simulation is satisfied [in Gallese's model] (Goldman & Sripada, 2005, p. 208).

If this is a necessary condition, it is clearly not a sufficient one, as Goldman (2006, p. 131 ff.) realizes, because on this minimal definition and without something further, it is not clear what would motivate me to ascribe the state that I was undergoing to someone else. Activation of the mirror system does not, on its own, necessarily involve mind reading or attribution to another agent. What one would have to add in order to get to the point of making this a simulation of the other's action, intention, or mental state are precisely the pretense and instrumental aspects that Goldman and Sripada are discounting.[8]

Furthermore, and importantly for implicit ST, if simulation were as automatic as mirror neurons firing, then it would seem that we would not be able to attribute a state different from our own to someone else. But we do this all the time. I can understand that the woman in front of me is enthusiastically and gleefully reaching to pick up a snake at the same time that I am experiencing revulsion and disgust about that very possibility. It seems straightforward to say that I see what she is doing and that I see she is doing it with enthusiasm and glee, but that my own feelings are quite

[6] It may seem contradictory to claim in the previous paragraphs that perception is enactive, or as Noë says, "perception is action," and in this argument to claim that the activation of the resonance system is the result of a passive elicitation, so that the motor aspect of perception does not involve our action, but is a case of us being affected by the other. I think that a fuller account of enactive perception has to be able to accommodate this passive, affective aspect of perception (see Gallagher, 2005).

[7] My thanks to an anonymous referee for this objection.

[8] This motivates Goldman's move to a hybrid model. Implicit mirroring may be considered a minimal condition for simulation, but for the full simulation that constitutes mind reading (the attribution of mental states to another) one requires higher-level processes of "classification" (which he characterizes as introspective, although not necessarily conscious introspection; 2006, p. 245 ff.) and "projection." Although pretense is not required for low-level mirroring, it is one possible way to realize high-level simulation and mind reading (p. 49).

different. Neither my neural states nor my feelings/cognitions match hers, and it thus seems that I can understand her actions and emotions (which are completely different from mine) without even meeting the minimal necessary condition for simulation, that is, matching my state to hers.

One might want to add to this minimally necessary condition for simulation the further stipulation that the simulating process must be "concretely similar" to the simulated process, where "concretely similar" means that there is a fine-grained (though not necessarily one-to-one) isomorphism for each step involved in the respective processes (Fisher, 2006). But this makes the problem worse if we are conceiving of the simulation as happening on a neuronal level. If my neural/mental/emotional state has to match the other person's neural/mental/emotional state, it is difficult to understand how we could attribute to someone else a mental or emotional experience that is different from our own.

In regard to these minimal definitions of simulation, also consider the difficulties involved if we were interacting with more than one other person, or trying to understand others who are interacting with each other. Is it possible to simulate the neural/mental/emotional states of two other people at the same time, or quickly enough, if in fact our simulations must be such that we instantiate, undergo, or experience, that very state, and/or that such simulations must be concretely similar? There would be an impossible amount of cognitive work or subpersonal matching required to predict or to understand the interactions of several people if the task involves simulating their mental states, especially if in such interpersonal interactions the actions and intentions of each person are affected by the actions and intentions of the others (see Morton, 1996, for a similar point).[9]

The woman reaching for the snake is a good example with which to make one more clarification. The term "simulation" is sometimes used by neuroscientists to refer to certain motor control processes for action planning. Efference copy sent through forward control mechanisms, for example, is said to constitute a simulation of an intended movement in order to compare it with an ongoing movement to predict its success. The brain runs this simulation to make fast nonconscious corrections to keep the action on track. This use of the term is entirely independent of ST, and the objections I am raising here do not apply to it.[10] In the context of motor control the term "emulation" is sometimes substituted for "simulation", and I'll follow that practice here.

In the case of seeing someone reach to pick up a snake, it seems possible that my understanding of the other's actions and their enthusiasm for those actions is initiated purely in enactive perceptual processes that would include (non-simulative) activation of shared representations. Those same shared representations may further generate, or may help to constitute a motoric emulation of my own action of picking up a snake, which leads to my sense of revulsion and disgust at the very possibility. This emulation, however, is not an implicit simulation that contributes to my understanding of the other person; it is rather something that helps to constitute my own feel-

[9] The requirement that the simulation has to be concretely similar also raises problems for the instrumental and pretense conditions even for the explicit version of ST. If our simulation has to be concretely similar to the simulated state for it to be considered a simulation, assuming explicit instrumental control of our simulation process, how will we know how to run or control our simulation unless we already know in some detail what the other's state is like. And how do we come by that knowledge? If the answer is through simulation, then we have an infinite regress. In regard to explicit pretense, Fisher (2006), who models simulation as a reasoning process, rejects this aspect as inconsistent with simulation being concretely similar, for if we simulate a reasoning process, we are really reasoning, and not just pretending to.

[10] Some theorists, however, have appealed to these motor emulation processes as possible mechanisms involved in the simulation of another's action (e.g., Gallese, 2001; Hurley, 2005, pp. 181–188; see Iacoboni, cited in Millikan, 2005, p. 188, note 2). Our own motor system comparators are activated to simulate and thereby anticipate the other's action. The brain could be said to predict the other person's actions in this way. On this account the perception of the other's action is automatically informed by a subpersonal simulation; perception of action involves a loop through the motor control comparator. Can ST adopt this model of simulation? The problem, again, is that the pretense condition is not met; there is no "*as if* it were I" involved, and in that regard it fails to be the kind of simulation required by ST. If indeed the subpersonal emulation is neutral in regard to whose action is at stake, then it can be only a representation of an intentional action in my motor system, but not a representation of my own motor action *as if it were* the other's. It is not at all clear that, as Gordon (2005, p. 96) suggests, the neurons respond "*as if* I were carrying out the [other's] behavior" in any sense in which the "as if" registers subpersonally. Even assuming a "Who system," a specification in my motor system that the action belongs to another is not equivalent to the specification "*as if* I were carrying out the action." If this is a simulation of intentional action, it is, nonetheless, not the kind of simulation that ST needs; it may be nothing more than motor priming or emulation, or what Hurley calls mirroring (2005, p. 184).

ings about this action, which turn out to be very different from the ones I understand to belong to the other person. Just this contrasting situation, between her enthusiastic action and my repulsion, may motivate a more explicit hermeneutical process. That is, depending on circumstances, I may explicitly frame her action in a narrative that makes sense of it (e.g., this woman is, after all, the curator of the zoo's snake house, she is trained to handle snakes, and does it all the time), or if there is some puzzling element involved (e.g., this woman is my wife who I know to be afraid of snakes), I may even engage in one of those rare instances of explicit simulation (e.g., assuming I am not in a position to ask her what she is doing, I might try to imagine what she is thinking—perhaps that this is a toy snake, etc.).

CONCLUSION

It is not clear why we should think of the activation of resonance systems as a simulation process of the sort required by ST. This is not to deny that resonance processes are at work in our perception of the other person. Moreover, the nature of the resonance processes involved in such encounters makes our perception of other conspecifics different from our perception of objects and instruments. But it does not make our perception and understanding of others the result of an implicit simulation. In effect, simulation is a personal-level concept that cannot be legitimately applied to subpersonal processes. As a personal-level concept it signifies a relatively rare and not at all pervasive or default way of understanding others.

I have suggested that implicit ST is only one, and not necessarily the best, interpretation of the significance of motor resonance systems. I have argued that implicit resonance processes are not simulations in any sense that is useful for ST. Furthermore, if implicit ST were a good account of our primary and pervasive ability to understand others, it would count as an argument against explicit ST, since explicit simulations would be redundant in this case. Likewise, however, if our default mode of understanding others were based on explicit simulation, then the claims of implicit ST about the adequacy of motor resonance processes would be wrong. Goldman's view of implicit motor resonance processes is that they do not constitute simulations of a sort that would be sufficient to do the full job, but do generate some background information that is useful to initiate the explicit simulation process. I have argued, in contrast to both explicit and implicit ST, that implicit motor resonance processes are important enactive processes that contribute to the constitution of the perceptual access that we have to the intentions of others.

I do not claim, however, that we get a full account of human intersubjectivity in the idea that we have *perceptual* access to the intentions of others. Perceptual access to the other person's contextualized bodily movements, gestures, facial expressions, and so forth does give us a partial sense of what is going on with them, what they mean and what they feel. This, together with our interactions with others in pragmatic and social contexts, where those contexts and situations enrich our understanding even further, gives us a relatively reliable, but still relatively elemental understanding of them. There is much more to say about the role of language and narrative competency in a fuller account of intersubjectivity (see Gallagher, 2006; Gallagher & Hutto, in press; Hutto, 2003, 2004, in press; Lewin, 2005). Even in that larger story, however, the ToM approaches that emphasize either simulation, or the role of folk psychology as background theory, have a minimal role to play in our normal and everyday interactions.

REFERENCES

Baird, J. A., & Baldwin, D. A. (2001). Making sense of human behavior: Action parsing and intentional inference. In B. F. Malle, L. J. Moses, & D. A. Baldwin (Eds.), *Intentions and intentionality: Foundations of social cognition* (pp. 193–206). Cambridge, MA: MIT Press.

Baldwin, D. A., & Baird, J. A. (2001). Discerning intentions in dynamic human action. *Trends in Cognitive Science*, *5*(4), 171–178.

Bernier, P. (2002). From simulation to theory. In J. Dokic & J. Proust (Eds.), *Simulation and knowledge of action* (pp. 33–48). Amsterdam: John Benjamins.

Blakemore, S.-J., & Frith, C. D. (2003). Self-awareness and action. *Current Opinion in Neurobiology, 13*, 219–224.

Bruner, J. (1986). *Actual minds, possible worlds*. Cambridge, MA: Harvard University Press.

Currie, G., & Ravenscroft, I. (2003). *Recreative minds*. Oxford, UK: Oxford University Press.

Decety, J. (2005). Perspective taking as the royal avenue to empathy. In B. F. Malle & S. D. Hodges (Eds.), *Other minds: How humans bridge the divide between self and other* (pp. 135–149). New York: Guilford Press.

deVignemont, F. (2004). The co-consciousness hypothesis. *Phenomenology and the Cognitive Sciences, 3*(1), 97–114.

Dokic, J., & Proust, J. (2002). Introduction. In J. Dokic & J. Proust (Eds.), *Simulation and knowledge of action.* (pp. vii–xxi). Amsterdam: John Benjamins.

Evans, G. (1982). *The varieties of reference.* Oxford, UK: Oxford University Press.

Fadiga, L., Fogassi, L., Pavesi, G., & Rizzolatti, G. (1995). Motor facilitation during action observation: A magnetic stimulation study. *Journal of Neurophysiology, 73*, 2608–2611.

Fisher, J. C. (2006). Does simulation theory really involve simulation? *Philosophical Psychology, 19*(4), 417–432.

Fogassi, L., & Gallese, V. (2002). The neural correlates of action understanding in non-human primates. In M. I. Stamenov & V. Gallese (Eds.), *Mirror neurons and the evolution of brain and language* (pp. 13–35). Amsterdam: John Benjamins.

Gallagher, S. (2001). The practice of mind: Theory, simulation, or interaction? *Journal of Consciousness Studies, 8*(5–7), 83–107.

Gallagher, S. (2004). Understanding interpersonal problems in autism: Interaction theory as an alternative to theory of mind. *Philosophy, Psychiatry, and Psychology, 11*(3), 199–217.

Gallagher, S. (2005). A new movement in perception: Review of Alva Noë's *Action in Perception* (book review). *Times Literary Supplement,* 9 September 2005.

Gallagher, S. (2006). The narrative alternative to theory of mind. In R. Menary (Ed.), *Radical enactivism: Intentionality, phenomenology, and narrative* (pp. 223–229) Amsterdam: John Benjamins.

Gallagher, S., & Hutto, D. (in press). Primary interaction and narrative practice. In Zlatev, J. Racine, T. Sinha, C. & Itkonen, E. (Eds.), *The shared mind: Perspectives on intersubjectivity.* Amsterdam: John Benjamins.

Gallagher, S., & Zahavi, D. (in press). *The phenomenological mind.* London: Routledge.

Gallese, V. (2001). The "shared manifold" hypothesis: From mirror neurons to empathy. *Journal of Consciousness Studies, 8*, 33–50.

Gallese, V. (2003). The roots of empathy: The shared manifold hypothesis and the neural basis of intersubjectivity. *Psychopathology, 36*, 171–180.

Gallese, V. (2005). "Being like me": Self–other identity, mirror neurons and empathy. In S. Hurley & N. Chater (Eds.), *Perspectives on imitation I* (pp. 101–118). Cambridge, MA: MIT Press.

Gallese, V., & Goldman, A. (1998). Mirror neurons and the simulation theory of mind-reading. *Trends in Cognitive Sciences, 2*, 493–501.

Georgieff, N., & Jeannerod, M. (1998). Beyond consciousness of external events: A "Who" system for consciousness of action and self-consciousness. *Consciousness and Cognition, 7*, 465–477.

Goldman, A. (1989). Interpretation psychologized. *Mind and Language, 4*, 161–185.

Goldman, A. (1992). In defense of simulation theory. *Mind and Language, 7*, 104–119.

Goldman, A. (2002). Simulation theory and mental concepts. In J. Dokic & J. Proust (Eds.), *Simulation and knowledge of action* (pp. 1–19). Amsterdam: John Benjamins.

Goldman, A. (2005a). Mirror systems, social understanding and social cognition. *Interdisciplines.* (Retrieved 12 January, 2007 from: http://www.interdisciplines.org/mirror/ papers/3)

Goldman, A. (2005b). Imitation, mind reading, and simulation. In S. Hurley & N. Chater (Eds.), *Perspectives on imitation* II (pp. 80–81). Cambridge, MA: MIT Press.

Goldman, A. (2006). *Simulating minds: The philosophy, psychology and neuroscience of mindreading.* Oxford, UK: Oxford University Press.

Goldman, A. I., & Sripada, C. S. (2005). Simulationist models of face-based emotion recognition. *Cognition, 94*, 193–213.

Gordon, R. M. (2004). Folk psychology as mental simulation. In N. Zalta (Ed.), *The Stanford encyclopedia of philosophy.* (Retrieved 12 January, 2007 from: http://plato. stanford.edu/archives/fall2004/entries/folkpsych-simulation/)

Gordon, R. M. (2005). Intentional agents like myself. In S. Hurley & N. Chater (Eds.), *Perspectives on imitation* I (pp. 95–106). Cambridge, MA: MIT Press.

Grèzes, J., & Decety, J. (2001). Functional anatomy of execution, mental simulation, and verb generation of actions: A meta-analysis. *Human Brain Mapping, 12*, 1–19.

Hurley, S. L. (1998). *Consciousness in action.* Cambridge, MA: Harvard University Press.

Hurley, S. L. (2005). Active perception and perceiving action: The shared circuits model. In T. Gendler & J. Hawthorne (Eds.), *Perceptual experience.* New York: Oxford University Press.

Hutto, D. (2003). Folk-psychological narratives and the case of autism. *Philosophical Papers, 33*(3), 345–361.

Hutto, D. (2004). The limits of spectatorial folk-psychology. *Mind and Language, 19*, 548–573.

Hutto, D. (in press). *Folk psychological narratives.* Cambridge, MA: MIT Press.

Jackson, P. L., & Decety, J. (2004). Motor cognition: A new paradigm to study self other interactions. *Current Opinion in Neurobiology, 14*, 259–263.

Jacob, P. (2002). The scope and limits of mental simulation. In J. Dokic & J. Proust (Eds.), *Simulation and knowledge of action* (pp. 87–109). Amsterdam: John Benjamins.

Jeannerod, M., & Pacherie, E. (2004). Agency, simulation, and self-identification. *Mind and Language, 19*(2), 113–146.

Lewin, P. (2005). Understanding narratively, understanding alterity. *Human Studies, 28*, 375–383.

Meltzoff, A. N. (1995). Understanding the intentions of others: Re-enactment of intended acts by 18-month-old children. *Developmental Psychology, 31*, 838–850.

Merleau-Ponty, M. (1964). *The primacy of perception and other essays on phenomenological psychology.* Evanston, IL: Northwestern University Press.

Millikan, R. G. (2005). Some reflections on the theory theory-simulation theory debate. In S. Hurley & N. Chater (Eds.), *Perspectives on imitation* II (pp. 182–188). Cambridge, MA: MIT Press.

Morton, A. (1996). Folk psychology is not a predictive device. *Mind*, *105*, 119–137.

Noë, A. (2004). *Action in perception*. Cambridge, MA: MIT Press.

Rizzolatti, G., Fadiga, L., Gallese, V., & Fogassi, L. (1996). Pre-motor cortex and the recognition of motor actions. *Cognitive Brain Research*, *3*, 131–141.

Rizzolatti, G., Fogassi, L., & Gallese, V. (2000). Cortical mechanisms subserving object grasping and action recognition: A new view on the cortical motor functions. In M. S. Gazzaniga (Ed.), *The new cognitive neurosciences* (pp. 539–552). Cambridge, MA: MIT Press.

Ryle, G. (1949). *The concept of mind*. New York: Barnes & Noble.

Saxe, R. (2005). Against simulation: The argument from error. *Trends in Cognitive Sciences*, *9*(4), 174–178.

Scheler, M. (1954). *The nature of sympathy* (Trans. P. Heath). London: Routledge & Kegan Paul. (Originally published in 1923 as *Wesen und formen der sympathie*. Bonn, Germany: Verlag Friedrich Cohen)

Shoemaker, S. (1968). Self-reference and self-awareness. *Journal of Philosophy*, *65*, 555–567.

Trevarthen, C. (1979). Communication and cooperation in early infancy: A description of primary intersubjectivity. In M. Bullowa (Ed.), *Before speech*. Cambridge, UK: Cambridge University Press.

Trevarthen, C. (1980). The foundations of intersubjectivity: Development of interpersonal and cooperative understanding of infants. In D. R. Olson (Ed.), *The social foundation of language and thought: Essays in honor of Jerome S. Bruner* (pp. 316–341). New York: Norton.

Trevarthen, C., & Hubley, P. (1978). Secondary intersubjectivity: Confidence, confiding and acts of meaning in the first year. In A. Lock (Ed.), *Action, gesture and symbol: The emergence of language*. San Diego, CA: Academic Press.

Wittgenstein, L. (1958). *The blue and brown books*. Oxford, UK: Basil Blackwell.

Subject index

For Product Safety Concerns and Information please contact our EU representative GPSR@taylorandfrancis.com Taylor & Francis Verlag GmbH, Kaufingerstraße 24, 80331 München, Germany